Control Engineering

For
Simon, Jamie and Nick

Clouds are not spheres,
mountains are not cones,
coastlines are not circles,
bark is not smooth,
nor does lightning travel in a straight line.

Benoit B. Mandelbrot

(and a control system is far more complex than the mathematics describing its behaviour)

Control Engineering

E.A. Parr
BSc, CEng, MIEE, MInstMC

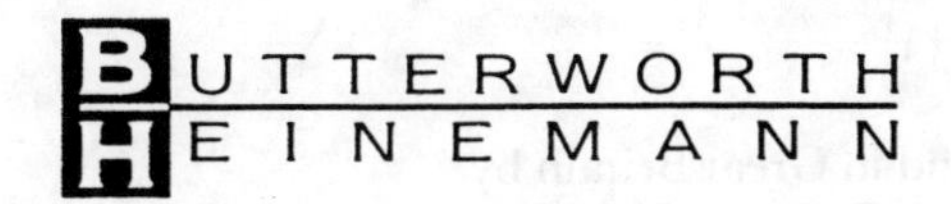

Butterworth-Heinemann
Linacre House, Jordan Hill, Oxford OX2 8DP
A division of Reed Educational & Professional Publishing Ltd

A member of the Reed Elsevier plc group

OXFORD BOSTON JOHANNESBURG
MELBOURNE NEW DELHI SINGAPORE

First published 1996

British Library Cataloguing in Publication Data

A catalogue record for this book
is available from the British Library

ISBN 0 7506 2407 8

Library of Congress Cataloguing in Publication Data

A catalogue record for this book
is available from the Library of Congress

Printed and bound in Great Britain by
Hartnolls Limited, Bodmin, Cornwall

Contents

Preface

Process control engineers tend to fall into two distinct categories. On the one hand are the practical engineers who view a plant as a collection of interesting technology which can only be understood by people who get their hands dirty. At the other extreme are academic engineers who view a plant as a collection of mathematical models which can best be analysed and understood at a desk, or better still at a computer terminal. In practice, both types of engineers tend to be surprised at the success of the other.

Most books on the mathematics of process control tend to be very dry and divorced from the realities of actual plants. This book has been written from my background of thirty years in process control, and aims to give a readable introduction to control theory. It is based on facts and techniques that I have found useful over the years.

I have tried to mix theory and relevant practical details (such as controller tuning) to provide a book which is useful and not just a collection of formulae and techniques needed to pass examinations.

The final chapter is concerned with computer simulation, and several analysis programs are given. It should, perhaps, be said that these programs are for instructional use and should not be used for the design of real plant ('not for navigation use', as it says on sail training charts). Use them, hack them and change them, but take care if real money and safety rely on them.

E.A. Parr
Minster on Sea
Kent

Preface

Chapter 1
Introduction to process control

1.1 Introduction

An industrial manufacturing plant is a complex social/economic/technical system that can be represented by Fig. 1.1. With an input of capital, materials, energy and human effort, products are produced which are sold to give profits for the investors and wages for the employees. It is in the interest of everyone concerned in the enterprise for the system to run as efficiently and economically as possible.

There are few, if any, processes that can be left to run themselves and most will require some form of supervision or control. Figure 1.2 is typical of most processes; operators give commands to some form of control equipment which translates these commands to action in the plant. The control equipment also conveys information from the plant for display to the operators in some convenient form. Although Fig. 1.2 is conceptually simple, it can be used as the basis for studying the most complex operation.

A process control engineer is responsible for the provision of the central block of Fig. 1.2 in such a way that the plant runs in the most

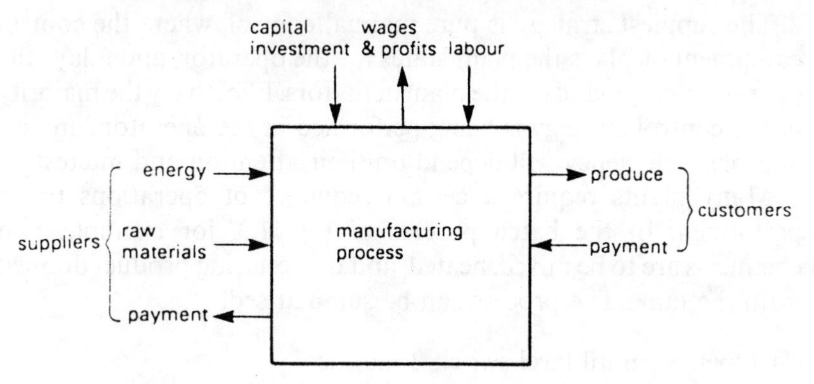

Fig. 1.1 Manufacturing as a social/economic system.

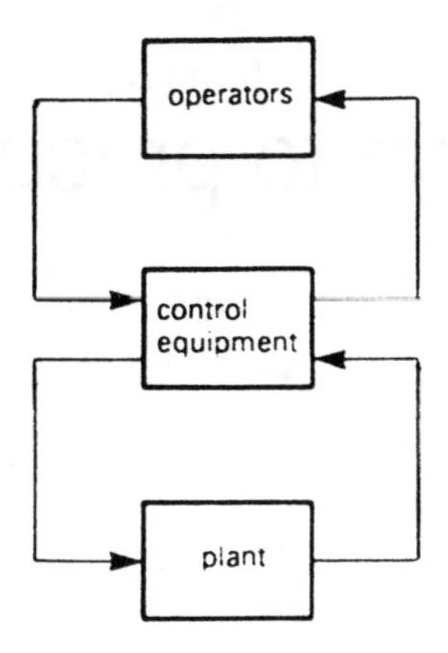

Fig. 1.2 Representation of a process control system.

economic manner consistent with safety. The latter consideration is, of course, of paramount importance, but economic operation is more subtle. The best theoretically obtainable control may require an impossibly large capital investment, or necessitate fast actuator movement which incurs premature wear and delays caused by plant failure. The good engineer should also be an economist and deal in acceptable, rather than perfect, control.

1.2. Control strategies

1.2.1. Types of control

It is possible to identify several forms that the control equipment of Fig. 1.2 can take. In reality, of course, this categorisation is an oversimplification of any plant system.

The simplest strategy is pure manual control, where the control equipment displays the plant states for the operator, and relays the operators' commands to the plant actuators. Effectively the majority of the control strategy is being performed in the operators' minds, and plant efficiency will depend on their attention and interest.

Many plants require a certain sequence of operations to be performed. In the batch process of Fig. 1.3, for example, two chemicals are to be mixed, heated, and the resulting product drained from the tank. The process can be summarised:

(1) Open V_1 until level switch 2 makes;
(2) Open V_2 until level switch 3 makes;

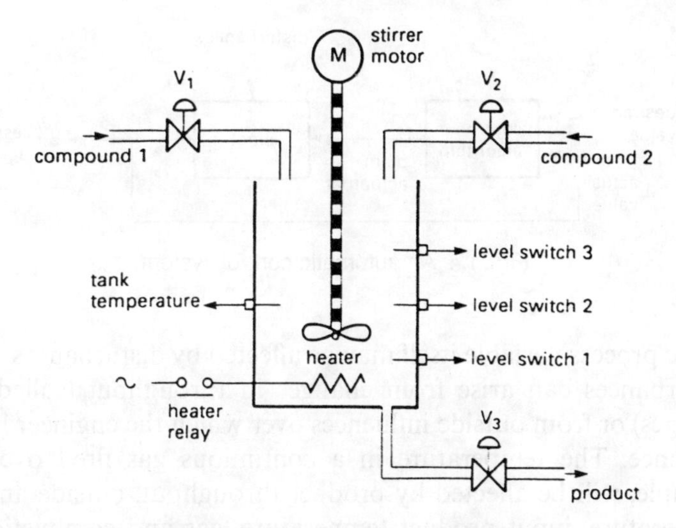

Fig. 1.3 Sequencing system.

(3) Mix for 90 secs;
(4) Heat until temperature reaches 75 °C. Wait until temperature cools to 25°C;
(5) Open V_3 until level switch 1 opens;
(6) Go back to step (1).

Not surprisingly, this type of control is known as sequencing, and is commonly performed by relays, digital logic or programmable controllers. They are designed by combinational or event driven logic techniques.

In many plants, process variables (e.g. temperature, pressure, flow, position, chemical analysis) are required to follow, or hold, some desired value. This control is achieved by manipulating plant actuators. Temperature control of a kiln, for example, could be achieved by manipulating the gas/air flow into a burner. It would be possible to control a process manually, but it is usually more efficient to use some form of automatic control.

The basis of an automatic control system is shown in Fig. 1.4. A control algorithm looks at the desired value, the actual value and possible outside influences affecting the plant, and on the basis of these observations adjusts the plant actuators to bring the process variable to the desired value.

The control algorithm has to cope with two circumstances. The desired value may be changing continuously (as in a position controlled telescope or an oven following a heating/cooling curve),

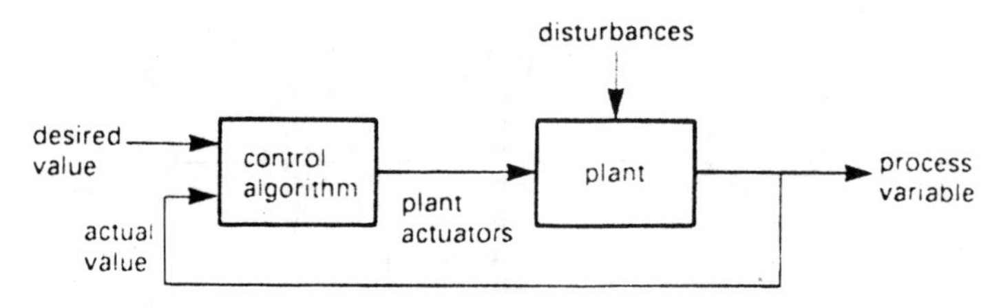

Fig. 1.4 An automatic control system.

or the process variable itself may be affected by disturbances. These disturbances can arise from changes in throughput (called load changes) or from outside influences over which the engineer has no influence. The temperature in a continuous gas fired oven, for example, will be affected by product throughput, outside ambient temperature, input product temperature, gas and combustion air pressure plus a multitude of other influences.

The majority of this book is concerned with the choice and design of suitable control strategies and the analysis of the subsequent behaviour of the plant.

1.2.2 Bang/bang servo

The simplest control strategy is the bang/bang servo shown controlling the temperature of an oven in Fig. 1.5a. The actual temperature is subtracted from the desired temperature to give an error signal. This error signal is passed to a comparator with hysteresis which controls a power control relay. If the temperature is too low, the relay will be energised and the heater turned on; if the temperature is too high the heater will be turned off.

The control system thus maintains the desired temperature by cycling power to the heater. The hysteresis in the comparator is necessary to prevent high speed chatter in the relay when the desired temperature is achieved.

Figure 1.5b shows what happens after a change in desired value. It can be seen that in the steady state the process variable oscillates about the desired value, with period and amplitude determined by the hysteresis and the characteristics of the process. Decreasing the hysteresis reduces the amplitude of the oscillations and decreases the period (i.e. the oscillations become faster) but in any practical system there is a limit below which the increasingly rapid actuator movement leads to excessive wear and early failure.

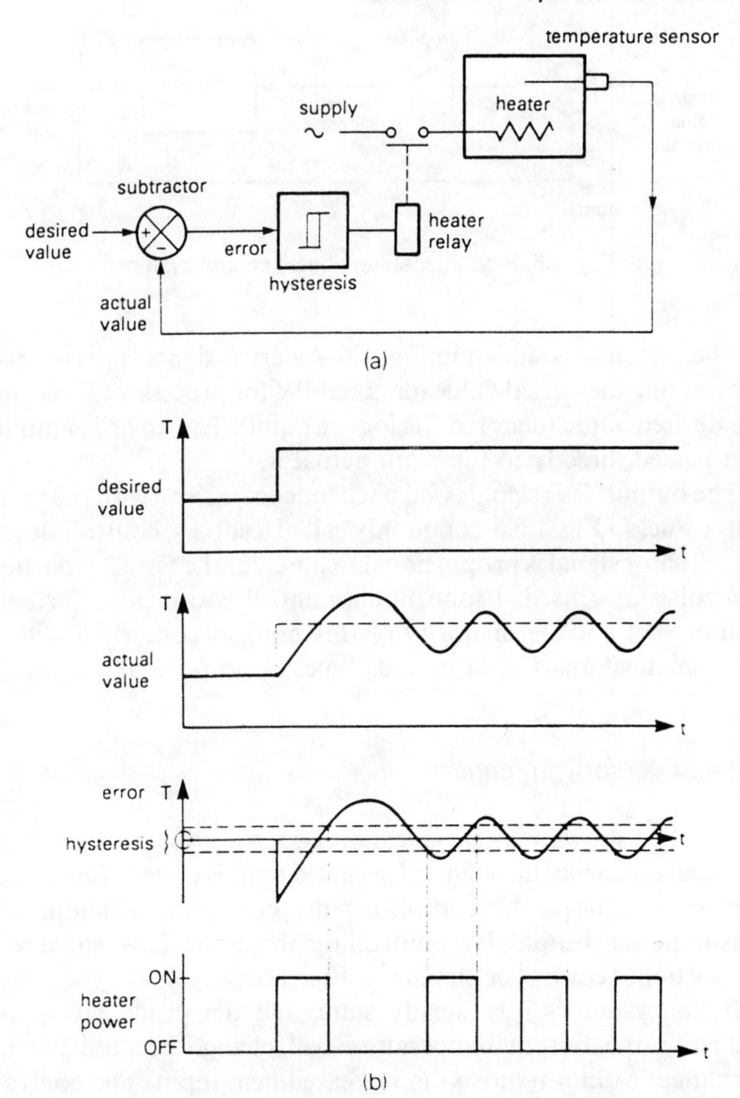

Fig. 1.5 A bang/bang servo system, (a) System block diagram; (b) System operation.

1.2.3. Proportional (feedback) control

In Fig. 1.5 the actuator can only be full on or full off, and in the steady state this leads to continuous oscillations. What is intuitively required is a large corrective action when the error is large and a small corrective action when the error is small. This necessitates proportional plant actuators which can give a controlled response over their full range.

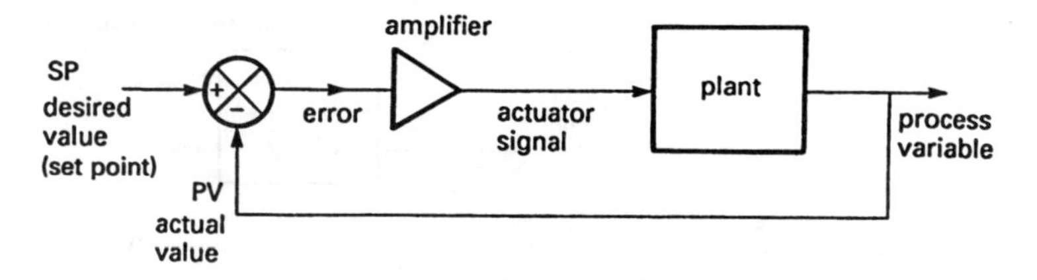

Fig. 1.6 Proportional (feedback) control system.

The principle is shown in Fig. 1.6. An error signal is produced by subtracting the actual value (denoted PV for process variable) from the desired value (denoted SP for set point). The error is amplified and passed directly to the plant actuator.

The output (PV) signal is fed back for comparison with the desired (SP) value, so Fig. 1.6 is commonly called feedback control. Because the actuator signal is proportional to the error the term proportional control is also used. Proportional control is discussed further in Section 1.3, and the majority of this book is concerned with the mathematical analysis of various types of proportional control.

1.2.4. Feedforward control

Figure 1.7a shows a fairly typical process, a product is being cooled by a heat exchanger through which chilled brine is passed. Temperature control is achieved by controlling the brine flow; a temperature sensor in the output leg controlling the brine flow valve via a proportional controller similar to Fig. 1.6.

If the system is in a steady state, and the product flow rises suddenly, the output temperature will change because the heat exchanger cannot remove the increased heat input. The controller will compensate and eventually return the output temperature to its correct level, but because proportional controllers can only react to an error after it has occurred, the temperature will rise transiently as Fig. 1.7b.

In Fig. 1.7c the control system has been modified by the inclusion of a product flow sensor, and the actuator signal now has two components. The first is from the temperature controller as before, and the second is the flow signal corresponding to product throughput. A change in product flow will now cause a change in brine flow before the output temperature is affected.

A plant model in the form of a simple scaling block is included so

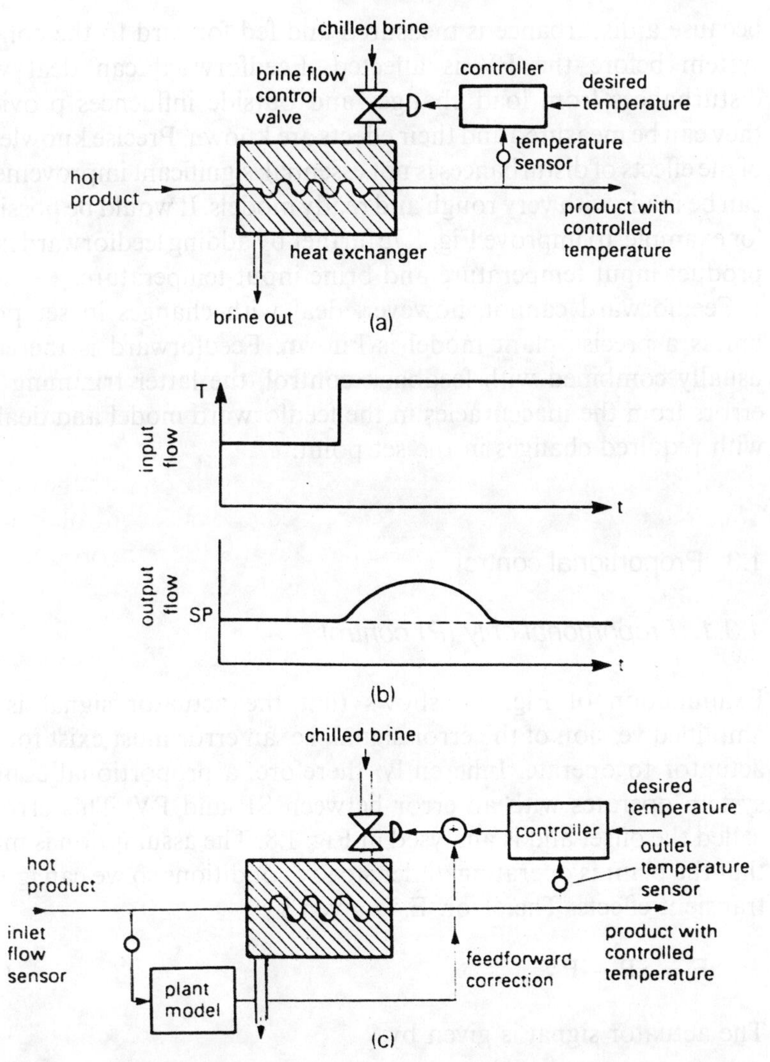

Fig. 1.7 Feedforward system, (a) Simple system with feedback control; (b) Effect of a change in input flow; (c) Improved performance with feedforward.

the relationship between product flow and brine flow is correct. If the scaling block has exactly the right value, the temperature controller will only have to deal with set point changes. In practice, a perfect match between scaling amplifier and plant is not possible (the brine control valve will probably have a non-linear flow/position characteristic, for example) and the output temperature will tend to change with flow. This disturbance will be dealt with by the temperature controller.

The arrangement of Fig. 1.7c is known as feedforward control because a disturbance is measured and fed forward to the control system before the PV is affected. Feedforward can deal with disturbances from load changes and outside influences provided they can be measured and their effects are known. Precise knowledge of the effects of disturbances is not essential; significant improvements can be made with very rough and ready models. It would be possible, for example, to improve Fig. 1.7c further by adding feedforward from product input temperature and brine input temperature.

Feedforward cannot, however, deal with changes in set point unless a precise plant model is known. Feedforward is therefore usually combined with feedback control, the latter trimming out errors from the inaccuracies in the feedforward model and dealing with required changes in the set point.

1.3. Proportional control

1.3.1. Proportional only (P) control

Examination of Fig. 1.6 shows that the actuator signal is an amplified version of the error signal, i.e. an error must exist for the actuator to operate. Inherently, therefore, a proportional control system operates with an error between SP and PV. This error is called the offset and is analysed in Fig. 1.8. The assumption is made that the plant is operating under steady conditions so we can ignore transient effects. The error, E, is given by:

$$E = SP - PV \tag{1.1}$$

The actuator signal is given by:

$$A = K(SP - PV) \tag{1.2}$$

where K is an adjustable gain.

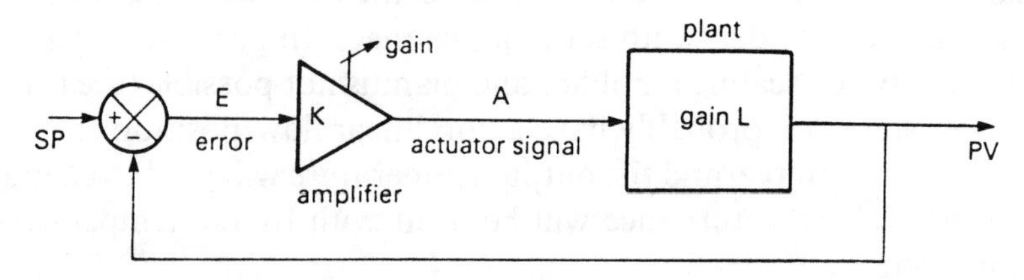

Fig. 1.8 Steady state representation of a feedback system.

Assuming the process is linear (a topic discussed further in Section 1.4) the plant output signal is:

$$PV = L.A \tag{1.3}$$

$$PV = L.K.(SP - PV) \tag{1.4}$$

$$PV(1 + L.K) = L.K.SP \tag{1.5}$$

or
$$PV = \frac{L.K.SP}{(1 + L.K)} \tag{1.6}$$

The product L.K is often called the open loop gain, as it is the gain that would be seen if the loop were broken at any point and a signal injected. If we denote the open loop gain by G (i.e. G = L.K) then:

$$PV = \frac{G.SP}{1 + G} \tag{1.7}$$

and the error is found by substitution into equation 1.1

$$E = \frac{SP}{1 + G} \tag{1.8}$$

It can be seen that for large values of G, the value of PV approaches SP, and the offset error, given by equation 1.8, approaches zero. If G is 20, for example, the value of PV is 0.95 SP and the error is about 4·7%.

The implication of equations 1.7 and 1.8 is that the open loop gain G should be as large as possible. Usually the plant gain is fixed so, to reduce offset error, K should be made as high as possible. Unfortunately open loop gain cannot be increased indefinitely without the system becoming unstable, a topic discussed further in Section 1.5.

Equations 1.1 to 1.8 were derived for a change in set point. A similar set of equations could be developed for disturbances from load changes or outside influences. Proportional only control will again compensate partially for disturbances, but an offset error will occur, the size of which decreases for increasing open loop gain.

The amplifier gain of a proportional controller is often referred to as the 'proportional band' or PB which is expressed in percent. It is an indication of the range over which the controller operates, and can be seen by reference to Fig. 1.9. Here an amplifier of gain K is used to amplify an error signal. Assume both input and output voltage range is the same, ± 15 V say.

The input signal range that causes the output to saturate is

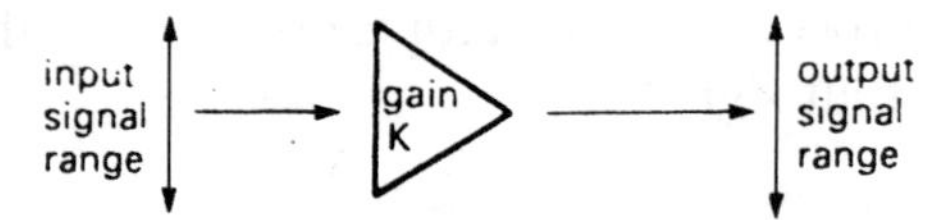

Fig. 1.9 Relationship between gain and proportional band.

therefore ±15/K V, signals outside this range cause no further change in output. The amplifier thus has an input proportional band of ±15/K V or 100/K expressed as a percentage of full input range. If input and output have the same range, the PB is 100/K%; a gain of 5 corresponds to a PB of 20%, for example.

If the input and output swings are unequal (or of different quantities, as could occur, for example, on a voltage to pneumatic pressure amplifier) the PB is the percentage of input range over which the output does not saturate.

1.3.2. Proportional plus integral (P + I) control

Figure 1.10 shows an, albeit manual, way in which the offset caused by set point changes can be overcome. The operator has an additional control which, historically, is labelled 'reset'. After a change of set point, the operator waits for transient effects to die away, then slowly adjusts the reset control until the observed error is zero. The offset error from the change in set point has been removed, but offsets from disturbances will still be present.

The circuit of Fig. 1.10 is called proportional with manual reset control. An obvious improvement would be the addition of automatic reset. This is provided by the circuit of Fig. 1.11. The signal to the control amplifier has two terms, a straight proportional signal as before, and a signal which is a scaled time integral of the error. The arrangement is called, for obvious reasons, a P + I controller.

Suppose a steady offset error is present. The integral signal will ramp, causing the actuator signal to change and reduce the error. When the error is zero, the integral signal will be steady. Intuitively, therefore, the effect of the integrator will, in the steady state, bring the error to zero.

Similar considerations will show that offsets caused by disturbances will also be reduced over a period of time. Integral action acts to reduce (SP − PV) to zero, but the time taken to remove the error depends on the settings of the gain K and M. It may be thought odd that the proportional and integral terms are added before the gain

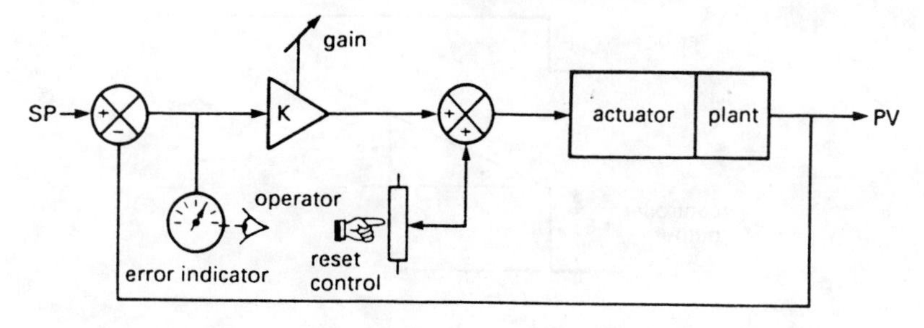

Fig. 1.10 Removal of offset with operator adjusted reset control.

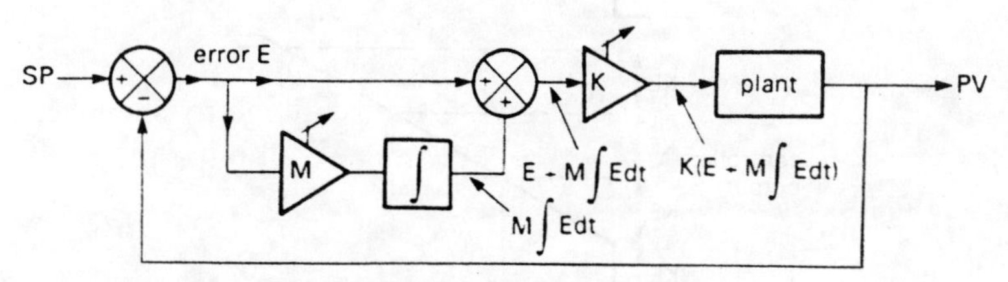

Fig. 1.11 Removal of offset with integral action.

setting amplifier, but as will be seen in later circuits this arrangement simplifies mathematical analysis.

The action of a P + I controller is summarised in Fig. 1.12. This shows the result of a step error of E%. In a closed loop system, of course, the error would reduce, but a step response is easy to analyse. The initial response of the controller will be a step of height K.E% from the proportional term. The integral term will then cause the controller output to ramp up with a slope of KME% per unit time.

The 'gain' of the integral term has the units of inverse time, and can be specified in two ways. The first notes that for a step error input the integral action will repeat the proportional step KE at fixed time intervals. This is termed the 'integral time', usually denoted by T_i. It can be seen that $M = 1/T_i$, so the controller output is given by:

$$V_o = K\left(E + \frac{1}{T_i}\int E\,dt\right) \tag{1.9}$$

T_i has the units of time, and can be given in seconds or minutes.

The second way of specifying integral action specifies how often the initial proportional step is repeated by the integral action per unit time (usually minutes are used). An integral time of 20 seconds corresponds to 3 repeats per minute as shown in Fig. 1.12b. It can be seen that increasing the repeats per minute or decreasing the integral time will speed up the integral contribution.

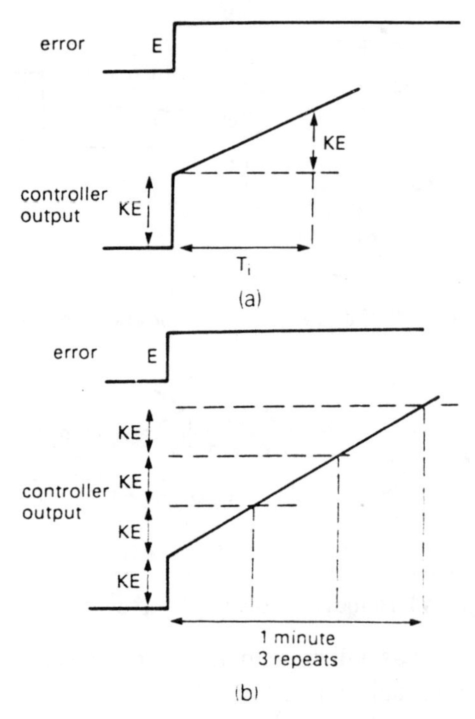

Fig. 1.12 Definition of controller integral action, (a) Definition in terms of integral time T_i; (b) Definition in terms of repeats/minute.

For the fastest response to a set point change or a disturbance a short integral time is required. Decreasing the integral time, however, tends to destabilise the system.

1.3.3. Derivative (rate) action

If we consider how an operator manually controls a plant, it is possible to identify an additional factor that can be usefully added to a proportional controller. In Fig. 1.13a the voltage from a steam driven generator is being controlled manually. The operator observes the voltage, and adjusts the steam valve to keep the voltage steady.

Suppose a sudden load is applied to the system. This will cause the generator speed and voltage to fall. The inertia of the system will, however, stop, the voltage falling instantly, and if left uncorrected the voltage would fall over a period of time as Fig. 1.13b. If the operator applied proportional and integral action alone, the voltage would be brought back to the correct level eventually, but the inertia has

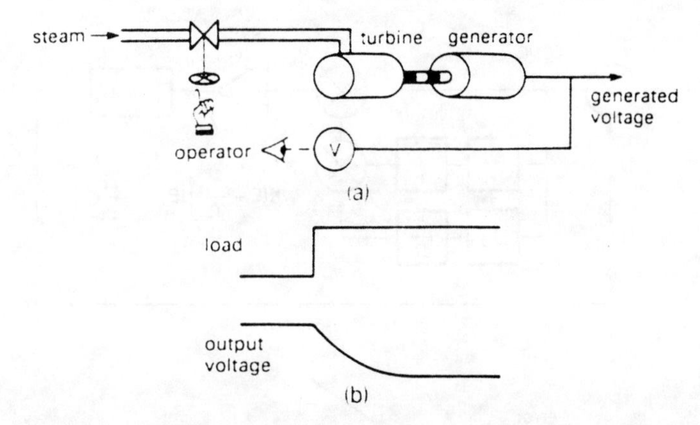

Fig. 1.13 System requiring operator action based on rate of change of error, (a) Manually controlled generation system; (b) Effect of load change.

extended the duration of the droop as both P and I actions are dependent on the magnitude of the error.

The operator, however, can anticipate what action is needed by looking at how *fast* the error is changing, and adding more corrective actions when the error is changing rapidly. If the error is increasing, the actuator signal is increased beyond that supplied by the P and I terms. This anticipates, to some extent, the control action required and compensates for the inertia induced time lag.

When the error is reducing, a similar anticipation will reduce the actuator signal such that the error comes to zero as fast as possible without overshoot.

The operator has anticipated the error signal and compensated to some extent for plant inertia, by adding corrective action based on the rate of change of error. This is known as derivative, or rate, action and can be achieved with the block diagram of Fig. 1.14a which combines proportional, integral and rate action. The actuator signal is given by

$$V_o = K\left(E + \frac{1}{T_i}\int E\,dt + T_d\frac{dE}{dt}\right) \tag{1.10}$$

where K and T_i were defined earlier for equation 1.9 and T_d is a scaling factor for the rate of change signal. This scaling factor has the dimensions of time, and is usually given in minutes or seconds according to the speed of the process. It is known as the derivative time or the rate time.

The derivative time can be visualised as Fig. 1.14b, where a ramp

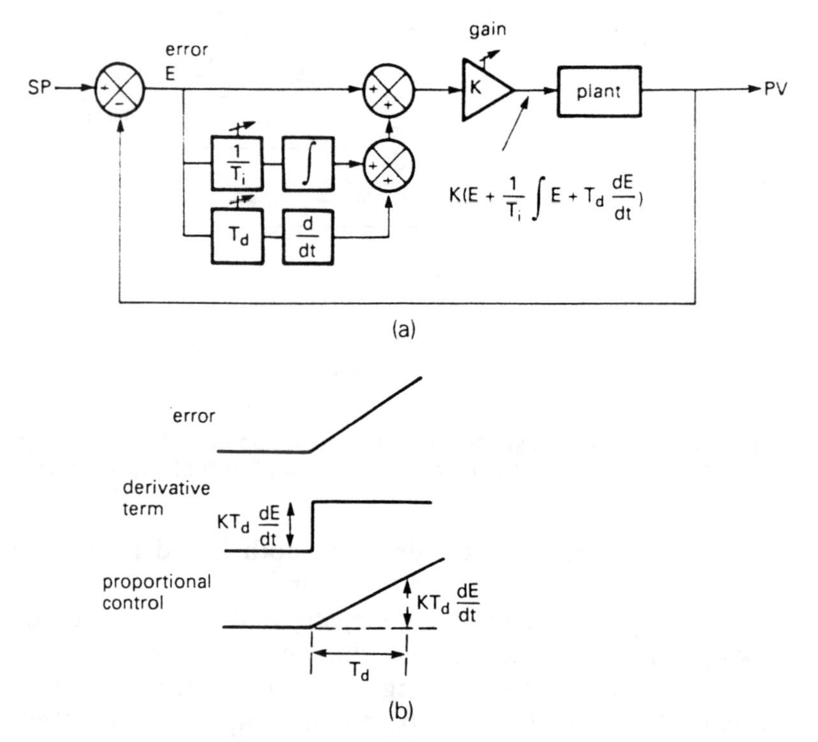

Fig. 1.14 The addition of derivative action, (a) Block diagram of three term controller; (b) Definition of derivative (rate) time T_d.

error signal is applied to the controller. The derivative term will contribute a step to the actuator signal, and the proportional term a ramp, as shown. The derivative time T_d is the time taken for the proportional signal to equal the derivative signal.

It can be seen that the longer the value to T_d the more contribution the rate of change of error will make to the output actuator signal.

Derivative action can cause problems in some circumstances. Consider the effect of applying a step change of set point to Fig. 1.14a. This will be seen as an infinitely fast rate of change of error and result in a large, fast, and possibly damaging, actuator signal. For this reason, many controllers implement rate action as Fig. 1.15, where a rate of change of PV (as opposed to error) signal is used. With a steady set point this behaves exactly as Fig. 1.14a, but does not 'kick' the actuator on a change of set point. If the controller is, however, required to follow a continually changing set point (as in ratio or cascade control described later) the form of Fig. 1.14a is preferred.

Derivative action can also be problematical where the measurement

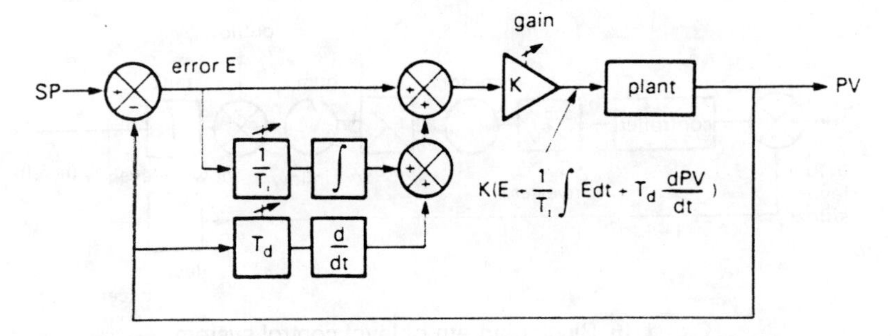

Fig. 1.15 Derivative action based on rate of change of PV rather than error.

of the process variable is inherently noisy. A typical example is level control where ripples and liquid resonance tend to give a continually fluctuating level signal even in the steady state. If derivative action is used, these rapid fluctuations will be seen as rapid changes of error, causing rapid unnecessary actuator movements and premature wear.

1.3.4. Three term (P+I+D) control

Equation 1.10 describes an actuator signal which has three components:

(1) proportional to error;
(2) proportional to the time integral of error;
(3) proportional to the rate of change of error.

The strategy of equation 1.10 is therefore called three term or P + I + D control, and is widely used in process control.

There are three adjustable constants in equations 1.10, the gain K, the integral time T_i, and the derivative time T_d. The setting of these is crucial to the performance of the controlled plant. The theory and practice of tuning a controller is discussed in later sections.

1.4. Plant modelling

1.4.1. Introduction

The first stage in designing a suitable control strategy is the determination of the plant characteristics. Usually these can be considered fixed and not subject to modification by the control engineer, although, of course, the replacement of any item which has

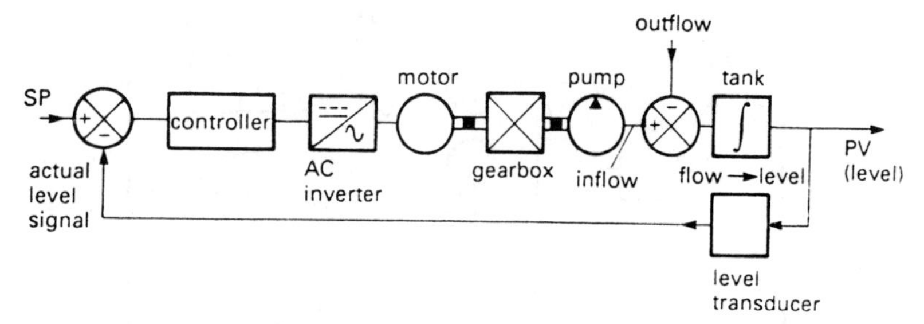

Fig. 1.16 Block diagram of level control system.

a characteristic that is detrimental to the plant performance should be considered.

Most plants can be considered as a series or parallel connection of blocks whose characteristics can be described by mathematical relationships. Figure 1.16 is a block diagram of a level control system which has to compensate for changes in outflow by controlling the speed of a pump delivering make up liquid to the tank.

The blocks need defining for both static and dynamic conditions. Usually the static, or steady state conditions, are easy to describe. The AC inverter in Fig. 1.16, for example, could go from 20 to 100 Hz for an input voltage of 0–10 V, and cause the motor speed to vary from 600 to 3000 rpm. Similar relationships can be derived for the gearbox ratio, between pump outflow and input shaft speed.

The 'steady state' relationship between tank level and flows is not quite so simple. The outflow and inflow are subtracted to give a net flow which can be into, or out of, the tank. This net flow causes the level to change at a steady rate given by

$$\text{rate of change of level} = \frac{(\text{inflow} - \text{outflow})}{A}\ \text{m/sec} \qquad (1.11)$$

where A is the tank cross-sectional area, and both inflow and outflow are given in cubic metres per second. If differing units are used (e.g. flow in litres/minute and level in metres) a scaling factor will also be needed.

A static definition for the level transducer can easily be obtained from the manufacturers' data sheets; an output signal of 0–10 V for a level range of 5–15 m, say.

The examples above show that static relationships do not necessarily follow the form $Y = AX$. With the exception of items such as the tank or non-linear elements (described in Section 1.5), most can be defined by $Y = AX + B$ where A is the sensitivity (1 volt per metre for the level transducer) and B an offset (an input offset of 5 metres for the level transducer).

It will be noted that the steady state open loop gain is the product of the steady state gain of the individual elements, and is dimensionless and independent of the units chosen to define the individual blocks. For Fig. 1.16 the open loop gain is:

$$\underset{\text{Controller}}{\frac{\text{volts}}{\text{volts}}} \times \underset{\text{Inverter}}{\frac{\text{Hz}}{\text{volts}}} \times \underset{\text{Motor}}{\frac{\text{rpm}}{\text{Hz}}} \times \underset{\text{Gearbox}}{\frac{\text{rpm}}{\text{rpm}}} \times \underset{\text{Pump}}{\frac{\text{flow}}{\text{rpm}}} \times \underset{\text{Tank}}{\frac{\text{level}}{\text{flow}}} \times \underset{\text{Transducer}}{\frac{\text{volts}}{\text{level}}}$$

Note that if the flow was expressed in gal/min rather than l/min. terms 5 and 6 would both change and the loop gain would remain unchanged.

To analyse how the plant responds to changing loads, disturbances or set points, it is also necessary to define how the outputs of the blocks respond to changing inputs. This is rarely as simple as defining steady state relationships.

Blocks can be considered as being linear or non-linear elements. A linear element is one which, when subjected to a sine wave input, produces a sine wave output of the same frequency, albeit of different amplitude and possibly shifted in phase as Fig. 1.17a. A non-linear element produces a non-sinusoidal output when driven with sine wave input. Figure 1.17b is therefore a non-linear element. The remainder of this section considers linear elements, non-linear elements being described in Section 1.5.

1.4.2. Simple gain

The output of some elements can be considered to follow input changes exactly. A well-engineered gearbox is a typical example. Such elements can be defined as having simple gain which can be greater, or less, than unity.

1.4.3. First order lag

The output of a first order lag follows an exponential response to a step change in input. Figure 1.18 represents a simple oven whose temperature is controlled by an electric heater. Heat loss through the walls is proportional to the temperature difference between the oven interior and the ambient temperature. Heat input is determined by the power fed to the heater element.

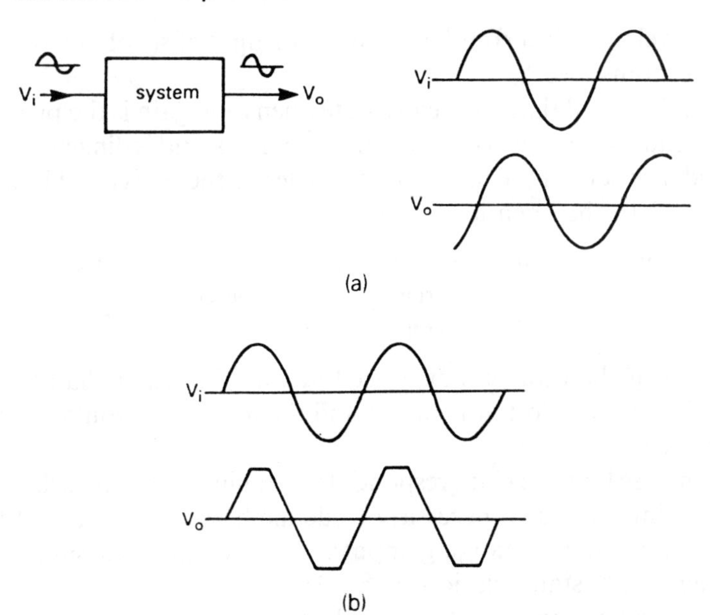

Fig. 1.17 Linear and non-linear systems, (a) A linear system driven by a sine wave produces a sine wave output of the same frequency (with possibly different amplitude and shifted in phase); (b) A non-linear system does not produce a sine wave output for a sine wave input.

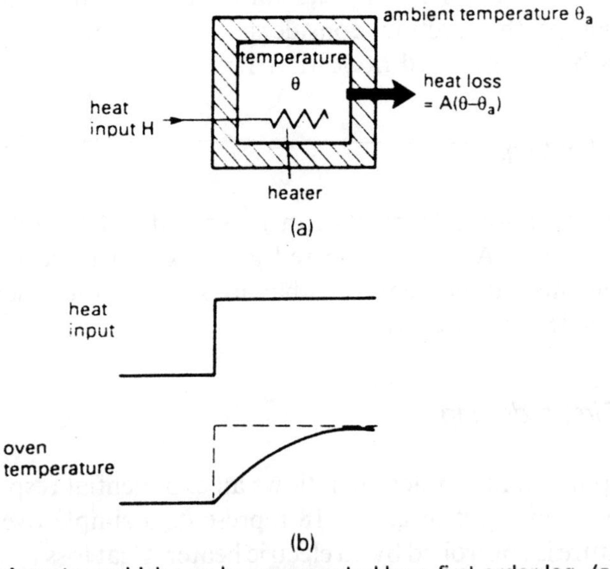

Fig. 1.18 A system which can be represented by a first order lag, (a) Block diagram of oven; (b) System response to step change in heat input.

$$\text{heat loss} = A(\theta - \theta_a) \tag{1.12}$$

where A is a constant, and θ_a is the ambient temperature

$$\text{heat input} = H \tag{1.13}$$

The oven temperature will stabilise when these are equal, i.e.

$$\theta = \frac{H}{A} + \theta_a \tag{1.14}$$

These equations describe the steady state conditions.

The rate of change of temperature is proportional to the net heat input into the oven, i.e. the difference between the heat input and heat loss.

$$\frac{d\theta}{dt} = B(H - A(\theta - \theta_a)) \tag{1.15}$$

where B is a constant.

$$\frac{d\theta}{dt} + B.A.\theta = B.H + B.A.\theta_a \tag{1.16}$$

In the steady state $d\theta/dt$ is zero and this reduces to equation 1.14.

The solution of equation 1.16 is of the form:

$$\theta = \theta_s(1 - e^{-t/T}) \tag{1.17}$$

where θ_s is the steady state temperature ($H/A + \theta_a$) and T, called the time constant, is 1/A.B.

Equation 1.17 predicts an exponential rise for a step change of heat input as Fig. 1.18b. This time constant determines the rate at which the temperature rises; reaching approximately 63% of the final value in one time constant. The rise can be tabulated:

Time (in terms of T)	% of final value
T	63
2T	86
3T	95
4T	98
5T	99

For most practical purposes, the final value is reached in about five time constants.

First order lags are characteristic of some forms of storage or capacity, and as such are common in elements involving temperature or inertia. Most temperature transducers, for example, behave as

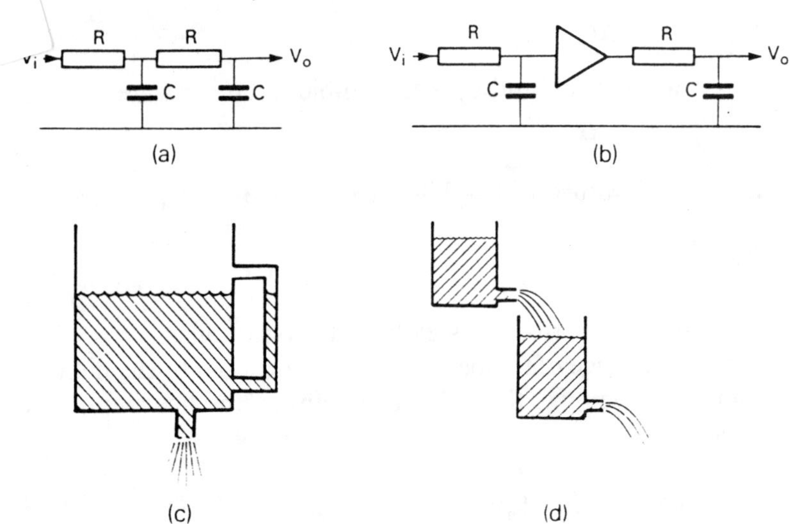

Fig. 1.19 Interacting (coupled) and non-interacting lags, (a) Two interacting (coupled) lags; (b) Two non-interacting lags; (c) An example of interacting lags, a tank level sight glass; (d) An example of non-interacting lags, two independent tanks.

first order lags with time constants ranging from 0.1 s to around 30 s dependent on size.

Series connected lags can be interacting, as Fig. 1.19a, or effectively non-interacting as Fig. 1.19b. For small deviations, the level in a tank with a free flowing outlet can be considered to respond to input flow as a first order lag (the true response is non-linear, but for small deviations the assumption is reasonable). The sight glass of Fig. 1.19c is an example of interacting lags, whereas the separate tanks of Fig. 1.19d form a pair of non-interacting lags.

1.4.4. Second order systems

Many mechanical systems can be considered to behave as Fig. 1.20a. This arrangement has an input displacement applied to one end of the spring with an output displacement taken from the load. Consider what happens as a sinusoidal displacement is applied at various frequencies.

At low frequencies the system will have unity gain with the output following the input. At high frequencies the system will exhibit low gain with the output hardly moving. There may, however, be some intermediate frequency where the system will exhibit resonance, and

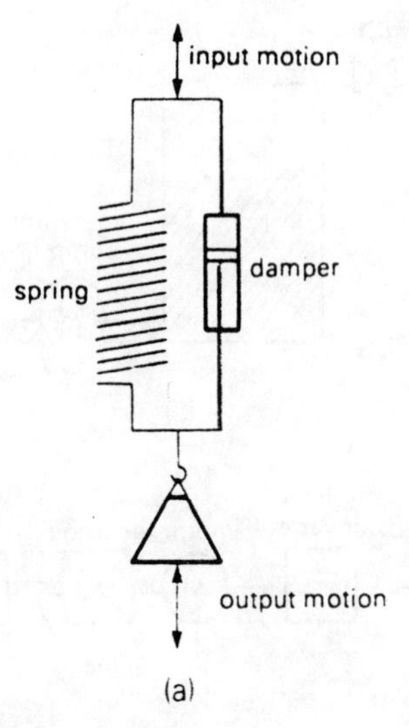

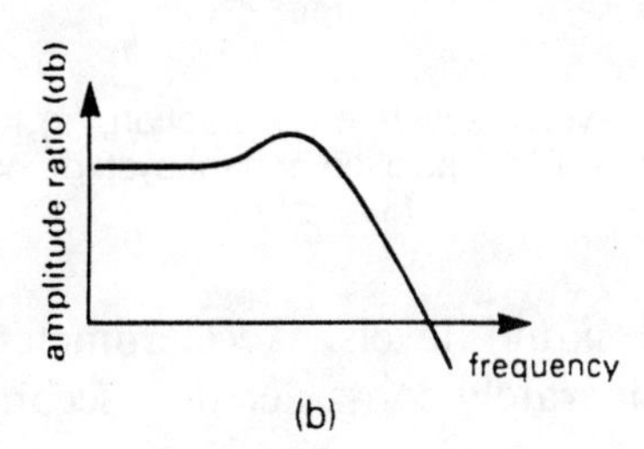

Fig. 1.20 Second order systems, (a) Example of a second order system; (b) Frequency response of second order system.

the output amplitude will be greater than the input amplitude. The system is said to exhibit resonance and has a gain/frequency response as Fig. 1.20b. The range of frequencies over which the gain is greater than unity is determined by the spring constant and the characteristics of the viscous damper.

Blocks behaving as Fig. 1.20 are said to be second order systems, and as will be seen later, can be represented by a second order differential equation of the form

$$\frac{d^2x}{dt^2} + 2b\omega_n\frac{dx}{dt} + \omega_n^2 x = \omega_n^2 x_i \tag{1.18}$$

where ω_n is called the natural frequency and b the damping factor.

Resonance is found in mechanical structures (e.g. position controls),

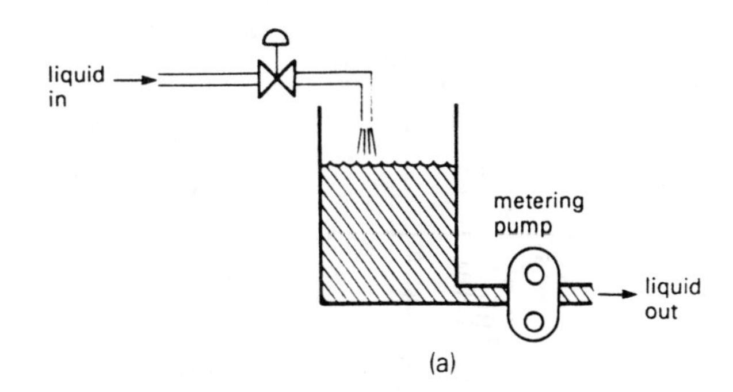

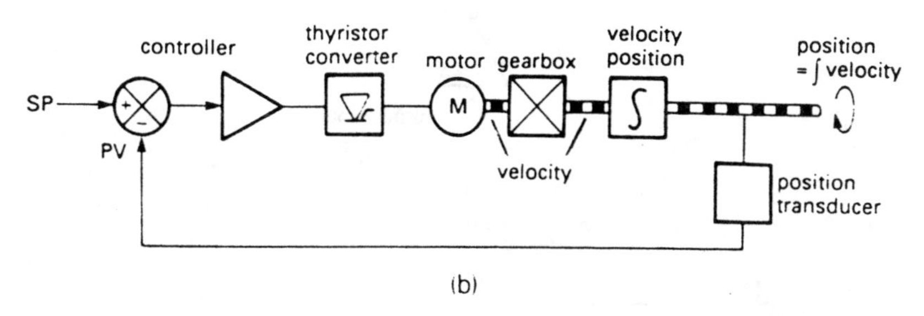

Fig. 1.21 Examples of systems with integral action, (a) In a tank, level is the integral of net flow/area; (b) In a position control system, position is the integral of velocity.

gas pressures and liquid levels. Accelerometers and vibration transducers are deliberately constructed as second order devices.

1.4.5. Integral action

In Fig. 1.21a the outflow from a tank is determined by a metering pump and is independent of tank level. The inflow is controlled by a level control system. Because the outflow is independent of level, any mismatch between inflow and outflow (however small) will cause the level to increase or decrease indefinitely. If V_i is the volumetric inflow and V_o the volumetric outflow, and h the level in the tank, the rate of change of level will depend on the tank cross-sectional area A and the difference between inflow and outflow, i.e.

$$\frac{dh}{dt} = \frac{1}{A}(V_i - V_o) \tag{1.19}$$

which gives us

$$h = \frac{1}{A}\int(V_i - V_o)dt + B \tag{1.20}$$

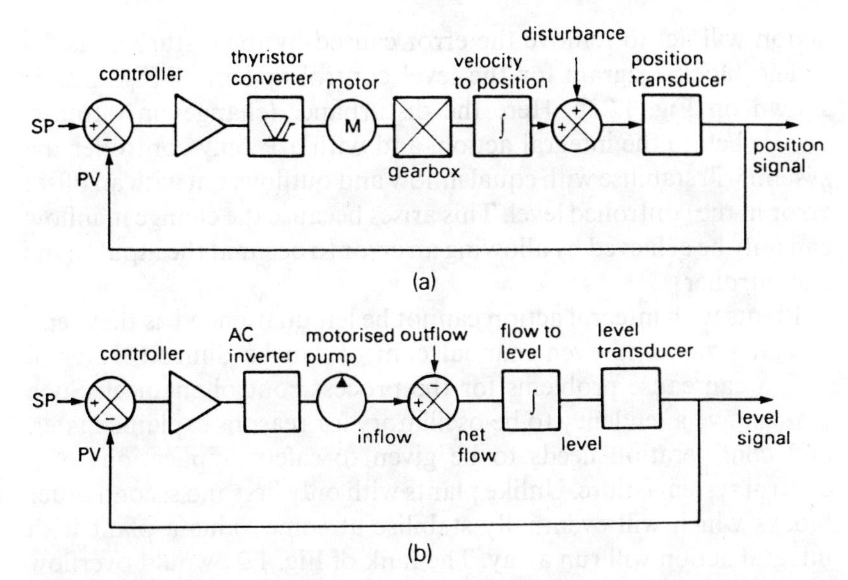

Fig. 1.22 The effect of integral action on disturbances, (a) The system will remove errors which occur from disturbances arising after the integral action; (b) Disturbances occurring before the integral action cause an offset.

where B is an initialisation constant.

Equation 1.20 shows that the level varies as the time integral of the difference between inflow and outflow, any fixed difference leading to a steady change of level. As such, it can be said that the tank can be represented as an integral action block.

Many items can be represented by integral action. Figure 1.21b is a position control system which could be a steerable radio telescope, a steel rolling mill roll gap control or a gyro stabilised gun on a warship. The position is the integral of the velocity of the gearbox output shaft. Similarly the flow through a motorised valve is the integral of the motor shaft velocity. (Motorised valves are commonly used where it is desirable for a valve to hold the last position in a fault condition.)

As integral action in a controller removes offset error, it is reasonable to expect a plant with integral action to operate with zero error when P only controllers are employed. This assumption is true if the disturbances occur after the integral action.

Figure 1.22a represents a position control system where the output position is disturbed by outside influences. It could, for example, be a radio telescope affected by side wind loading. Examination of the block diagram will show that the plant integral action will act to remove the error caused by the disturbances.

The block diagram for the level control system of Fig. 1.21 is shown on Fig. 1.22b. Here the disturbance (changes in outflow) occurs before the integral action, and with a P only controller the system will stabilise with equal inflow and outflow, but with an offset error in the controlled level. This arises because the change in inflow can only be achieved by allowing an error to occur at the input to the P controller.

Plants with integral action cannot be left unattended as they tend to run away, and even manual control can be difficult. Integral action can cause problems for the process control engineer. Such plants have a tendency to be oscillatory for reasons explained later, and consideration needs to be given to safety implications of a control system failure. Unlike plants with only first and second order blocks which will eventually stabilise at some value a plant with integral action will run away. The tank of Fig. 1.21 would overflow or drain, for example, if the controller failed or the flow control valve jammed.

1.4.6. Transit delay

Figure 1.23a is a system feeding bulk powder to a process. The powder is conveyed at a constant speed on a conveyor belt driven at fixed speed by an AC motor. The powder feed rate is determined by the speed of a vibrating feeder, and controlled by a weigher which weighs a short length of belt a distance d metres downstream of the feeder.

If the belt velocity is V metres per second, it will take d/V secs for any change in feed rate to be seen at the weigher. This time is called a transit delay, or dead time, and obviously limits how fast controlled changes can be made to the system.

The output from a transit delay is identical to the input but delayed by a fixed time as Fig. 1.23b. A transit delay is a linear element as a sine wave input will give a sine wave output of the same frequency but shifted in phase.

Transit delays are a function of time and distance. Common elements exhibiting this property are liquid flow controllers where the control valve and primary flow sensor are separated, systems with pneumatic control, analytical sampling in chemical process control where long piping runs exist between the plant and the analyser, and computer based systems which effectively introduce a

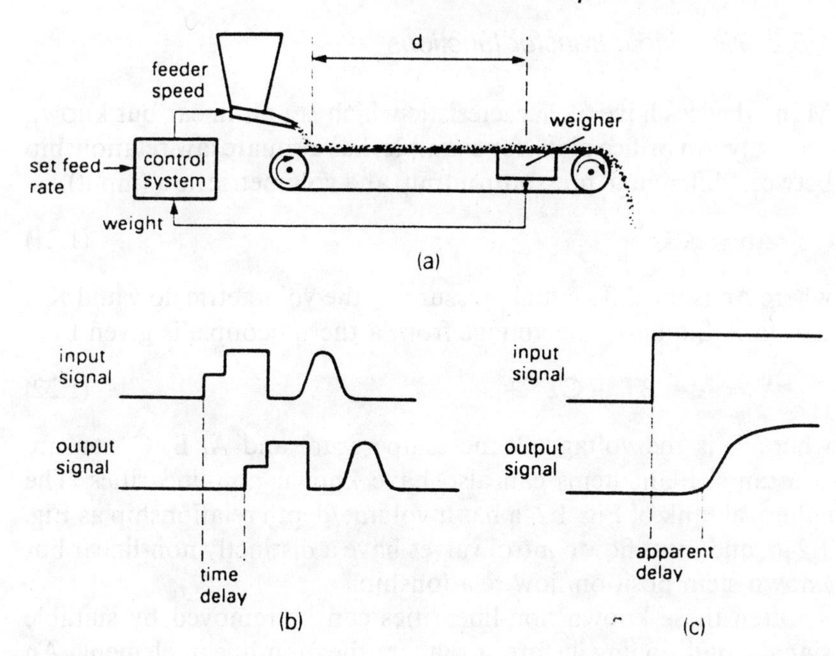

Fig. 1.23 The transit delay, (a) A feed system with a transit delay; (b) The effect of a transit delay; (c) Cascaded first order lags approximating to a transit delay.

transit delay by observing the process intermittently at a predetermined scan rate.

Multiple first order lags have a step response similar to Fig. 1.23c which approximates to a transit delay in series with a single first order lag.

Where a plant exhibits a transit delay it is frequently this element that determines the best performance and stability that can be achieved. A plant with both integral action and transit delay can be quite difficult to control.

1.5. Non-linear elements

1.5.1. Introduction

Section 1.4 describes linear blocks which can be used to describe the behaviour of parts of a plant. In practice, however, most plants contain at least one, and usually several, non-linear elements which cannot be easily described by mathematical models. This section describes the more common non-linearities. Methods of dealing analytically with non-linearities are dealt with in Section 5.7.

1.5.2. Non-linear transfer functions

Many devices have a characteristic which is non-linear but known precisely. An orifice plate, for example, has a square law relationship between differential pressure (output) and volumetric flow (input), i.e.

$$\Delta p = K.Q^2 \tag{1.21}$$

where Δp is the differential pressure, Q the volumetric flow and K a constant. Similarly the voltage from a thermocouple is given by

$$V = A + BT + CT^2 + \ldots \tag{1.22}$$

where V is the voltage, T the temperature and A, B, C, etc. are constants. Plant items can also have known non-linearities. The spherical tank of Fig. 1.24a has a volume/depth relationship as Fig. 1.24b, and most flow control valves have a distinctly non-linear but known stem position/flow relationship.

Often these known non-linearities can be removed by suitable signal conditioning before, or after, the non-linear element. An orifice plate, for example, can give a linear flow signal if it is followed by a square root circuit.

An alternative solution is to allow the plant only to operate over a limited region where it is reasonably linear. Examination of Fig. 1.24 shows that the tank volume/level relationship is reasonably linear about $\pm 20\%$ of the mid-position.

1.5.3. Saturation

To some extent every block is non-linear because each will eventually reach some physical limit which prevents its output increasing or decreasing indefinitely. The output of a controller, say, is limited to

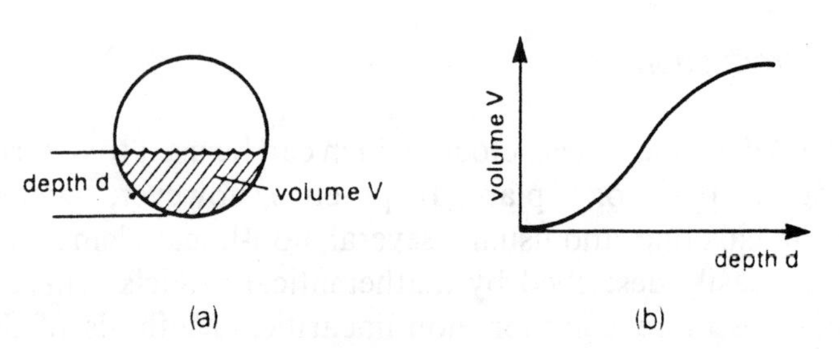

Fig. 1.24 A non-linear system, (a) Spherical tank; (b) Volume/depth relationship.

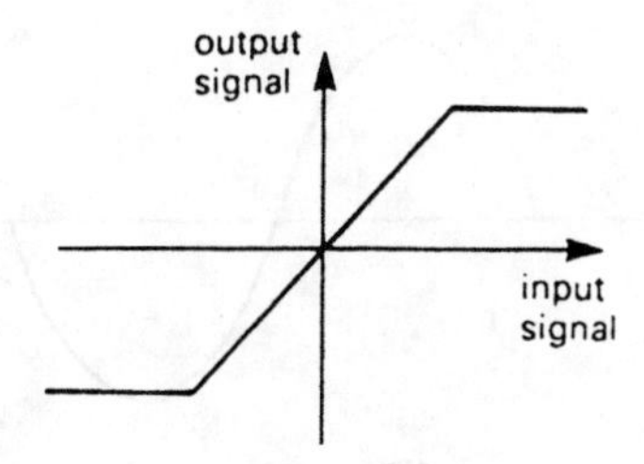

Fig. 1.25 Saturation.

the range 0–10 V, a tank will overflow or drain dry, a flow valve is limited to 5 cm stem travel and so on.

Saturation can be represented by Fig. 1.25. It is usually convenient to deal with saturation of an element by two blocks; one describing the 'pure' element, followed by a unity gain saturation block. This shows that saturation has no effect within the saturation limits.

The effect of saturation is to effectively reduce the gain at high amplitudes, slowing the plant response to disturbances. As the control system has lost control of the plant, saturation can lead to integral windup, a problem discussed later in Section 4.2.5.

1.5.4. Velocity limiting

The speed of a DC motor must be limited to prevent centrifugal forces damaging the motor and couplings. Such a velocity limited motor can follow slow positional changes requiring speeds below the limit, but will lag when called to perform high speed changes as Fig. 1.26.

Similar velocity limitations occur in pneumatic and hydraulic actuators which can accurately follow low speed signals but lag high speed signals.

A related non-linearity is acceleration limiting, which is commonly imposed to reduce strains on gearboxes and couplings.

1.5.5. Hysteresis (or backlash)

Poorly machined gears as Fig. 1.27a or loose fitting linkages as Fig. 1.27b exhibit backlash. This manifests itself as a differing input/output relationship according to the direction of movement of the input shaft. Stiction and friction cause a similar effect. This can be

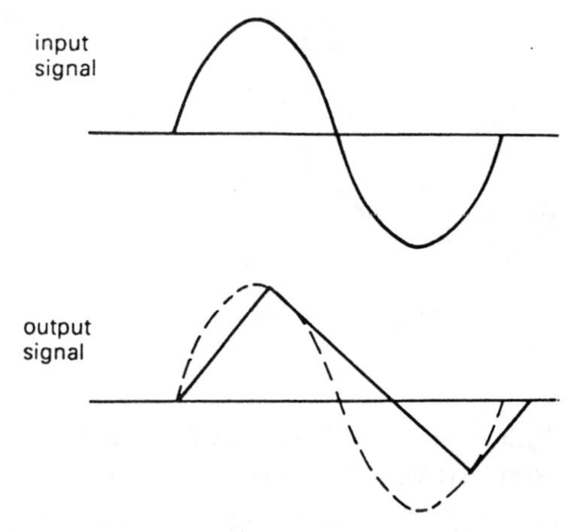

Fig. 1.26 The effect of velocity limiting.

visualised as Fig. 1.27c, and is known in control engineering as hysteresis or dead band.

Hysteresis can be a source of problems and often manifests itself as a 'dither' about the set point as the control system hunts in the dead band. It can also be self-reinforcing, as dither in either Fig. 1.27a or b will lead to more wear and more backlash.

Hysteresis can be designed out by careful manufacturing techniques such as spring loaded gears and pretensioned linkages. In systems which do not exhibit overshoot, a unidirectional approach can be employed as Fig. 1.27d. Movements to A, B, C are all the same direction. The movement to D is a reversal, so a deliberate overshoot is introduced by driving briefly to X, then approaching D in the same direction as the A, B, C movements.

1.5.6. Dead zone

Dead zone has the response of Fig. 1.28, this should not be confused with dead band of Section 1.5.5. Dead zone is often deliberately introduced into a controller characteristic after the error subtraction to prevent response to small errors. A typical application is a level control system in a surge tank where the level is allowed to vary within limits without corrective action being taken, and it is undesirable for the control system to attempt to correct for ripples and hydraulic resonance.

Dead zone can also occur with two mode control as Fig. 1.28b.

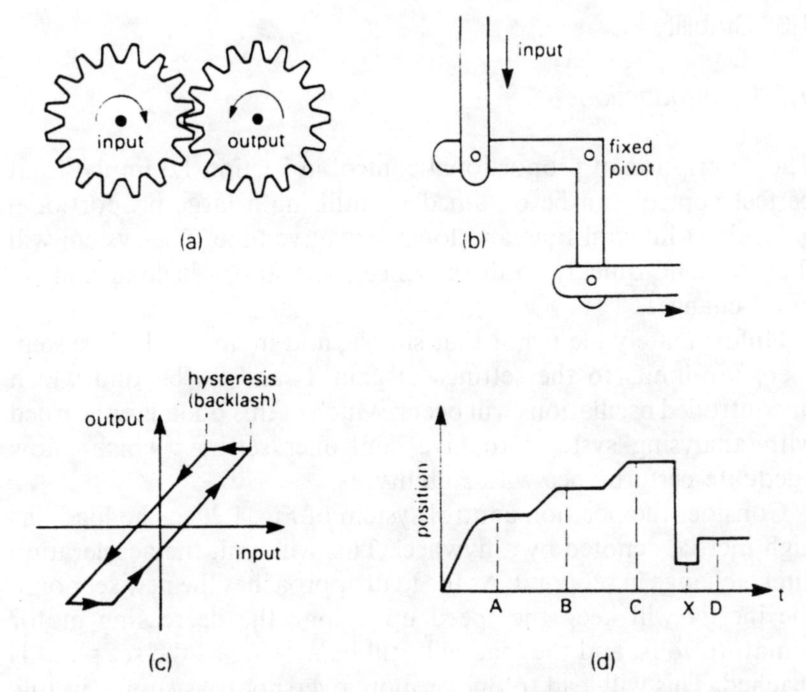

Fig. 1.27 The effects of hysteresis and backlash, (a) Poorly machined gears exhibit backlash; (b) Mechanical linkages exhibit backlash; (c) Representation of backlash; (d) Overcoming hysteresis with a unidirectional approach.

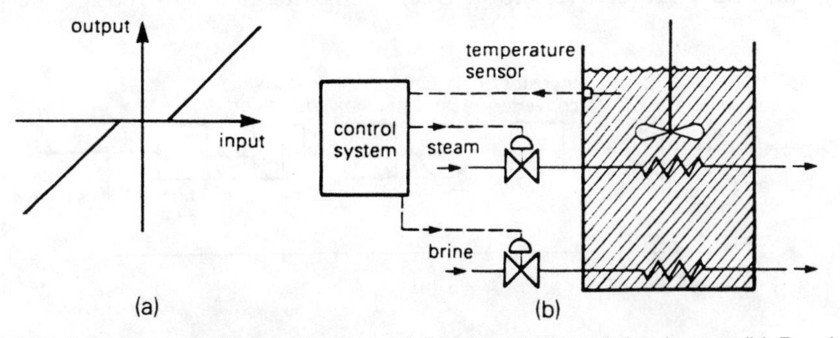

Fig. 1.28 The effect of dead zone, (a) Representation of dead zone; (b) Dead zone deliberately introduced to heating/cooling system.

Here the temperature of a chemical reaction is to be controlled. As the reaction is exothermic heat can be added by a steam line or removed by chilled brine. A dead zone is deliberately introduced to prevent both steam and brine being present together. Dead zones at changeover points also arise when an extended range of process variables is covered by two or more transducers.

1.6. Stability

1.6.1. Introduction

The description of proportional control in Section 1.3 implies that perfect control can be obtained by utilising a large proportional gain, short integral time and long derivative time. The system will then respond quickly to disturbances, alterations in load and set point changes.

Unfortunately life is not that simple, and in any real life system there are limits to the settings of gain T_i and T_d beyond which uncontrolled oscillations will occur. Much of this book is concerned with analysing systems to give controller settings which allow adequate performance with stability.

Consider the position control system of Fig. 1.29a. The load has high inertia, denoted by a flywheel. This will limit the acceleration after a change in set point. As the load approaches the new set point, the inertia will keep the speed up despite the decreasing motor armature volts, and the load will still be moving as the set point is reached. This will lead to one or more overshoots as shown in Fig. 1.29b. Increasing the gain will speed up the initial acceleration (assuming no element saturates) but will lead to a higher running speed and a larger overshoot. Decreasing the gain will reduce the

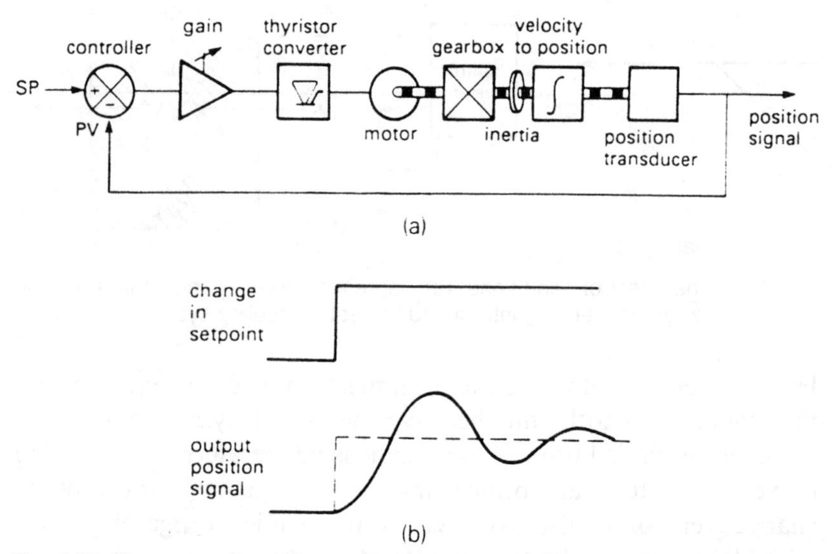

Fig. 1.29 Response of position control system, (a) Position control system with high inertia; (b) Response to a step change in set point.

overshoot, but give a slower initial response. Like many engineering systems, the setting of the controller is a compromise between conflicting requirements.

1.6.2. Definitions and performance criteria

It is often convenient (and not too inaccurate) to consider that a closed loop system behaves as a second order system, with an angular frequency ω_n and a damping factor b as equation 1.18. It is then possible to identify five possible performance conditions, shown for a set point change and a disturbance in Fig. 1.30a and b.

An unstable system exhibits oscillations of increasing amplitude. In practice the oscillations will continue until some element saturates or fails. A marginally stable system will exhibit constant amplitude oscillations; a theoretical condition which is almost impossible to achieve in practice, but is of importance in selecting controller gains.

An underdamped system will be somewhat oscillatory, but the amplitude of the oscillations decreases with time and the system is stable. (It is important to appreciate that oscillatory does not necessarily imply instability.) The rate of decay is determined by the damping factor. An often used performance criteria is the 'quarter amplitude damping' of Fig. 1.30c which is an underdamped response with each cycle peak one quarter of the amplitude of the previous. For many applications this is an adequate, and easily achievable, response.

An overdamped system exhibits no overshoot and a sluggish response. A critical system marks the boundary between underdamping and overdamping and defines the fastest response achievable without overshoot.

For a simple system Fig. 1.30a and b can be related to the gain setting of a P only controller; overdamped corresponding to low gain with increasing gain causing the response to become underdamped and eventually unstable.

It is impossible for any system to respond instantly to disturbances and changes in set point. Before the adequacy of a control scheme can be assessed, a set of performance criteria is usually laid down by production staff. Those defined in Fig. 1.31 are commonly used. These assume the closed loop response is similar to a second order system.

The 'rise time' is the time taken for the output to go from 10% to 90% of its final value, and is a measure of the speed of response of the

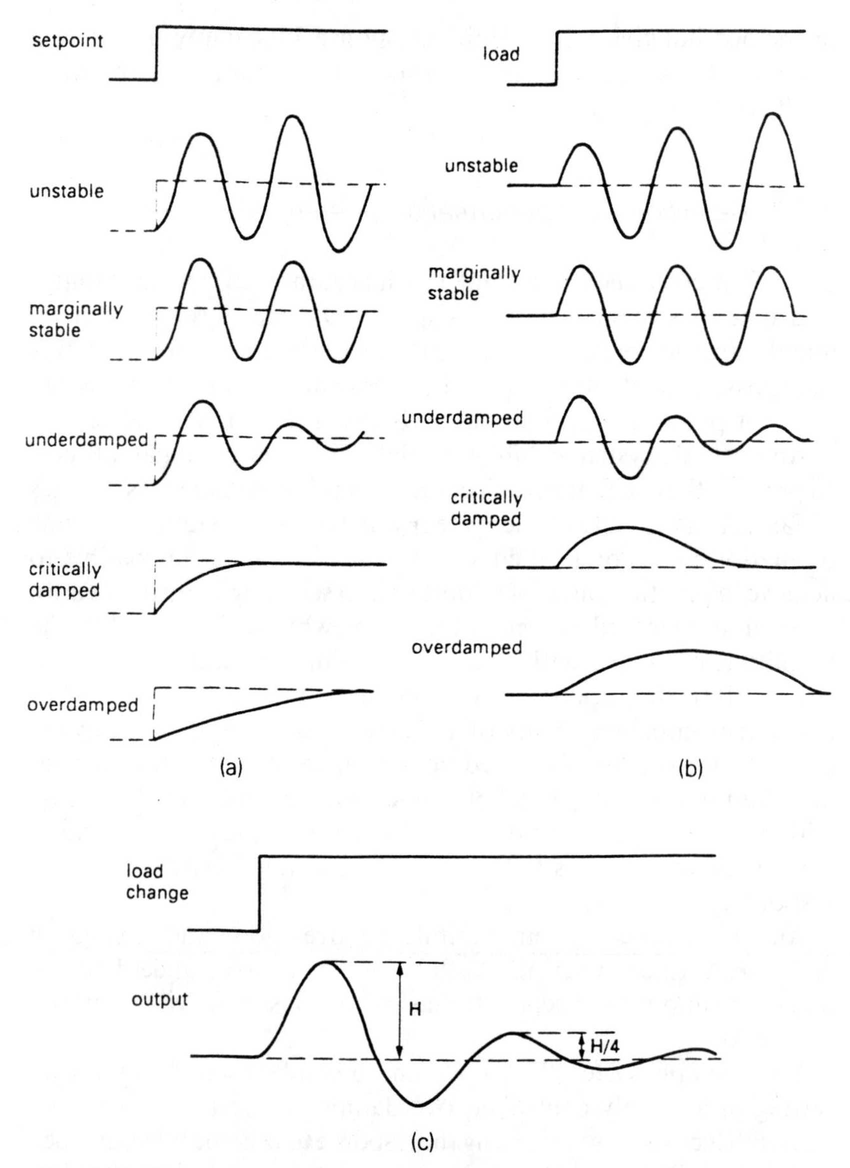

Fig. 1.30 Various forms of system response, (a) Step change in set point; (b) Step change in load; (c) Quarter amplitude damping.

system. The time to achieve 50% of the final value is called the 'delay time'. This is a function of, but not the same as, any transit delays in the system. The first overshoot is usually defined as a percentage of the corresponding set point change, and is indicative of the damping factor achieved by the controller.

As the time taken for the system to settle completely after a change

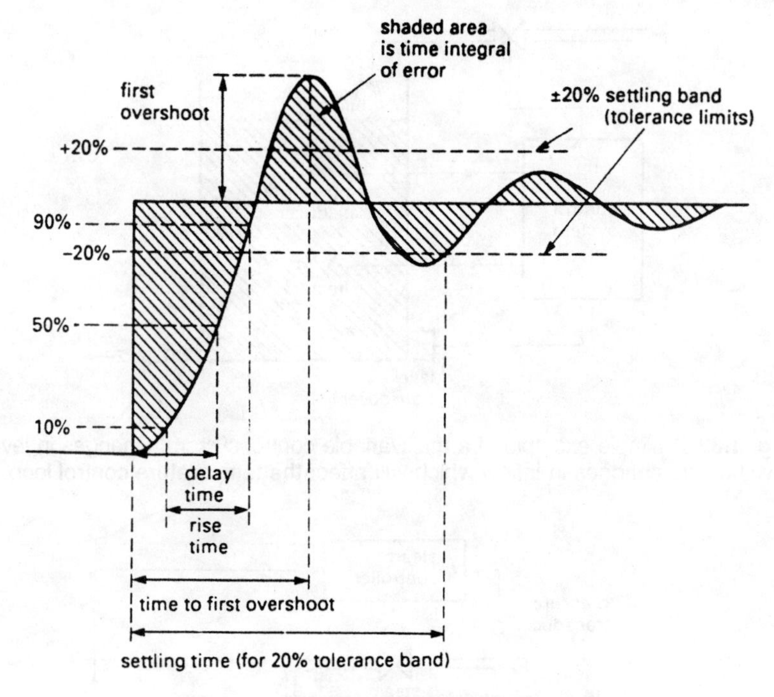

Fig. 1.31 Definitions of system response.

in set point is theoretically infinite, a 'tolerance limit' or 'maximum error' is usually defined. This is usually expressed as a percentage of a set point change; once the plant is within the tolerance limit it is considered to have achieved the set point. A typical value for the tolerance limit is 5%. The settling time is the time taken for the system to enter, and remain within, the tolerance limit.

The shaded area is the integral of the error and can be used as an index of performance. Note that for a system with a standing offset (as occurs with a P only controller) the area under the curve will increase with time and not converge to a final value. Stable systems with integral action control have error areas that converge to a finite value. The area between the curve and the set point is called the integrated absolute error (IAE) and is an accepted performance criterion.

An alternative criterion is the integral of the square of the instantaneous error. This weights large errors more than small errors, and is called integrated squared error (ISE). It is used for systems where large errors are detrimental, but small errors can be tolerated.

The performance criteria above were developed for a set point

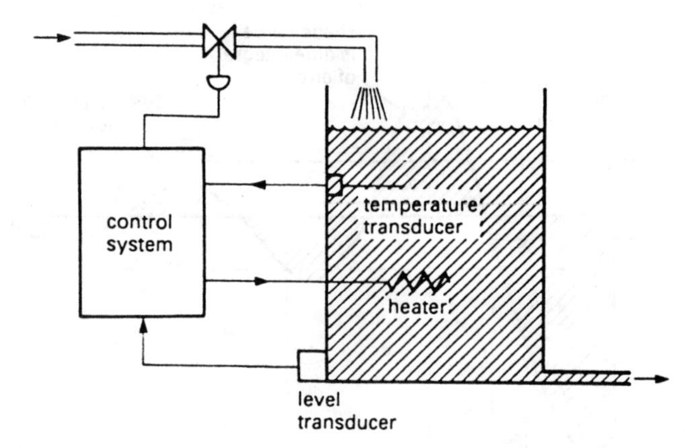

Fig. 1.32 A simple example of a multivariable control system. Changes in level will cause changes in inflow which will affect the temperature control loop.

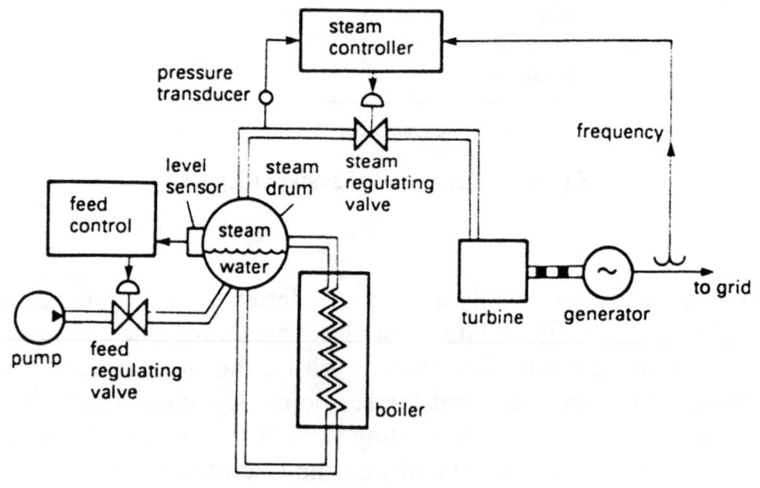

Fig. 1.33 An example of multivariable control; a power station boiler.

change. Similar criteria can be developed for disturbances and load changes.

1.7. Complex systems

Previous sections have implied that plants consist of a combination of single, unrelated control loops. This is rarely so. There are a variety of complex systems which can occur.

Interaction between loops is common. Figure 1.32 shows a typical problem where the level and temperature of liquid in a tank are

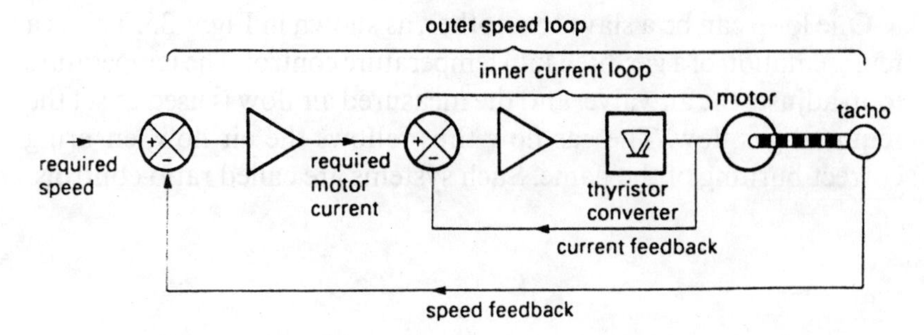

Fig. 1.34 Motor speed controller, an example of cascade control.

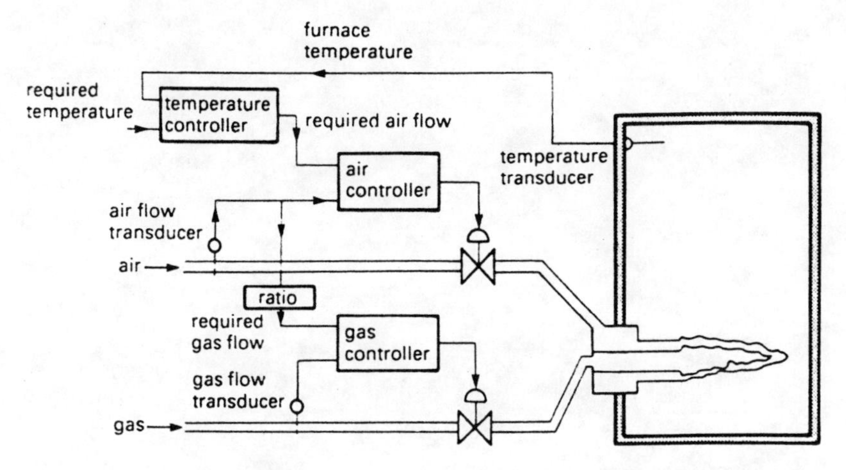

Fig. 1.35 An example of ratio and cascade control, a gas fired furnace.

controlled. The level control will interact detrimentally with the temperature control. Such plants are called multivariable control systems. Interactions can be difficult to analyse. Figure 1.33 is a simplified representation of a boiler feed water system. Here there are control loops for steam pressure, drum level and turbine speed. These all interact in apparently unrelated ways. A sudden demand for steam can lower the drum pressure and cause a large percentage of the drum liquid to flash to steam with a sudden drop in level. The cooling effect of increased make-up water flow can cause an apparent fall in drum level as levitation of the water in steam bubbles momentarily ceases.

Loops can exist within loops as Fig. 1.34. This is called cascade control. Speed controls based on thyristor drives utilise an outer speed control loop, inside which is a motor current loop. It is essential that the inner loop(s) have a faster response than the outer loop(s).

One loop can be a slave to another, as shown in Fig. 1.35. This is a representation of a gas-fired kiln temperature control. The temperature loop adjusts the air valve, and the measured air flow is used to set the required gas flow. The gas flow thus follows the air flow, ensuring correct burning of the flame. Such systems are called ratio controls.

Chapter 2
Analytical methods and system modelling

2.1. Introduction

Chapter 1 introduced the idea of plant modelling, that is producing mathematical descriptions of the various items that comprise the plant under study. From these individual models of plant components, a complete mathematical model can be constructed and used to deduce the theoretical plant behaviour.

The performance of a plant is normally studied to determine:

(a) Its steady state performance, i.e. how it behaves in the long term in the absence of disturbances and when all transients have died away. Usually it is required that the process variables attain the set point values in the long term. The expression 'long term' depends on the plant; for a motor speed control loop the long term is a few seconds. For a large boiler temperature control the long term may be several tens of minutes.
(b) Its stability, i.e. whether the plant returns to a stable condition, runs away or bursts into uncontrolled oscillation when disturbed.
(c) Its transient response, which is how it responds to changes in set point, throughput and outside disturbances.

It is important to appreciate the limitations of modelling. Mathematical models generally assume a plant is linear (a type discussed later) and ignore effects such as saturation and hysteresis which can often dominate plant behaviour. A reasonable plant model can, however, often give good insight into a plant and serve as a basis for further investigations.

2.2. Differential equations

2.2.1. A continuous casting plant

Figure 2.1 shows a plant used to cast liquid steel into solid billets. Liquid steel from a ladle is admitted to a tundish (used to give a constant head and distribute the steel to several casting lines). The steel passes to a water cooled mould, where the steel starts to solidify. The steel is extracted from the mould by electrically driven rolls. At the exit from the mould, the core of the steel is still liquid, surrounded by a thin solidified skin. Spray water is used to cool the steel further so that it is totally solidified by the time it reaches the cutting torches.

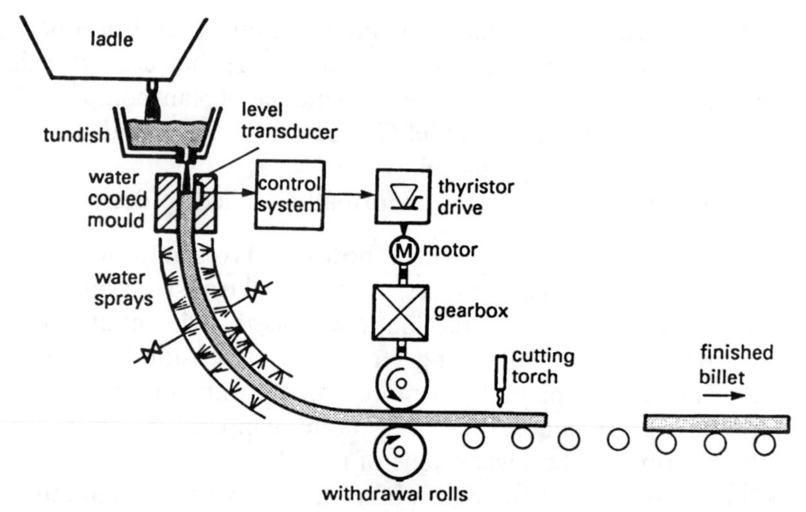

Fig. 2.1 Steel casting system.

Starting the operation is obviously 'interesting', but once running it is a continuous process, with liquid steel being converted to solid billet. It is obviously crucial that steel is extracted from the mould at the same rate as steel enters. If the withdrawal rate is too low, the steel will gradually rise in the mould and overflow (causing considerable damage to the plant and a very real hazard to personnel in the area). If the withdrawal rate is too high the steel level will progressively fall, until the point is reached where the solid skin has not formed at the mould exit and molten steel flows out of the bottom of the mould. The resultant mixing of liquid steel and water causes steam explosions which again lead to plant damage and possible human injury.

There are many control loops on this plant but we shall consider the method of balancing the steel inflow and outflow from the mould. A level sensor is used to measure the steel level in the mould, and this is used to control the speed of the withdrawal rolls. If the tundish nozzle blocks, say, the steel flow will fall and the level in the mould will fall. The roll speed is then automatically reduced to compensate for the new flow rate. If the blockage is cleared, the flow will rise causing the level to rise and the rolls to speed up.

In order to see how this system might behave, we need to produce models for each component. Figure 2.2 shows a block diagram of the relevant items.

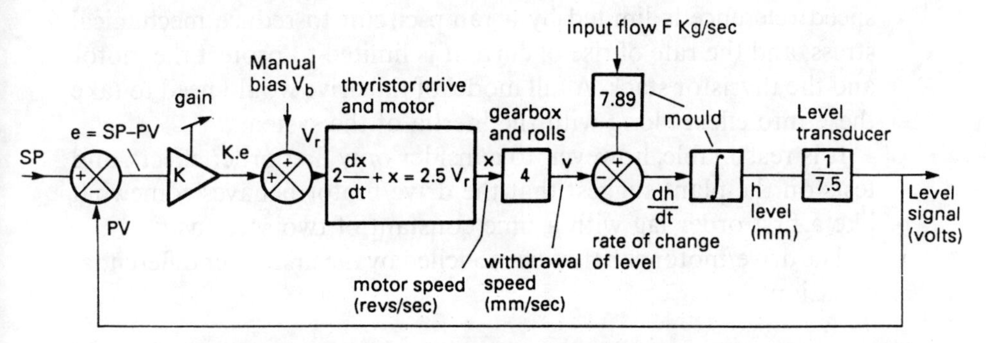

Fig. 2.2 Block representation of casting system.

The level is measured by a radioactive source and ionisation chamber and gives a 0–10 V signal over a 0–75 mm range. The bottom of the measuring range we will assume is a level datum, so a level of 10 mm is measured from the bottom of the sensor 'window'. The level sensor we will assume acts instantaneously, so it can be considered as an amplifier block of gain 1/7·5 V/mm.

The required level is a 0–10 V signal set on a potentiometer. Normally this is set around 5 V corresponding to the centre of the level sensor measuring range. The error signal (SP − PV) is amplified by a controller with adjustable gain K, to give a motor reference signal.

A bias is added to the controller output. This serves two purposes. The output from the controller can be positive or negative dependent on whether the actual steel level is above or below the set point. In practice, of course, it would be highly undesirable to have the motor reverse direction and push the solidified steel back up into the mould if the level fell too far. The bias, set to a speed corresponding to the nominal casting speed, ensures that the motor speed reference remains positive at all times.

The system proposed in Fig. 2.2 is essentially proportional only control. This leads to an error offset, as explained in Section 1.3.1. This bias control is operator accessible and can be used to trim out the error offset such that in the steady state the error is zero.

The motor will be required to decelerate the mass of semi-solid steel as well as accelerate it, so a full four quadrant thyristor drive is needed. The motor/drive is set up such that a 0–10 V reference gives a speed range of 0–1500 rpm or 0–25 rps. The steady state gain is thus 2·5 rps/V.

The motor/drive is itself a complex control system with a tacho speed control loop and a current control loop. The rate of rise of speed reference is limited by a ramp circuit to reduce mechanical stress, and the rate of rise of current is limited to protect the motor and the thyristor stack. A full model of the drive would need to take these into effect along with the inertia of the system.

It is reasonable, however, to consider only first order effects, and tests on the plant suggest that the drive/motor behaves somewhat like a first order lag with a time constant of two seconds.

The drive/motor can thus be modelled by the first order differential equation

$$2\frac{dx}{dt} + x = 2{\cdot}5\,V_r \qquad (2.1)$$

where x is the motor speed (in revs per second) and V_r is the speed reference (in volts). V_r is the sum of the amplified error signal and the offset adjustment. In the above equation, 2 represents the time constant and 2·5 the steady state gain.

The motor drives the steel via a gearbox and pinch rolls. The inertia of these is included in the assumed first order response of the motor above. It is therefore only necessary to calculate the gain of the gearbox and rolls from revs per second to linear speed in mm/sec.

The gearbox ratio is 550:1 and the nominal roll diameter is 700 mm. The linear speed v is simply

$$v = \frac{\pi \times 700 \times x}{550} \qquad (2.2)$$

$$= 3{\cdot}996x \text{ mm/sec}$$

As the roll diameters start off at about 705 mm when new, and are turned down on a lathe as they wear to a lower limit of about 695 mm, it is pointless to take the above relationship to three significant figures. We will assume

$$v = 4x \text{ mm/sec} \tag{2.3}$$

where x is the input shaft speed in revs per second, and v the steel linear speed in mm/sec. The top motor speed of 25 rps (1500 rpm) thus corresponds to a linear speed of 100 mm/sec. The nominal casting speed of 3 metre/min (50 mm/sec) corresponds to a motor speed of 12·5 rps, and a speed reference of 5v to the thyristor drive.

The gearbox/pinch rolls thus act as an amplifier of gain 4.

The mould has an area of 127 × 127 mm. The steel level is being lowered at a rate of v mm/sec by the withdrawal mechanism. The level is being raised by the incoming flow from the tundish. The actual rate of change of level will be the difference between these, i.e.

$$\frac{dh}{dt} = (\text{inflow rate of rise}) - (\text{outflow rate of fall}) \tag{2.4}$$

where h is the steel level in mm above our datum position. The rate of fall due to outflow is, of course, simply v from equation 2.3.

For an input flow rate of F kg/sec, the equivalent rate of rise in the mould is

$$\text{rise rate} = \frac{F}{A \times \text{density}} \tag{2.5}$$

where A is the mould cross-sectional area, 127 × 127 sq mm. The density of steel is 7·86 tonnes/cubic metre, so

$$\text{rise rate} = \frac{F}{127 \times 127 \times 7{\cdot}86 \times 10^{-6}} \tag{2.6}$$

$$= 7{\cdot}89F \text{ mm/sec}$$

when F is in kg/sec.

We can now write the mathematical equation for the mould

$$\frac{dh}{dt} = 7{\cdot}89F - v \tag{2.7}$$

where h is the steel level in mm above the level sensor datum.

We can now derive an expression for the whole system. For simplicity we shall assume constant flow rate from the tundish, and consider the response to changes in set level. Although the system is designed to compensate to changes in flow, practically the set point is easier to change than the flow. In most systems, the response observed when the loop is 'kicked' by a set point change is similar to the response to a kick from a throughput or load change.

Combining the motor, controller and sensor models gives

$$2\frac{dx}{dt} + x = 2{\cdot}5\left(K\left(\frac{h}{7{\cdot}5} - SP\right) + V_b\right) \tag{2.8}$$

The mould equation 2.7 gives

$$v = 7{\cdot}89F - \frac{dh}{dt} \tag{2.9}$$

Including the gearbox/rolls gives

$$x = \frac{1}{4}\left(7{\cdot}89F - \frac{dh}{dt}\right) \tag{2.10}$$

$$\text{or } x = 1{\cdot}97F - 0{\cdot}25\frac{dh}{dt} \tag{2.11}$$

For constant F

$$\frac{dx}{dt} = -0{\cdot}25\frac{d^2h}{dt^2} \tag{2.12}$$

Substituting back into equation 2.8 gives

$$-0{\cdot}5\frac{d^2h}{dt^2} + 1{\cdot}97F - 0{\cdot}25\frac{dh}{dt} = \frac{Kh}{3} - 2{\cdot}5\ K.SP + 2{\cdot}5\ V_b \tag{2.13}$$

Rearranging

$$0{\cdot}5\frac{d^2h}{dt^2} + 0{\cdot}25\frac{dh}{dt} + \frac{Kh}{3} = 1{\cdot}97F + 2{\cdot}5\ K.SP - 2{\cdot}5\ V_b \tag{2.14}$$

Multiplying by two puts this into the standard form

$$\frac{d^2h}{dt^2} + 0.5\frac{dh}{dt} + 0{\cdot}67\ K.h = 3{\cdot}94F + 5\ K.SP - 5\ V_b \tag{2.15}$$

The simplified mould level control system thus behaves as a second order differential equation, but what does this tell us?

2.2.2. Steady state response of casting level control

The behaviour of a closed loop system has two components; a transient portion that is the response of the system after a disturbance and a steady state portion that is the response after the transients have died away.

In the steady state, dx/dt, d^2x/dt^2 and higher terms must all be zero, so we can rewrite equations 2.15:

$$0{\cdot}67\ K.h = 3{\cdot}94F + 5\ K.SP - 5\ V_b \tag{2.16}$$

$$\text{or } h = 5{\cdot}91\frac{F}{K} + 7{\cdot}5SP - \frac{7{\cdot}5\ V_b}{K} \tag{2.17}$$

Let us suppose the system is set up with K = 1, a set point of 5v (i.e. 37·5 mm) and a bias voltage of 5 V. Equation 2.17 simplifies to:

$$h = 5{\cdot}91.F\ mm \tag{2.18}$$

For a nominal casting rate of 6 kg/sec, simple substitution gives a level of 37 mm, just under the mid-point of the mould. If the casting rate increases by 20% to 7·2 kg/sec, the mould level will rise to 43 mm; the increase in height being necessary to give the increased error signal, and hence an increased motor speed to match the increased flow.

This change in process variable with a disturbance is typical of proportional only control. Closed loop control has stabilised the level by matching inflow and outflow, and reduced, but not eliminated, the effects of a disturbance.

The controller gain plays a crucial roll in determining the offset. Let us assume the set point stays at 5 V (37·5 mm), the bias voltage stays at 5 V, and the gain is reduced to 0·3. Simple arithmetic in equation 2.17 gives the new level as 31 mm at the nominal casting rate of 6 kg/sec. This is low in the mould, so the operator would restore the level to the middle of the mould by reducing the bias voltage. A bias voltage of 4·7 V will restore the level to 37 mm.

Assume now the flow increases by 20% as before. The level will rise to 61 mm before inflow and outflow are balanced.

A 20% change in flow, therefore, produces about a 7 mm change in level for unity gain, but a 24 mm change with a gain of 0·3. Not surprisingly, the lower the gain, the less well the system compensates for changes in flow. A gain of 5, for example, will keep the deviation for a 20% flow change to less than 2 mm.

The moral would seem to be use a very high gain. Unfortunately life is not that simple. A high gain will lead to large speed changes for small changes in level, and the continual acceleration and deceleration will stress, and probably fracture, the thin steel skin at the exit from the mould. This practical consideration is one limit on how high a gain can be used.

A more fundamental limit, however, comes from a consideration of the transient response of the system. To find the transient response we must solve the differential equation represented by equation 2.15.

2.2.3. Solution of differential equations

A linear differential equation has the form

$$a_n \frac{d^n x}{dt^n} + a_{n-1} \frac{d^{n-1} x}{dt^{n-1}} + \ldots + a_1 \frac{dx}{dt} + a_o x = f(t) \quad (2.19)$$

where $a_o - a_n$ are constants. The left-hand side of the equation represents the response of the system, and the right-hand side, called the driving function, the signal being applied to, or disturbing the system.

We have already, intuitively, deduced that the solution of a differential equation has two parts; a transient part which (hopefully) dies away with time, and a steady state part which is the system response in the 'long term'. The total system response is the sum of these two parts, i.e.

$$x(t) = x_t(t) + x_s(t) \quad (2.20)$$

where x_t is the transient response and x_s is the steady state response. The transient response is called, in mathematical texts, the complementary function and the steady state response the particular integral.

We have been calculating the steady state responses of the mould level control system to changes in flow or set point in the previous section by setting dx/dt and higher terms to zero. The transient response depends on the setting of the controller gain. Possible transient and steady state responses are shown in Fig. 2.3. In each case, the total response is the sum of the transient and steady state responses.

The section below gives methods for determining the solution of differential equations. The techniques are given without rigorous proof (and are rather dubious from a strict mathematical viewpoint). The techniques are based on the underlying principle that there are two functions which replicate on differentiation. These are the trigometrical functions sin(x) and cos(x), plus the exponential function e^{Ax}:

$$\frac{d(\sin x)}{dt} = \cos x, \quad \frac{d(\cos x)}{dt} = -\sin x, \quad \frac{de^{Ax}}{dt} = Ae^{Ax} \quad (2.21)$$

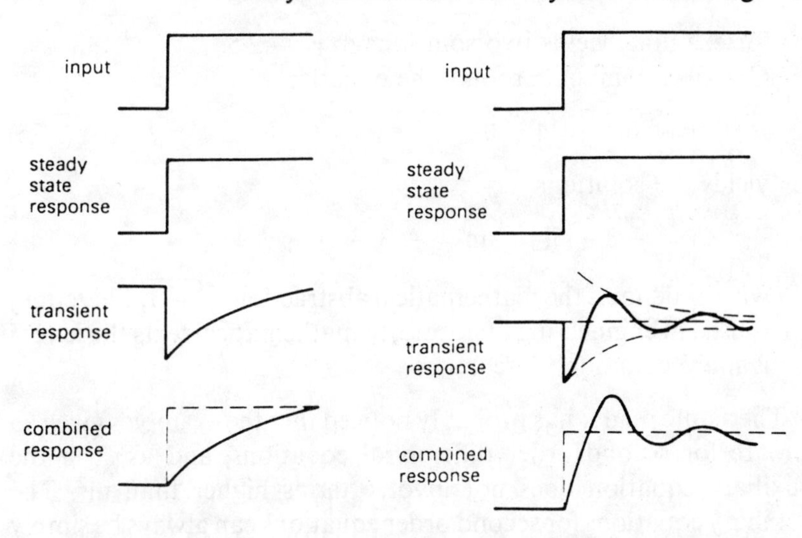

Fig. 2.3 Combined response as the sum of transient and steady state responses.

2.2.3.1. Transient response

The transient response is determined by replacing the differential equation by an algebraic equation (called the auxiliary function). Each term $d^n x/dt^n$ is replaced by m^n. This algebraic equation is then solved for m, with the driving function set to zero.

For example, the differential equation

$$\frac{d^2x}{dt^2} + 5\frac{dx}{dt} + 6x = \sin 5t \tag{2.22}$$

represents a system being driven by a sine wave. Following the above procedure gives:

$$m^2 + 5m + 6 = 0 \tag{2.23}$$

which is the auxiliary equation corresponding to equation 2.22. Solving for m yields $m = -3$ and $m = -2$.

There are, in general, three possible cases for m, when the auxiliary equation is solved:

(a) Real unequal roots (of which equation 2.23 is typical).
(b) Pairs of equal roots. The equation

$$m^2 + 5m + 6{\cdot}25 = 0 \tag{2.24}$$

for example, yields two solutions $m = -2{\cdot}5$.

(c) Complex conjugate roots. The equation

$$m^2 + 6m + 34 = 0 \tag{2.25}$$

yields the solutions

$$m = -3 + 5j \qquad m = -3 - 5j$$

where j denotes the mathematical abstraction $\sqrt{-1}$. The letter j is used in engineering literature; in mathematical texts the letter i is more commonly used.

The astute reader has probably noticed that the examples given so far are for second order differential equations, and as such the auxiliary equation does not involve terms higher than m^2. The auxiliary equations for second order equations can always be simply solved, using the well-known procedure

for $Am^2 + Bm + C = 0$

$$m = -\frac{B \pm \sqrt{B^2 - 4AC}}{2A} \tag{2.26}$$

Third, and higher order, equations are dealt with in a similar manner, but their solution is more difficult. The equation

$$m^4 + 13m^3 + 77m^2 + 235m + 250 = 0 \tag{2.27}$$

factorises to

$$(m^2 + 6m + 25)(m^2 + 7m + 10) = 0$$

from which there are four solutions

$$m = -5, \quad m = -2, \quad m = -3 + 4j, \quad m = -3 - 4j$$

Few real life equations factorise so neatly, and computer based iterative routines need to be used. The unit MATHUTIL in Chapter 7 contains procedures for solving polynomials up to order five.

With the auxiliary equation solved, the transient response is built up from each solution as follows.

(1) For each real solution $m = \alpha$ a term $Ae^{\alpha t}$ appears in the transient response, where A is an integration constant determined by the initial conditions.

(2) For repeated real solutions $m = \alpha$, an expression

$$(A_1 + A_2t + \ldots + A_nt^{n-1})e^{\alpha t}$$

appears in the transient response where $A_1 - A_n$ are integration constants, and n is the number of repeated roots.

(3) For each pair of complex conjugate solutions $m = \alpha \pm j\omega$ an expression

$$e^{\alpha t}(A_1 \sin \omega t + A_2 \cos \omega t)$$

appears in the transient response. A_1 and A_2 are again integration constants.

The form of the transient response of equation 2.22 is thus

$$x = Ae^{-3t} + Be^{-2t}$$

where A and B are constants.

The transient response corresponding to equation 2.24 is

$$x = (A + Bt)e^{-2 \cdot 5t}$$

with A and B again constants.

For equation 2.25, the transient response has the form

$$x = (A \sin 5t + b \cos 5t)e^{-3t}$$

with A and B the usual constants.

Finally the transient response of the fourth order equation 2.27 has the form:

$$x = Ae^{-5t} + Be^{-2t} + (C \sin 4t + D \cos 4t)e^{-3t}$$

with A–D constants.

Without determining the value of the integration constants, we can make some observations of the form of the transient response.

A term of the form

$$x = Ae^{\alpha t} \tag{2.28}$$

represents an exponentially increasing response if α is positive as Fig. 2.4a. Such a response indicates an uncontrolled or unstable system, and should not occur in a well-ordered control system. If α is negative, an exponential decay occurs as Fig. 2.4b.

Terms of the form

$$x = (A \sin \omega t + B \cos \omega t)e^{\alpha t} \tag{2.29}$$

represent a sinusoid with an exponential envelope. Again, if α is positive the sinusoids will increase in amplitude as Fig. 2.4c; an undesirable characteristic for a control system. With negative α, the sinusoids die away indicating a stable, but oscillatory, system.

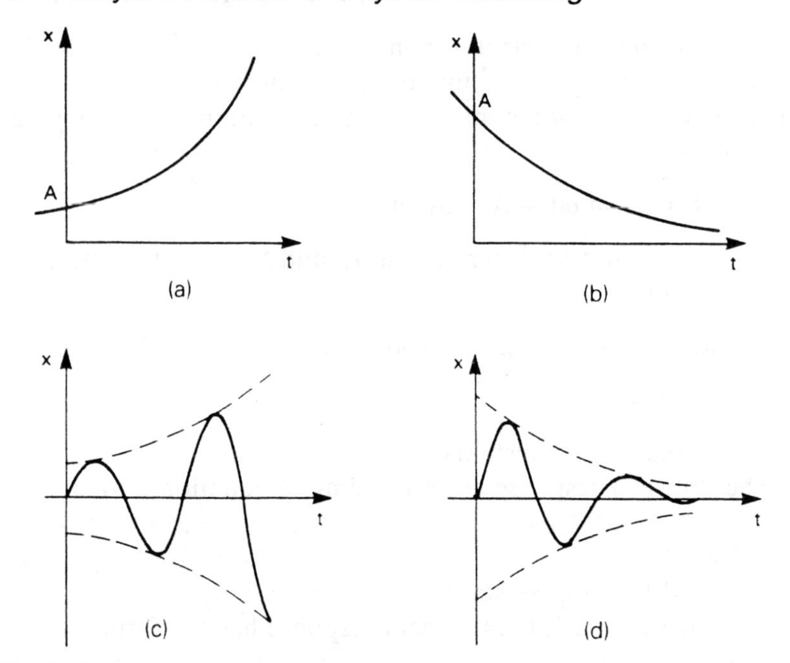

Fig. 2.4 Various possible forms of transient response, (a) $x=Ae^{\alpha t}$; (b) $x=Ae^{-\alpha t}$; (c) $x=Ae^{\alpha t}\sin\omega t$; (d) $x=Ae^{-\alpha t}\sin\omega t$.

Terms of the form

$$x = (A + Bt + \ldots)e^{\alpha t} \tag{2.30}$$

increase or decrease, in a similar way to Fig. 2.4a/b according to the sign of α.

2.2.3.2. Steady state response

The total solution of a differential equation depends on the driving function and the initial conditions. Common driving functions are:

(a) A constant or step ($f(t) = A$ for positive t).
(b) A ramp driving function ($f(t) = At$ for positive t). This function is important for tracking servos in applications such as gunnery, astronomy and radio telemetry.
(c) A sinusoid ($f(t) = A\sin\omega t$ for positive t). This function leads to frequency response analytical methods, described in later sections.

To determine the steady state response the D operator is used, which represent d/dt. $D^n x$ thus represents $d^n x/dt^n$.

The procedure to determine the steady state response varies

according to the driving function.

For a constant or step function, the equation

$$a_n \frac{d^n x}{dt^n} + a_{n-1} \frac{d^{n-1} x}{dt^{n-1}} + \ldots + a_1 \frac{dx}{dt} + a_0 x = A \qquad (2.31)$$

is re-written

$$(a_n D^n + a_{n-1} D^{n-1} + \ldots + a_1 D + a_0)x = A$$

D is equated to zero, giving

$$x = \frac{A}{a_0} \qquad (2.32)$$

This is equivalent to the earlier derivation of the steady state solution for the mould level control where we set dh/dt and higher terms to zero.

If a_0 is zero, there is no term in x in the differential equation. This corresponds to pure integral action plus other terms. The steady state solution is then derived as below.

First the equation is formed in terms of D as before

$$(a_n D^n + a_{n-1} D^{n-1} + \ldots + a_1 D)x = A$$

The factor D is taken out to give

$$(a_n D^{n-1} + a_{n-1} D^{n-2} + \ldots + a_1)Dx = A$$

$$\text{or } (a_n D^{n-1} + a_{n-1} D^{n-2} + \ldots + a_1)x = \frac{A}{D}$$

We now set D to zero on the left-hand side only giving

$$a_1 x = \frac{A}{D}$$

$$\text{or } x = \frac{A}{a_1 D}$$

If D represents d/dt, it is reasonable to assume that 1/D represents integration with respect to time (a step which probably strikes the reader as very suspicious, but one which can be justified with mathematics that need not concern us here). It follows that

$$x = \frac{At}{a_1} + B$$

where B is a constant determined by the initial conditions.

A ramp driving function can be represented by At, so a ramp driven differential equation can be represented by

$$(a_n D^n + \ldots + a_1 D + a_0)x = At$$

this can be rearranged as

$$x = \frac{At}{(a_0 + a_1 D + \ldots + a_n D^n)} \tag{2.33}$$

The right-hand denominator is arranged in a series of ascending powers of D using the binomial expansion

$$(1 + y)^{-1} = 1 - y + y^2 - y^3 \text{ etc.}$$

Rearranging the denominator gives

$$x = \frac{A}{a_0}\left(1 + \left(\frac{a_1}{a_0}D + \ldots + \frac{a_n}{a_0}D^n\right)\right)^{-1} t$$

When expanded, the D operators act on t (with terms higher than D producing zero) to give the steady state solution (another seemingly doubtful procedure, but one which can again be justified mathematically). For example

$$\frac{d^2x}{dt^2} + 5\frac{dx}{dt} + 6x = 3t \tag{2.34}$$

Using the D operator gives

$$(D^2 + 5D + 6)x = 3t$$

$$x = \frac{3t}{(D^2 + 5D + 6)}$$

$$\text{or } x = \frac{3}{6}\left(1 + \left(\frac{5}{6}D + \frac{1}{6}D^2\right)\right)^{-1} t$$

Ignoring terms higher than D gives

$$x = \frac{3}{6}\left(1 - \frac{5}{6}D\right)t$$

$$= 0{\cdot}5t - \frac{5}{12} \qquad \left(\text{since } Dt = \frac{d(t)}{dt} = 1\right)$$

The practical inference of this is that x follows the ramp but lags by a constant offset of 5/12 as shown in Fig. 2.5.

The solution for a ramp driving function is a special case of driving

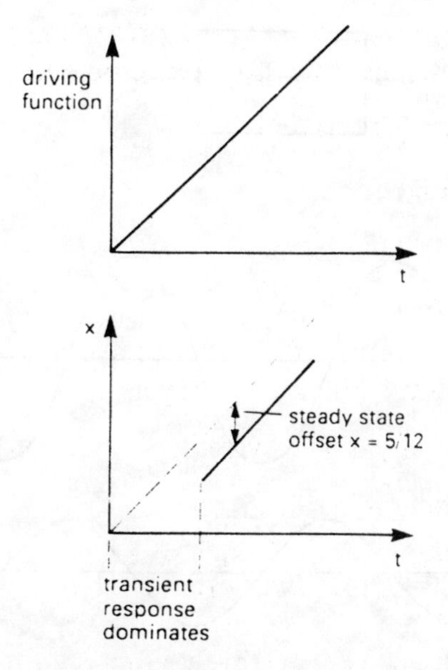

Fig. 2.5 System response to ramp driving function (constant velocity).

functions of the form At^p where p is a power. The expression At^2, for example, represents constant acceleration. The general case is solved as above but ignoring terms higher than D^p in the binomial expansion. Note, for example, that $Dt^2 = 2t$, and $D^2t^2 = 4$.

A sinusoidal driving function has a special place in control theory as it is the basis of frequency response methods described in later sections, and as such is worthy of some discussion.

Figure 2.6a represents a linear system (represented by a differential equation) being driven by a time varying function. A loose definition of a linear system is one whose output has the same form as the input, when driven by a sinusoid but possibly of different amplitude and shifted in time.

Figure 2.6b shows a sinusoidal driving function. For a linear system this must, by definition, produce a sinusoidal output. In Fig. 2.6b, at the frequency shown, the output is shifted by 0·75 s, and is half the amplitude of the input. For sinusoidal signals it is more convenient to represent time shifting as a phase shift. For Fig. 2.6b the sinusoidal period is 3 sec, so a 0·75 s shift represents a phase shift of 90° or $\pi/2$ radians.

The driving function has the form A sin ωt, with A denoting the amplitude and ω the angular frequency in radians per second. The output will have the form AK sin $(\omega t - \phi)$ where ω is the same angular frequency as in the driving function (by definition of a linear

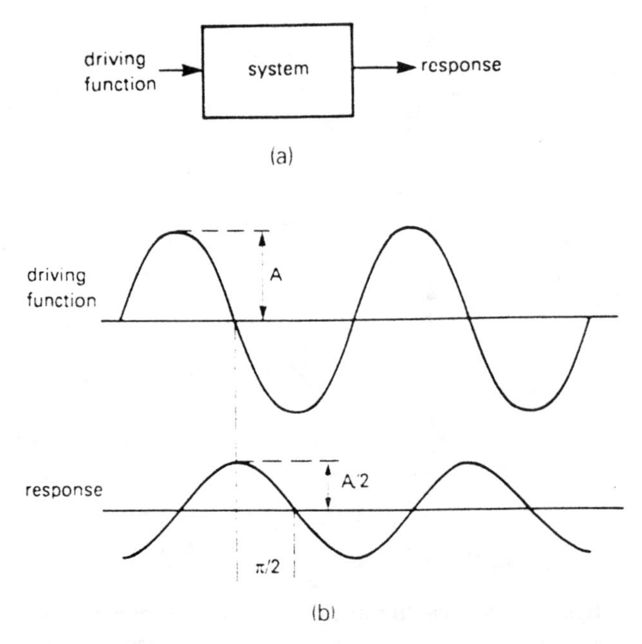

Fig. 2.6 System response to sinusoidal driving function, (a) Representation of system; (b) System response with gain of 0·5 and phase shift of $\pi/2$ radians (90°).

system), ϕ is the phase shift and K the steady state gain.

Determination of the steady state response to a sinusoidal driving function is thus a determination of K and ϕ for a particular ω. The easiest way to achieve this is via a seemingly unrelated diversion into complex numbers.

In Appendix A, it is shown that the function A sin $(\omega t - \phi)$ is equivalent to A exp $(j\omega - \phi)$. A system driven by a sinusoid can thus be represented in D form as

$$(a_n D^n + \ldots + a_1 D + a_0)x = Ae^{j\omega} \tag{2.35}$$

$$\text{or } x = \frac{Ae^{j\omega}}{(a_n D^n + \ldots + a_1 D + a_0)}$$

The next step is to replace D by $(j\omega)$, remembering that $j^2 = -1$, $j^3 = -j$, $j^4 = 1$ etc. The reader is asked to accept this apparently random substitution, it will be justified (but not proved) in section 2.3.7 later.

This substitution produces a complex number in the denominator of the form $(B + jC)$, which can be represented in polar form as

$\sqrt{B^2 + C^2}$ exp ($\tan^{-1}$ (C/B)). This last step is also justified in Appendix A. Call this M.exp (θ)

We now have

$$x = \frac{A \exp (j\omega)}{M \exp (\theta)} = \frac{A}{M} \exp (j\omega - \theta)$$

which is the required form where the gain Z is 1/M and θ is the phase shift.

For example, consider a system of equation 2.34 driven by a sinusoid of angular frequency 4 rads/sec and amplitude 3, i.e.

$$\frac{d^2x}{dt^2} + 5\frac{dx}{dt} + 6x = 3 \sin 4t \tag{2.36}$$

In D form this gives

$$(D^2 + 5D + 6)x = 3e^{4jt}$$

$$x = \frac{3e^{4jt}}{D^2 + 5D + 6}$$

Substituting 4j for D gives

$$x = \frac{3e^{4jt}}{16j^2 + 20j + 6}$$

$$= \frac{3e^{4jt}}{-10 + 20j}$$

The denominator is $-10 + 20j$, which becomes

$$\sqrt{100 + 400} \exp (j \tan^{-1}(20/-10)$$

The angle represented by $\tan^{-1}(20/-10)$ is denoted by ϕ on Fig. 2.7, which is, as shown,

$$\begin{aligned} \phi &= \pi - \tan^{-1}(20/10) \text{ radians} \\ &= 3{\cdot}14 - 1{\cdot}11 \\ &= 2{\cdot}03 \text{ radians} \end{aligned}$$

Thus, after a bit of routine, but lengthy, manipulation, we have

$$x = \frac{3e^{4jt}}{22{\cdot}4e^{2{\cdot}03j}}$$

$$\text{or } x = \frac{3}{22{\cdot}4} e^{(4jt - 2{\cdot}03j)}$$

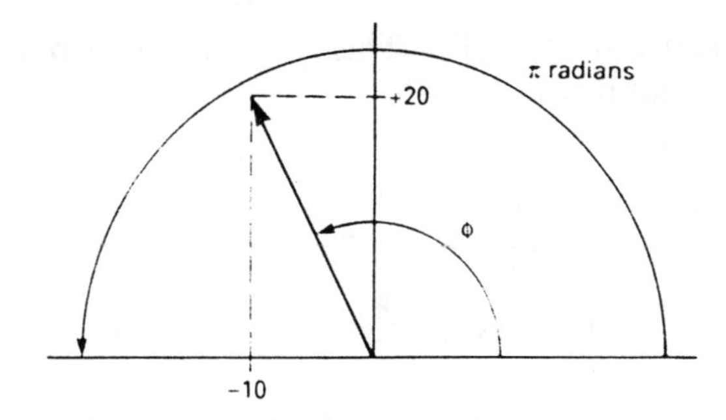

Fig. 2.7 Representation of ϕ.

This is in the standard form $Ae^{j(\omega t - \varphi)}$, so we can write

$$x = \frac{3}{22{\cdot}4} \sin (4t - 2{\cdot}03)$$

i.e. the system has a gain of 1/22·4 and a phase shift of 2·03 radians at angular frequency of 4 radians/sec.

2.2.3.3. The complete solution

The complete solution is simply the sum of the transient and steady state solutions. These will contain integration constants which are determined from the initial conditions.

For example, consider the response of the system described by the equation

$$\frac{d^2x}{dt^2} + 2\frac{dx}{dt} + 5x = f(t)$$

to a step of height 4.

The auxiliary equation is

$$m^2 + 2m + 5 = 0$$

which yields $m = -1 \pm 2j$.

The transient response therefore has the form

$$x_t = e^{-t}(A \sin 2t + B \cos 2t)$$

where A and B are constants.

The steady state solution is simply $x = 0{\cdot}8$ (found by putting dx/dt and higher terms to zero).

The complete solution is thus

$$x = 0{\cdot}8 + e^{-t}(A \sin 2t + B \cos 2t)$$

If the system is at rest at time $t = 0$, $x = dx/dt = d^2x/dt^2 = 0$. Substituting into the equation for $t = 0$ gives

$$0 = 0{\cdot}8 + e^0(0 + B)$$

therefore $B = -0{\cdot}8$

$$\text{so } x = 0{\cdot}8 + e^{-t}(A \sin 2t - 0{\cdot}8 \cos 2t)$$

Differentiating by parts yields

$$\frac{dx}{dt} = -e^{-t}(A \sin 2t - 0{\cdot}8 \cos 2t) + e^{-t}(2A \cos 2t + 1{\cdot}6 \sin 2t)$$

At $t = 0$

$$0 = -e^0(0 - 0{\cdot}8) + e^0(2A - 0)$$

$$\text{or } A = -0{\cdot}4$$

The complete solution is thus

$$x = 0{\cdot}8 - e^{-t}(0{\cdot}4 \sin 2t + 0{\cdot}8 \cos 2t)$$

steady state   transient response

This function is shown in Fig. 2.8 and, as can be seen, corresponds to a damped oscillatory response.

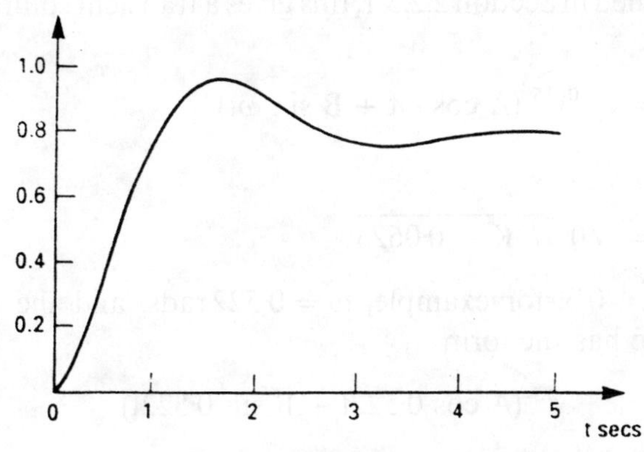

Fig. 2.8 Plot of $x = 0{\cdot}8 - e^{-t}(0{\cdot}4 \sin 2t + 0{\cdot}8 \cos 2t)$.

2.2.4. Transient behaviour of casting level control

The differential equation determining the behaviour of the casting level control system for a change in set point is given by equation 2.15. The driving function on the right-hand side is rather unwieldy, and it is more convenient to express it as *deviations* from a steady state. This allows us to remove the terms representing the flow and bias, and rewrite the equation as

$$\frac{d^2h}{dt^2} + 0{\cdot}5\frac{dh}{dt} + 0{\cdot}67\ Kh = 5\ Ks \tag{2.37}$$

where h and s are deviations from a steady state, and K is the controller gain. Let us assume the system is steady (i.e. $h = 0$, $s = 0$, $dh/dt = 0$, $d^2h/dt^2 = 0$) and we change the set point by 20 mm.

Using the techniques in Section 2.2.3, we first form the auxiliary equation

$$m^2 + 0{\cdot}5\,m + 0{\cdot}67\,K = 0$$

This has solutions

$$m = -\frac{0{\cdot}5 \pm \sqrt{0{\cdot}25 - 4 \times 0{\cdot}67\ K}}{2}$$

The transient behaviour is determined by the value of K, and there are three possible conditions:

(a) For $K > 0{\cdot}093$ the solution is

$$m = -0{\cdot}25 \pm j\sqrt{0{\cdot}67\ K - 0{\cdot}0625}$$

As outlined in Section 2.2.3.1, this gives a transient solution of the form

$$h_t = e^{-0{\cdot}25}\ (A \cos \omega t + B \sin \omega t)$$

where

$$\omega = \sqrt{0{\cdot}67\ K - 0{\cdot}0625} \tag{2.38}$$

with $K = 0{\cdot}5$, for example, $\omega = 0{\cdot}522$ rads, and the transient response has the form

$$h_t = e^{-0{\cdot}25}\ (A \cos 0{\cdot}522t + B \sin 0{\cdot}522t)$$

Note that increasing the gain increases ω.

(b) For $K = 0{\cdot}093$ the solution is

$$m - 0{\cdot}25 \qquad \text{(with double roots)}$$

From Section 2.2.3.1, this yields a transient form

$$h_t = (A + Bt)e^{-0{\cdot}25t}$$

This is known as critical damping, a response discussed later.

(c) For $K < 0{\cdot}093$ the solution yields two real roots

$$m = -0{\cdot}25 \pm Z$$

where $Z = \sqrt{0{\cdot}0625 - 0{\cdot}67\ K}$

This gives a transient form

$$h_t = Ae^{-0{\cdot}25+Z} + Be^{-0{\cdot}25-Z}$$

Let us assume $K = 1$. This predicts a damped oscillatory response because $K > 0{\cdot}093$.

Substitution in equation 2.38 gives $\omega = 0{\cdot}78$ and a transient response of the form

$$h_t = e^{-0{\cdot}25t}\,(A \cos 0{\cdot}78t + B \sin 0{\cdot}78t)$$

The steady state solution is found by setting dh/dt and higher terms to zero. A set point change of 20 mm corresponds to a voltage set point change of $20 \times 100/75$ V, so

$$0{\cdot}67\ Kh = \frac{5 \times K \times 20 \times 10}{75}$$

or $h = 20$ mm.

This implies that a set point change yields an equal level change, as the K term cancels out on both sides of the equation.

The complete solution for a 20 mm change in set point and $K = 1$ is thus

$$h = 20 + e^{-0{\cdot}25t}\,(A \cos 0{\cdot}78t + B \sin 0{\cdot}78t)$$

To find the integration constants A, B we insert the initial conditions. At time $t = 0$, $h = 0$, giving

$$0 = 20 + e^{0}\,(A \cos 0 + B \sin 0)$$

since $\sin 0 = 0$, and $\cos 0 = 1$, the constant $A = -20$, giving

$$h = 20 + e^{-0{\cdot}25t}\,(-20 \cos 0{\cdot}78t + B \sin 0{\cdot}78t)$$

differentiating by parts

$$\frac{dh}{dt} = -0{\cdot}25\, e^{-0{\cdot}25t}(-20 \cos 0{\cdot}78t + B \sin 0{\cdot}78t)$$

$$+ e^{-0{\cdot}25t}(15{\cdot}6 \sin 0{\cdot}78t + 0{\cdot}78B \cos 0{\cdot}78t)$$

At time $t = 0$

$$0 = -0{\cdot}25(-20 + 0) + (0 + 0{\cdot}78B)$$

$$\text{or } B = -6{\cdot}4$$

Giving the complete solution

$$h = 20 - e^{-0{\cdot}25t}(20 \cos 0{\cdot}78t + 6{\cdot}4 \sin 0{\cdot}78t)$$

This step response is shown on Fig. 2.9.

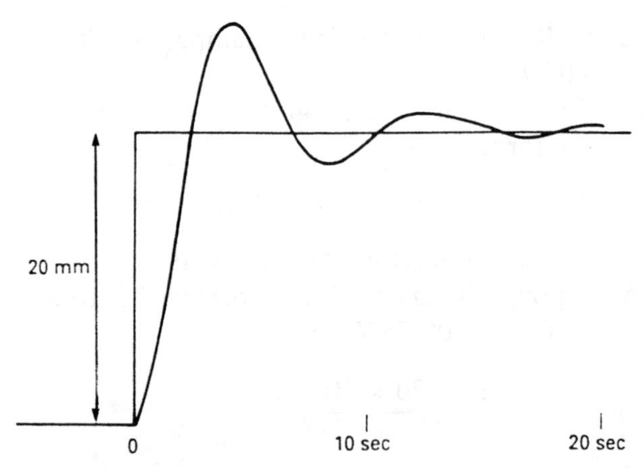

Fig. 2.9 Response of level control system to 20 mm set point change and gain of 1.

2.2.5. Shortcomings of the caster model

The control scheme described for the caster level control has several obvious problems. A high gain is needed to prevent changes in flow affecting the level, but gain greater than about 0·1 causes an oscillatory transient response. The step response for various gains is shown in Fig. 2.10. None of these are ideal.

Obviously what is needed is an integral term in the controller, to give P + I control which will always return the level to the set point (assuming a stable system).

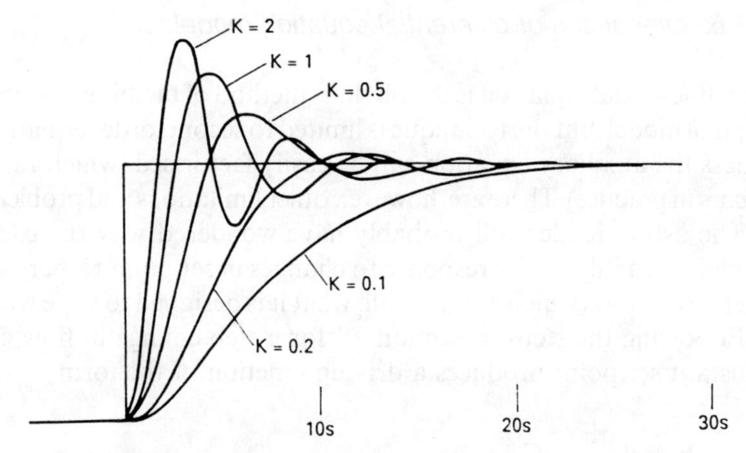

Fig. 2.10 Response of level control system for various gains.

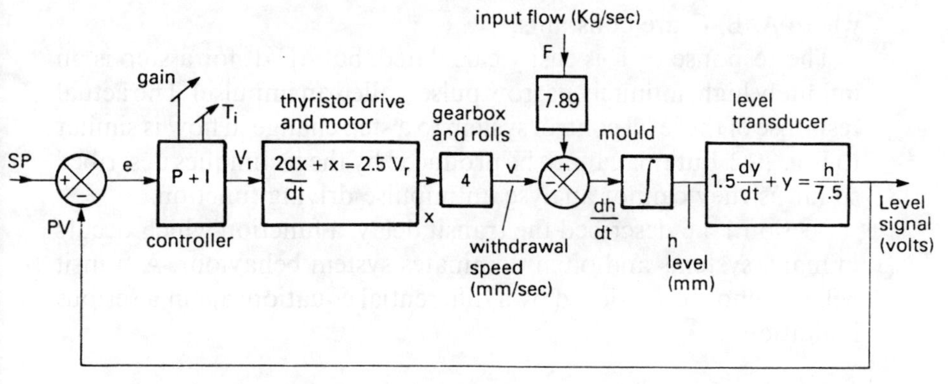

Fig. 2.11 A more accurate casting level control system.

A further practical problem occurs with the level measurement. The steel in the mould is very turbulent from the incoming steel flow and the vaporisation of the oil injected into the mould walls to lubricate the steel as it solidifies. In addition, the level transducer (a radioactive source and GM tube) is itself inherently noisy. The level signal is therefore passed through a first order filter with a 1·5 sec time constant.

Figure 2.11 therefore represents a more accurate model of the caster but what can we say of the predicted behaviour?

Unfortunately the more accurate model does not resolve easily to a simple differential equation that could be solved with the techniques described earlier. Terms in ds/dt and d^4h/dt^4 appear which cannot be dealt with simply. Further analytical tools are therefore needed.

2.2.6. Limitations of differential equation models

The differential equation is the obvious method of tackling a process control model, but the technique is limited to second order equations unless the auxiliary equation can be easily factorised (which rarely occurs in practice). There are, however, other limitations and problems.

The astute reader will probably have wondered why the caster model was analysed for response to changes in set point rather than changes in flow (which are, after all, what it is designed to cope with).

Following the steps in Section 2.1 for a step change in flow and constant set point produces a driving function of the form

$$AF + B\frac{dF}{dt} + C$$

where A, B, C are constants.

The response to F is easily calculated, but dF/dt for a step is an infinitely high, infinitely narrow pulse (called an impulse). The actual response of the level control system to a step change in flow is similar to Fig. 2.12, but this cannot be produced by the techniques described so far as they cannot analyse an impulse driving function.

Section 1.4.6 described the transit delay, a function which occurs in many systems and often dominates system behaviour. A transit delay cannot be modelled by a differential equation; again a serious limitation.

2.3. Frequency response methods

2.3.1. Introduction

Figure 2.13 represents a linear system being driven by a sine wave input. After the transients have died away, the output will also be a sine wave of the same frequency as the input, but of different amplitude and shifted in phase. In Fig. 2.13b, for example, the output amplitude is 0·75 the input amplitude, and is delayed by 60°. The techniques in Section 2.2.3.2 describe how to calculate the steady state response for a system modelled by differential equations and driven by a sine wave.

For any system, the gain (i.e. ratio of the output amplitude to the input amplitude) and the phase shift will vary with the driving frequency. Figure 2.14, for example, has a first order lag driven by a

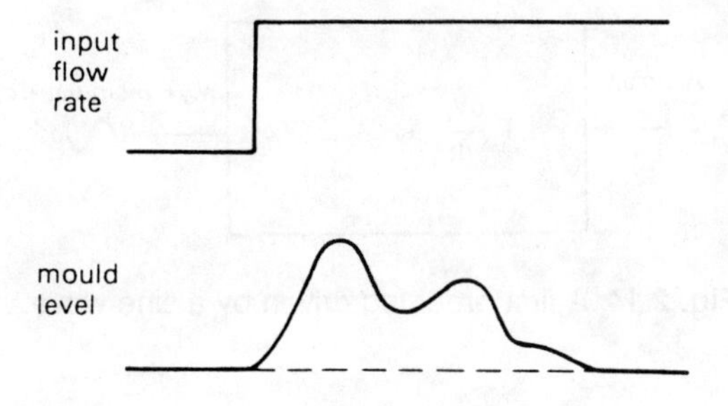

Fig. 2.12 Response of mould level system to load change.

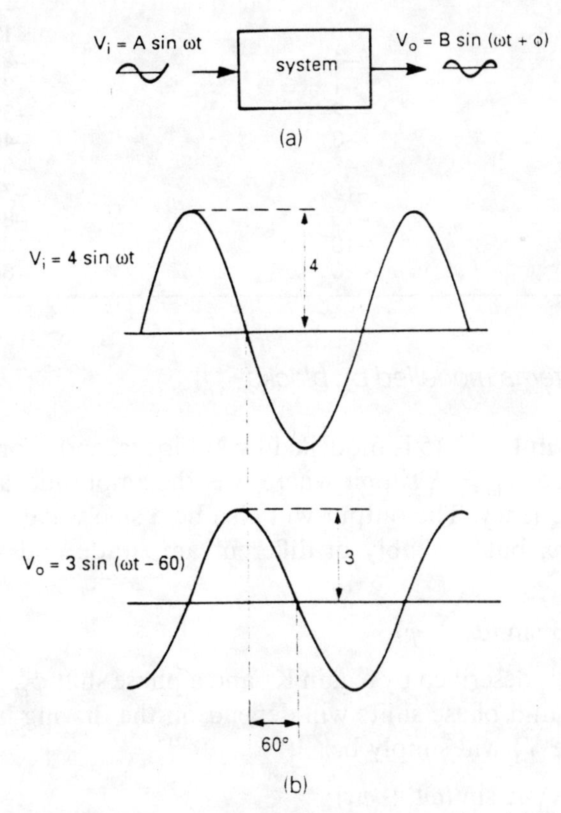

Fig. 2.13 System response to a sine wave driving function, (a) System driven by a sine wave; (b) System with gain of 0·75 and phase shift of 60°.

sine wave. The gain and phase shift at various frequencies is shown in Table 2.1. It would therefore seem reasonable that an analytical method could be developed based on knowledge of how a system responds to sine wave inputs.

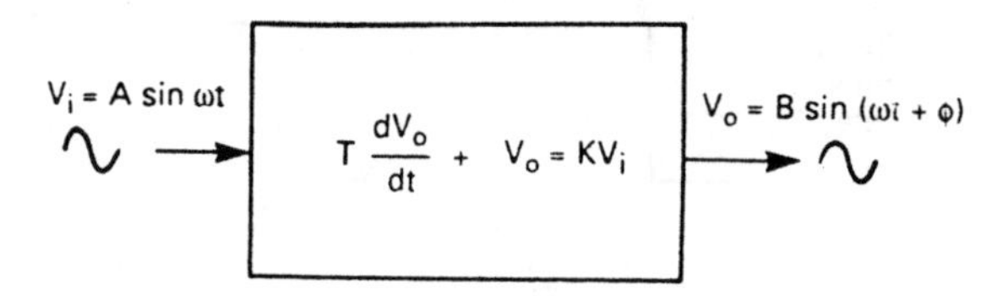

Fig. 2.14 A first order lag driven by a sine wave.

Table 2.1

Omega	*Gain (dB)*	*Phase shift*
0·1	0	−6
0·2	0	−12
0·4	−1	−22
0·6	−1	−31
0·8	−2	−39
1·0	−3	−46
2·0	−7	−64
4·0	−12	−76
6·0	−16	−81
8·0	−18	−83
10·0	−20	−85

2.3.2. Systems modelled by blocks

The system in Fig. 2.15 is modelled by N blocks, and is driven by a sine wave, i.e. $V_{in} = A \sin \omega t$ where A is the amplitude, and ω the angular frequency. The output will also be a sine wave of angular frequency ω, but probably of different amplitude and shifted in phase, i.e.

$$V_0 = B \sin (\omega t + \phi)$$

Each block is described by a gain K_n and a phase shift ϕ_n. Note that these gains and phase shifts will depend on the driving frequency. The voltage V_1 will simply be:

$$V_1 = K_1 A \sin (\omega t + \phi_1)$$

Similarly, V_2 will be given by multiplying V_1 by K_2, and adding a

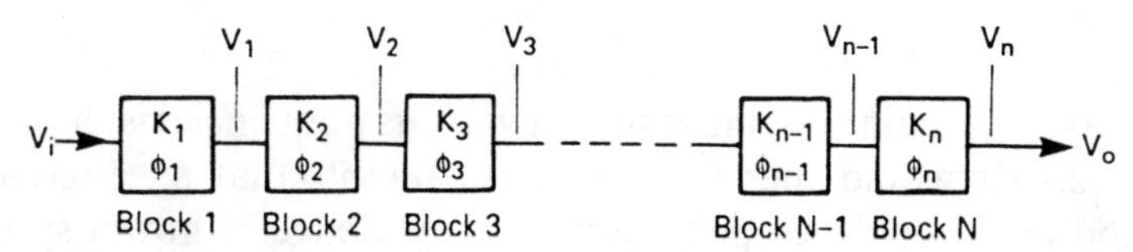

Fig. 2.15 Gains and phase shifts for a system modelled by blocks.

further phase shift ϕ_2, i.e.

$$V_2 = K_1K_2 \text{ A} \sin(\omega t + \phi_1 + \phi_2)$$

Extending this argument we arrive at

$$V_0 = K_1.K_2 \ldots K_n \text{ A} \sin(\omega t + \phi_1 + \phi_2 + \ldots + \phi_n) \qquad (2.39)$$

We can thus deduce that for a series arrangement of blocks, at any given frequency, the overall gain is the *product* of the individual gains, and the overall phase shift is the *sum* of the individual phase shifts.

The determination of the open loop frequency response of a complex system can thus be simplified considerably by considering each small subsystem separately and combining the individual gains and phase shifts.

The gain of a block is more conveniently expressed in decibels, a unit borrowed from acoustical engineering. The basic unit, the bel, is used to compare power, and is defined

$$\text{bel} = \log(\text{power } 1/\text{power } 2)$$

The bel is an inconvenient unit, and the decibel is more commonly used, where

$$\text{decibels} = 10 \log(\text{power } 1/\text{power } 2)$$

If the powers are expressed in terms of voltages, since power $\propto (\text{voltage})^2$, we can write

$$\text{decibels} = 20 \log(V_1/V_2) \qquad (2.40)$$

This last step is very dubious as it assumes equal load resistances, but equation 2.40 is commonly used to express block gains in control engineering. The use of decibels is even more unsound when applied to blocks such as current to pressure converters with different input and output quantities (4–20 mA in, 3–15 psi out, corresponding to a 'gain' of 20 log (12/16) = $-$ 2·5 dB). The total loop gain, however, is dimensionless (as explained in Section 1.4.1) so it is legitimate to express closed and open loop gains in dB even if the individual block gains are mathematically rather suspicious.

By substituting values into equation 2.40, it is easy to show that unity gain corresponds to 0 dB and a gain of two to 6 dB. Gains less than unity are negative when expressed in decibels, a gain of 0·5, for example, is equivalent to -6 dB.

Expressing gains in decibels allows large gains to be expressed easily (a gain of 1000, for example, is 60 dB), but the main advantage is that the gains of series blocks can be added when expressed in

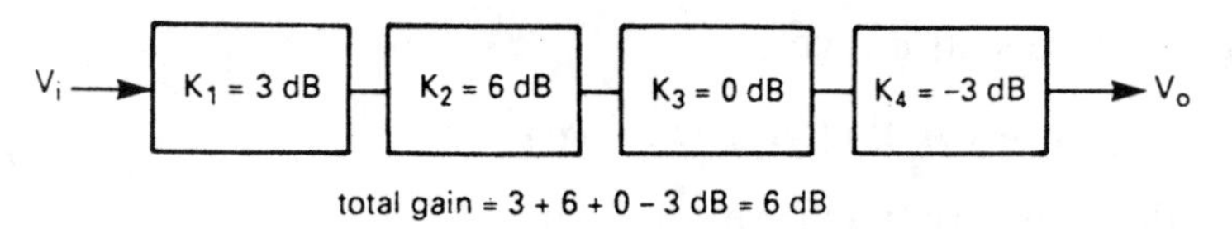

Fig. 2.16 The advantage of representing gain in decibels.

decibels. In Fig. 2.16, for example, the open loop gain is simply $(3 + 6 + 0 - 3)\text{dB} = 6\,\text{dB}$.

2.3.3. From open loop frequency response to closed loop frequency response

Given the open loop frequency response of a system at any frequency (i.e. the gain and phase shift) it is a straightforward, if somewhat laborious, operation to deduce the closed loop frequency response.

The behaviour of any system can be inferred by comparing its frequency response over a range of frequencies with blocks whose behaviour is known (e.g. first order and second order systems). Knowledge of the closed loop frequency response derived from the open loop response thus allows system performance to be predicted.

The system in Fig. 2.17 is being driven by a sinusoidally varying sine wave of angular frequency ω rads/sec. The error e and the output will also be sinusoids of angular frequency ω, but of different amplitude and shifted in phase. Suppose at frequency ω the system has a gain of 5 and a lag of 45°.

Assume that $e = \sin \omega t$ then

$$V_o = 5 \sin (\omega t - 45)$$

$$\text{But } e = V_i - V_o$$

$$\sin \omega t = V_i - 5 \sin (\omega t - 45)$$

$$\begin{aligned} \text{or } V_i &= \sin \omega t + 5 (\sin \omega t \cos 45 - \cos \omega t \sin 45) \\ &= \sin \omega t + 5 (0{\cdot}707 \sin \omega t - 0{\cdot}707 \cos \omega t) \\ &= 4{\cdot}5 \sin \omega t - 3{\cdot}5 \cos \omega t \end{aligned} \tag{2.41}$$

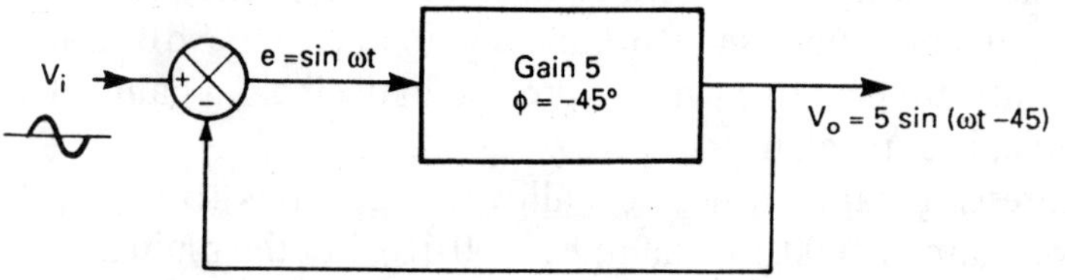

Fig. 2.17 Going from open to closed loop response.

An expression of the form A sin θ + B cos θ can be represented as

$$C(\sin\theta\cos\phi + \cos\theta\sin\phi)$$

where $\phi = \tan^{-1}(B/A)$

and $C = \sqrt{A^2 + B^2}$

but $\sin\theta\cos\phi + \cos\theta\sin\phi = \sin(\theta + \phi)$

Hence

$$A\sin\omega + B\cos\theta = \sqrt{A^2 + B^2}\,\sin(\theta + \tan^{-1}(B/A)) \quad (2.42)$$

Applying equation 2.42 to equation 2.41 gives

$$V_i = 5{\cdot}7\sin(\omega t - 37{\cdot}8)$$

The closed loop gain (V_o/V_i) is thus 5/5·7 = 0·88 and the closed loop phase shift is (−45 − (−37·8)) = − 7·2°.

Given the open loop gain and phase shift at any frequency, the closed loop gain and phase can be calculated in a similar way. Note that the closed loop response depends solely on the open loop gain and phase shift and the actual frequency does not enter into the calculation. Any system with an open loop gain of 5 and phase shift −45° at a particular frequency will have a closed loop gain of 0·88 and phase shift of −7·2°.

It would therefore seem reasonable to tabulate open/closed loop frequency responses in some way to allow the closed loop frequency response to be 'looked up' from the open loop response. This requirement is met by the Nichols chart, described in Section 2.3.6 and, to a lesser extent, by the M and N curves on the Nyquist diagrams described in Section 2.3.5.

2.3.4. Bode diagram

There are many ways of displaying the frequency response of a system for further analysis. The most obvious, but possibly the least useful, is to tabulate gain (in dB) and phase shift against frequency. Figure 2.18 represents a simple level control system. If this is analysed using the techniques described later, the open loop frequency response in Table 2.2 is obtained. This response will be plotted in various forms in this, and the following, sections.

The Bode diagram plots phase shift (in degrees) and gain (in decibels) on separate graphs against a logarithmic frequency scale (in

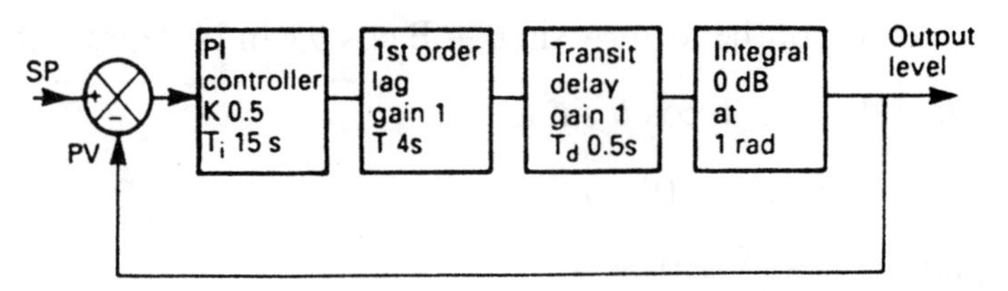

Fig. 2.18 Simple level control used for Bode, Nyquist and Nichols chart.

Table 2.2

Omega (rads)	*Gain (dB)*	*Phase shift (degrees)*
0·04	28	–160
0·06	22	–154
0·08	18	–150
0·10	15	–149
0·20	6	–153
0·30	1	–162
0·40	–3	–169
0·60	–10	–181
0·80	–15	–191
1·00	–18	–199
2·00	–30	–233

Hz or rads/sec whichever is more convenient). The use of a logarithmic frequency scale gives a wide frequency range and allows some simple 'rules of thumb' to be used when plotting. Log/lin graph paper is readily available for the plotting of Bode diagrams (e.g. Chartwell 5542). The system of Fig. 2.18 is plotted in Fig. 2.19.

2.3.5. The Nyquist diagram

In Fig. 2.20a, V_i is a sine wave applied to a linear block, giving an output signal V_o which will be a sine wave of the same frequency but differing in amplitude and phase. This system can be represented by a so-called 'phasor diagram' as Fig. 2.20b where the lengths of the arrows represent the amplitude and the angle between them the phase shift. The diagram drawn represents a gain of 0·75 and a phase *lead* of 30°.

A Nyquist diagram plots the locus of the phasor diagram as the frequency changes on a circular chart (e.g. Chartwell 4001). The distance from the origin represents the *linear* gain (i.e. *not* in decibels) and the angle the phase shift. In Fig. 2.21, for example, four points are plotted corresponding to (see table on p. 67):

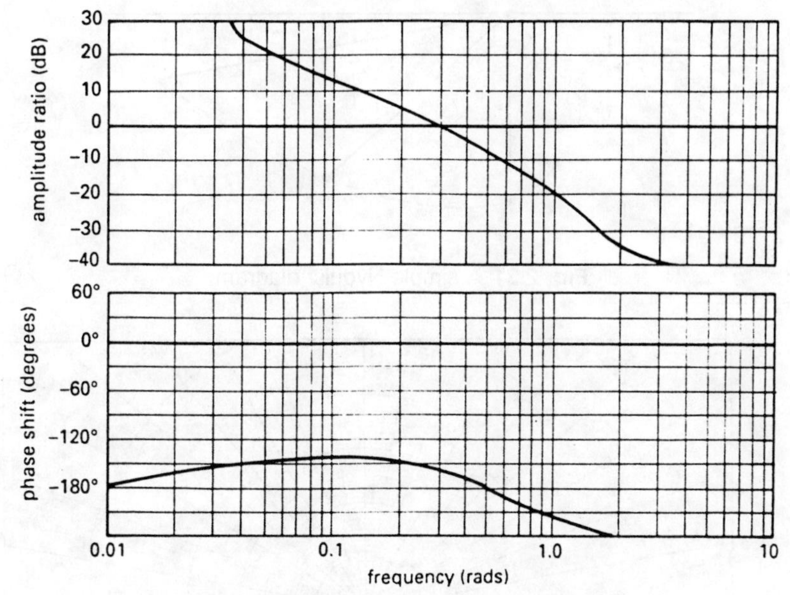

Fig. 2.19 System represented on Bode diagram.

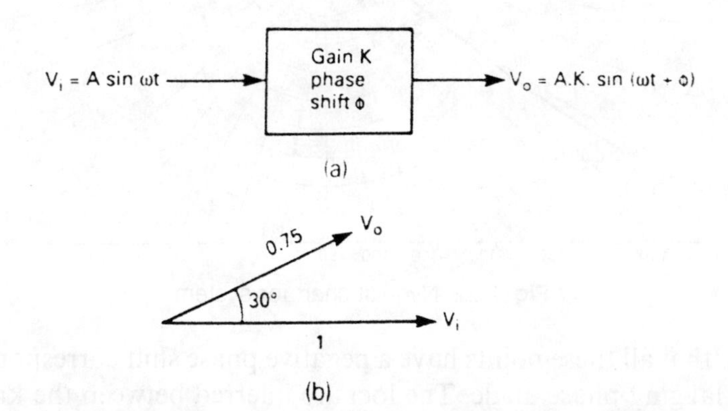

Fig. 2.20 Phasor representation, (a) Representation of system driven by a sine wave; (b) Phasor representation of gain of 0·75 and phase shift of 30°.

ω (rads/sec)	*Gain*	*Phase shift (degrees)*
1	3	−10
2	2	−40
4	1·5	−90
8	1	−130

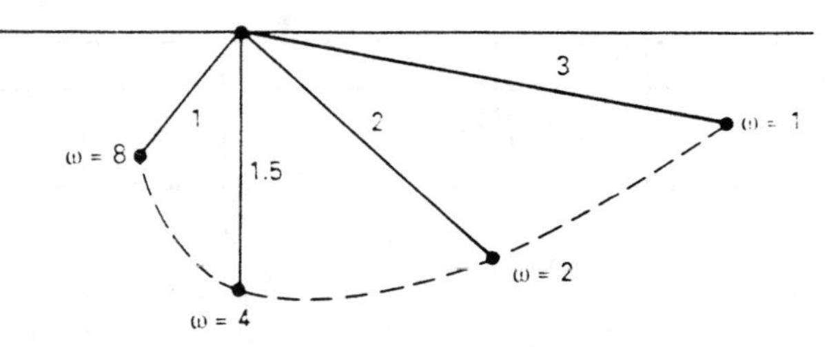

Fig. 2.21 A simple Nyquist diagram.

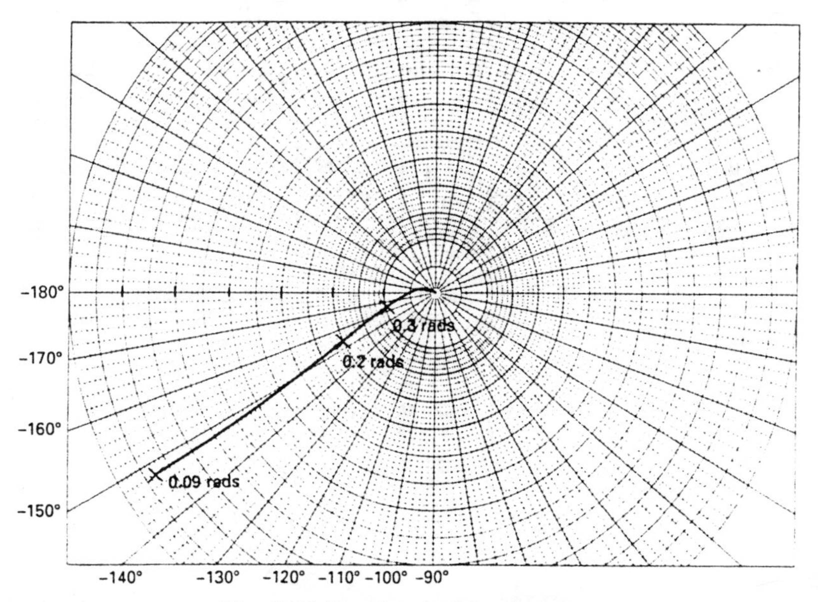

Fig. 2.22 Nyquist chart for system.

Note that all these points have a negative phase shift corresponding to a lagging phase angle. The loci are inferred between the known points, and shown dotted.

The system of Fig. 2.18 is shown plotted onto a Nyquist diagram in Fig. 2.22.

Figure 2.23 represents a system with unity feedback, i.e. the open loop gain is determined solely by the plant and the control elements. Most real life plants can be considered to be of this form if the assumption is made that the controlled variable is the output from the transducer rather than the actual quantity of interest on the plant. If the transducer has been correctly chosen this is a reasonable assumption.

Section 2.3.3 showed how closed loop gain and phase shift can be determined from open loop gain and phase shift. Loci of constant

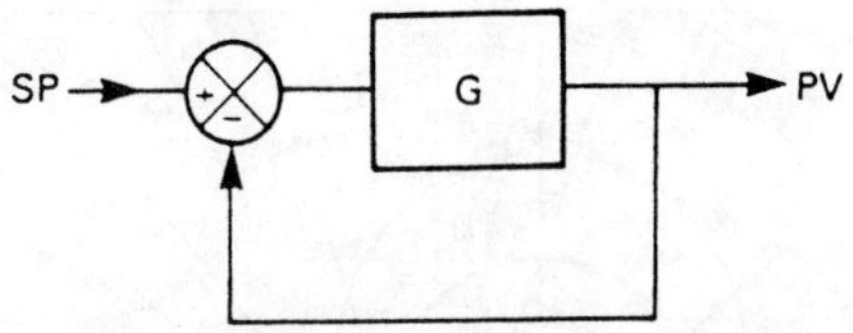

Fig. 2.23 A system with unity feedback.

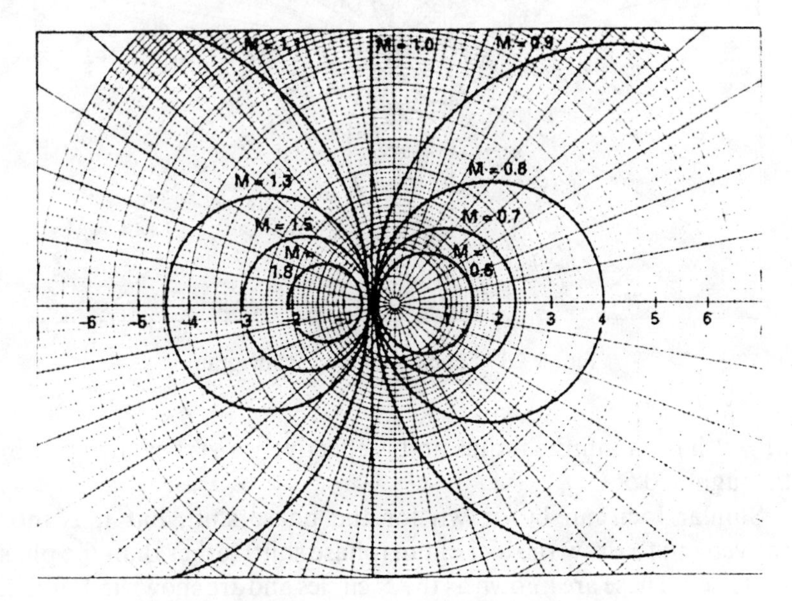

Fig. 2.24 M curves on Nyquist diagram.

gain and phase shift for a unity feedback system can be plotted onto a Nyquist diagram.

The gain loci, called the M circles, are shown in Fig. 2.24. These are circles of radius

$$R = \left| \frac{M}{1 - M^2} \right| \tag{2.43}$$

on centres

$$Y = 0$$

$$X = \frac{M^2}{1 - M^2} \tag{2.44}$$

where M is the closed loop gain.

For example, the circle corresponding to a closed loop gain of 1·5 has a radius of 1·2 centred on X = 1·8. Note that the M circle for

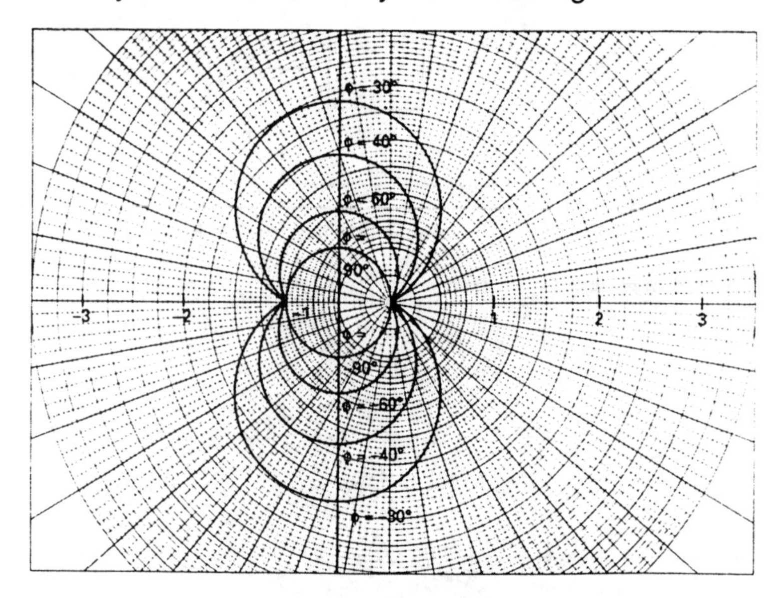

Fig. 2.25 N curves on Nyquist diagram.

M = 1 has an infinite radius and is effectively a vertical line passing through −0·5, 0.

Similar loci can be drawn for the phase shift, but it is more convenient to draw the loci of tan (phase shift) rather than the phase shift itself. These are known as the N circles and are shown in Fig. 2.25.

It can be shown that these too are circles, with radius

$$R = \left| \frac{\sqrt{N^2 + 1}}{2N} \right| \tag{2.45}$$

and centre

$$X = -0{\cdot}5 \tag{2.46}$$

$$Y = \frac{1}{2N} \tag{2.47}$$

For a phase shift of −45°, for example, N = tan(−45) = −1, giving R = 0·707, and centre X = −0·5, Y = −0·5.

By plotting the M and N circles onto a Nyquist diagram, and superimposing the frequency response curve of a system of interest, it is possible to 'read off' the corresponding closed loop response as shown in Fig. 2.26.

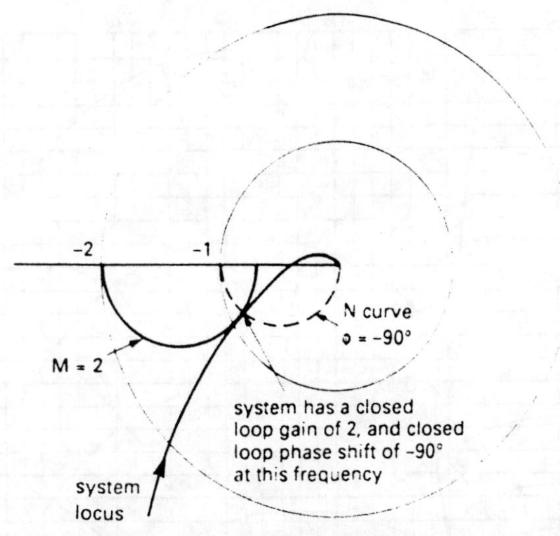

Fig. 2.26 Use of M and N curves to determine closed loop response.

2.3.6. The Nichols chart

The Nichols chart plots open loop gain, in decibels, against open loop phase shift. Figure 2.27 shows the system in Fig. 2.18 plotted onto a Nichols chart (Chartwell 7514).

The Nichols chart is a particularly effective method of displaying frequency response as it is easy to visualise the effect of adjusting controller parameters. Adjusting controller gain, for example, corresponds to a vertical shift of the frequency response curve.

M and N loci can be drawn on the Nichols chart, but these are not simple circles. A range of M curves is shown in Fig. 2.28a and N curves in Fig. 2.28b. It will be noted that the Chartwell graph paper used for Fig. 2.27 includes M and N curves, allowing closed loop frequency response to be read directly from a plot of open loop frequency response.

2.3.7. Determination of frequency response

A system modelled by differential equations can, in general, be represented by

$$a_n \frac{d^n x}{dt^n} + \ldots + a_2 \frac{d^2 x}{dt^2} + a_1 \frac{dx}{dt} + a_0 x = f(t)$$

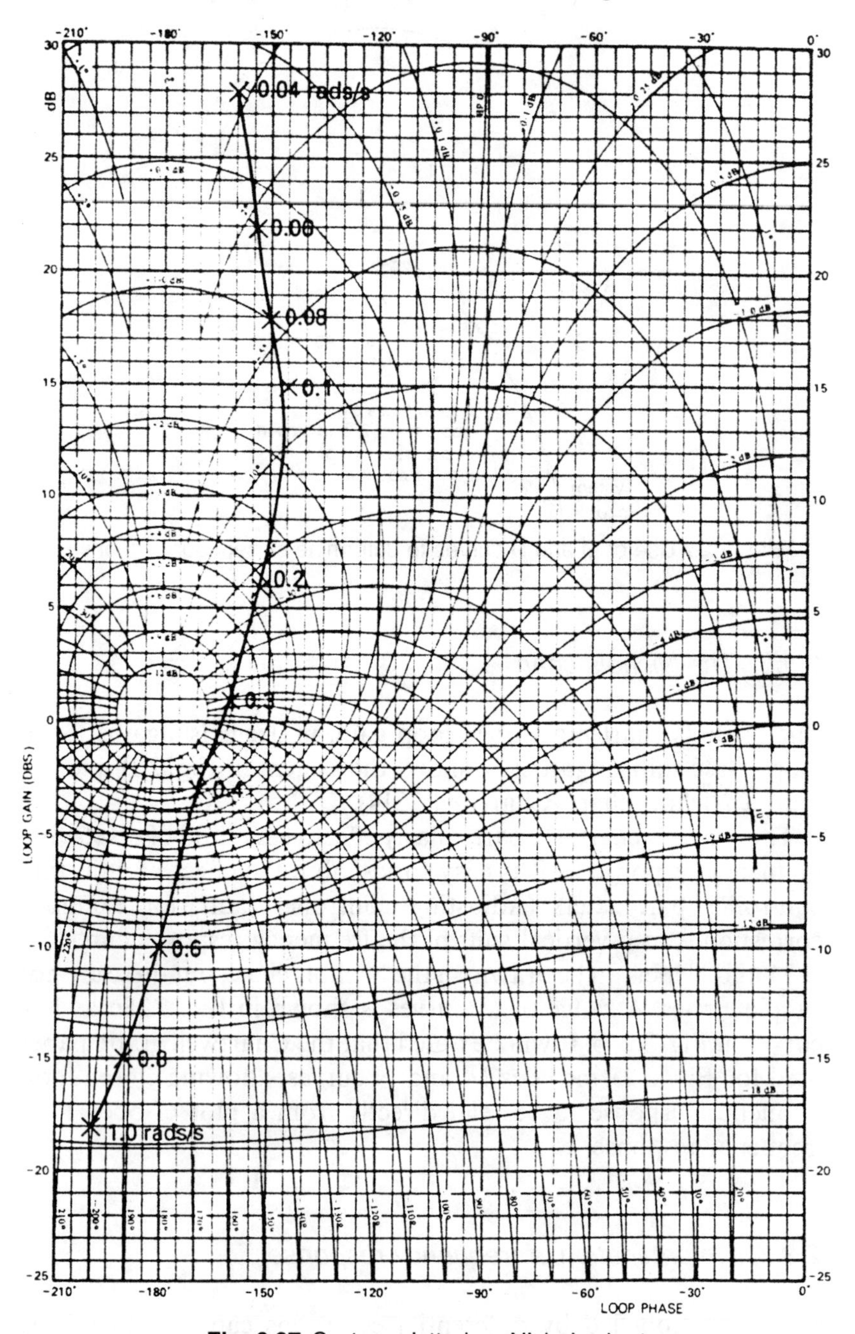

Fig. 2.27 System plotted on Nichols chart.

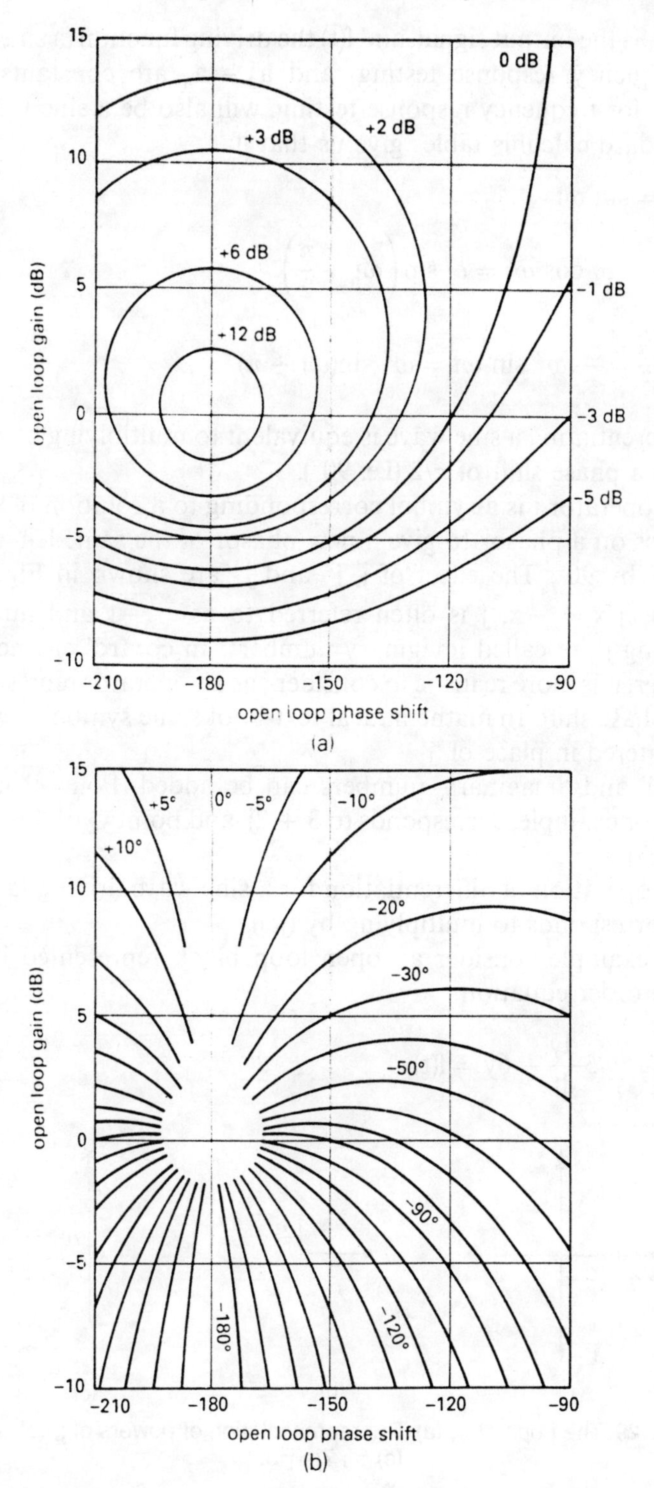

Fig. 2.28 M and N curves on Nichols chart, (a) M curves; (b) N curves.

where x is the output signal, and f(t) the driving function (a sine wave for frequency response testing) and $a_0 - a_n$ are constants. The output, for frequency response testing, will also be a sine wave.

Standard calculus tables give us that if:

$$x = \sin \omega t$$

$$\frac{dx}{dt} = \omega \cos \omega t = \omega \sin\left(\omega t + \frac{\pi}{2}\right)$$

$$\frac{d^2x}{dt^2} = -\omega^2 \sin \omega t = \omega^2 \sin(\omega t + \pi)$$

i.e. differentiating a sine wave is equivalent to multiplying by ω and adding a phase shift of $\pi/2$ (i.e. 90°).

The operator j is a symbol corresponding to a rotation of 90°. It operates on a phasor to give a new phasor of the same length but rotated by 90°. The effect of j, j^2 and j^3 are shown in Fig. 2.29. Because $j^2x = -x$, j is often referred to as $\sqrt{-1}$ and numbers involving j are called imaginary numbers. In control engineering, however, it is more realistic to consider j as an operator representing a 90° phase shift. In mathematical text-books, the symbol 'i' will be encountered in place of 'j'

'Real' and 'imaginary' numbers can be added. Point P in Fig. 2.29b, for example, corresponds to $3 + 2j$, and point Q on Fig. 2.29c to $5 + 7j + j^2 3$.

The operation of differentiation for a sine wave driving function thus corresponds to multiplying by $(j\omega)$.

For example consider an open loop block represented by the second order equation

$$\frac{d^2y}{dt^2} + 5\frac{dy}{dt} + 6y = f(t)$$

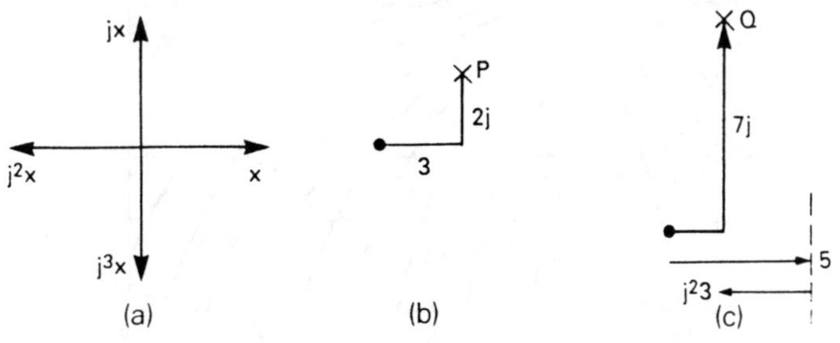

Fig. 2.29 The j operator, (a) The representation of powers of j; (b) $3+2j$; (c) $5+7j+j^2 3$.

This can be rewritten

$$((j\omega)^2 + 5j\omega + 6)Y = X$$

where Y is used to represent the phasor corresponding to the output sine wave, and X the phasor corresponding to the driving sine wave. This can be reorganised to give the system gain.

$$\frac{Y}{X} = \frac{1}{(j\omega)^2 + 5j\omega + 6}$$

Noting that $(j\omega)^2 = -\omega^2$ we get, by grouping real and imaginary parts

$$\frac{Y}{X} = \frac{1}{(6 - \omega^2) + j5\omega} \tag{2.48}$$

At this stage it is useful to remind ourselves of what we are trying to achieve. The determination of the frequency response at any frequency is simply the calculation of the gain (or amplitude ratio) and phase shift at that frequency.

If we have a complex number $C + jD$ as Fig. 2.30, the resultant magnitude Z is

$$Z = \sqrt{C^2 + D^2} \tag{2.49}$$

and the phase difference is

$$\phi = \tan^{-1}(D/C) \tag{2.50}$$

Any complex number

$$a + (j)^n b_n + (j)^{n-1} b_{n-1} + \dots + jb_0$$

can be represented in the form $A + jB$ by noting that $(j)^2 = -1$, $(j)^3 = -j$, $(j)^4 = 1$ and so on.

Equation 2.48 is of the form $1/(A + jB)$, and from the argument in the preceding paragraph it can be seen that the equation for any differential equation will ultimately end up in this form, where A and B both contain terms involving ω.

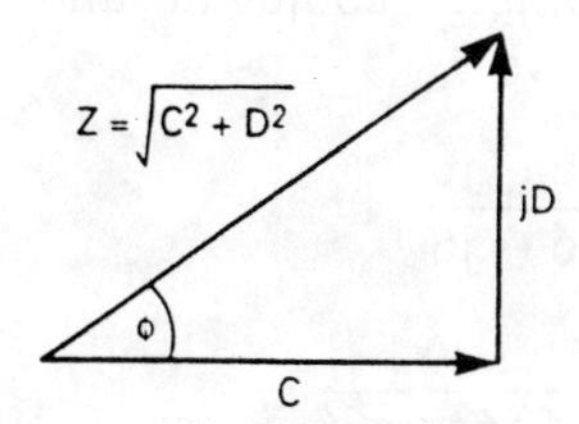

Fig. 2.30 Magnitude and phase shift of a complex number.

To determine the frequency response, we need to go from $1/(A + jB)$ to $C + jD$ at which point we can directly apply equations 2.49 and 2.50 to give the gain and phase shift.

Consider the multiplication

$$(a + jb)(a - jb)$$

Expanding gives

$$a^2 - jab + jab - j^2b = a^2 + b^2 \qquad \text{since } -j^2 = +1$$

$(a - jb)$ is called the complex conjugate of $(a + jb)$. Multiplying a complex number by its complex conjugate gives a real result.

Applying this to $1/(A + jB)$ gives us the result we require as follows

$$\frac{1}{A + jB} = \frac{1}{A + jB} \cdot \frac{A - jB}{A - jB}$$

$$= \frac{A - jB}{A^2 + B^2}$$

$$= C + jD$$

$$\text{where } C = \frac{A}{A^2 + B^2} \text{ and } D = -\frac{B}{A^2 + B^2}$$

After a rather lengthy diversion we can now return to our original equation for the system.

$$\frac{Y}{X} = \frac{1}{(6 - \omega^2) + j5\omega}$$

Let us calculate the frequency response for $\omega = 1{\cdot}2$ rads/sec. Substituting gives

$$\frac{Y}{X} = \frac{1}{(6 - 1{\cdot}44) + j6}$$

$$= \frac{1}{4{\cdot}56 + j6}$$

Multiplying by the complex conjugate and separating real and imaginary terms as above

$$\frac{Y}{X} = \frac{1}{4{\cdot}56 + j6} \cdot \frac{4{\cdot}56 - j6}{4{\cdot}56 - j6}$$

$$= \frac{4{\cdot}56}{4{\cdot}56^2 + 6^2} - \frac{j6}{4{\cdot}56^2 + 6^2}$$

$$= 0{\cdot}08 - j0{\cdot}11$$

The gain is

$$\sqrt{0{\cdot}08^2 + 0{\cdot}11^2} = 0{\cdot}136$$

and the phase shift is

$$\tan^{-1}(-0{\cdot}11/0{\cdot}08) = \tan^{-1}(-1{\cdot}375) = -54°$$

At a frequency of 1·2 rads/sec, therefore, the system has a gain of 0·136 and a phase shift of $-54°$. The gain and phase shift at any frequency could be obtained in a similar manner.

It is worth reviewing the steps to obtain the frequency response from a differential equation representing the open (or closed) loop behaviour of the system. Suppose the differential equation is of the form:

$$a_n \frac{d^n y}{dt^n} + \ldots + a_2 \frac{d^2 y}{dt^2} + a_1 \frac{dy}{dt} + a_0 y = f(t)$$

(a) Replace y by Y denoting an output phasor, and f(t) by X denoting the driving sinusoidal phasor.

(b) Replaced $d^n y/dt^n$ by $(j\omega)^n$ giving

$$(a_n (j\omega)^n + \ldots + a_2 (j\omega)^2 + a_1 (j\omega) + a_0) Y = X$$

(c) Rearrange to give the gain

$$\frac{\text{output}}{\text{input}} = \frac{Y}{X} = \frac{1}{(a_n (j\omega)^n + \ldots + a_1 (j\omega) + a_0)}$$

(d) Using the relationships $(j\omega)^2 = -1, (j\omega)^3 = -j\omega$ etc., reorganise the right-hand side of the equation into the form

$$\frac{Y}{X} = \frac{1}{A + jB}$$

When finding the amplitude ratio and phase shift for one specific frequency, it is normally more convenient to substitute for ω at this stage.

(e) Multiply top and bottom by complex conjugate $(A - jB)$ to convert to the form $C + jD$, i.e.

$$\frac{Y}{X} = \frac{1}{A + jB} \cdot \frac{A - jB}{A - jB} = C + jD$$

where $C = \dfrac{A}{(A^2 + B^2)}$ and $D = \dfrac{-B}{(A^2 + B^2)}$

(f) The gain is then simply $\sqrt{C^2 + D^2}$, and the phase shift $\tan^{-1}(D/C)$. The gain can also be expressed directly in terms of A, B as

$1/\sqrt{A^2 + B^2}$ since the magnitude of

$$\frac{A - jB}{A^2 + B^2} \text{ is } \sqrt{\frac{A^2 + B^2}{(A^2 + B^2)^2}}$$

Similarly the phase shift can be expressed directly as $\tan^{-1}(-B/A)$.

Using this technique, the frequency response of several common system building blocks will now be obtained.

2.3.8. First order lag

A first order lag can be represented by the equation

$$T\frac{dy}{dt} + y = f(t) \tag{2.51}$$

where f(t) is the driving function and T the time constant.

Following the steps above gives the relationship between output and input as

$$\frac{Y}{X} = \frac{1}{1 + (j\omega)T}$$

$$= \frac{1}{1 + \omega^2T^2} - j\frac{\omega T}{1 + \omega^2T^2}$$

The gain is thus

$$G = \frac{\sqrt{1 + \omega^2T^2}}{1 + \omega^2T^2} = \frac{1}{\sqrt{1 + \omega^2T^2}}$$

and the phase shift

$$\phi = \tan^{-1}(-\omega T)$$

These are plotted in Bode diagram of Fig. 2.31. This is drawn normalised to the so-called 'corner frequency' ω_c, where $\omega_c = 1/T$. It can be seen that the gain is unity (0 dB) at low frequencies, and falls at −20 dB per decade at high frequencies. At the corner frequency the gain is −3 dB.

The phase shift is 0° for low frequencies, and −90° at high frequencies. The phase shift is −45° at the corner frequency. Between $0{\cdot}1\omega_c$ and $10\omega_c$ the phase shift increases at −45° per decade.

Figure 2.31 also shows a useful straight line approximation, with 0 dB gain up to ω_c, and then a fall off at 20 dB/decade. The phase shift

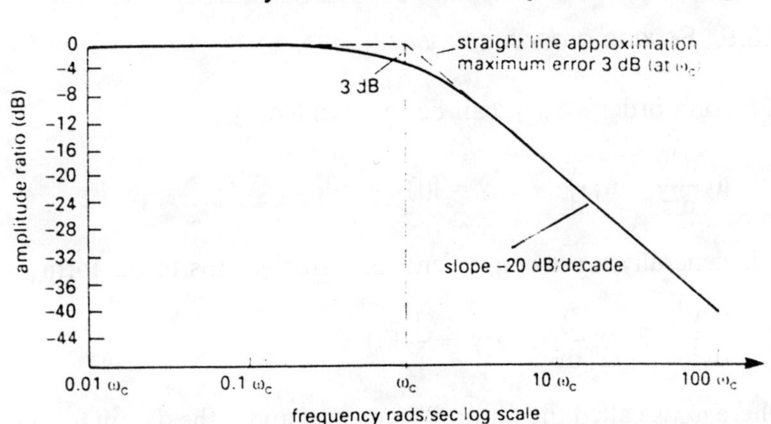

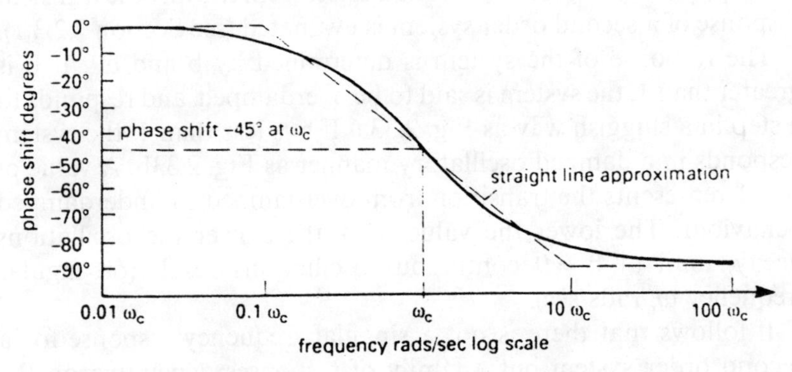

Fig. 2.31 Bode diagram of first order lag and straight line approximation.

is approximated by 0° below $0{\cdot}1\omega_c$ and $-90°$ over $10\omega_c$ and a straight line of slope $-45°$ per decade between these two points. The maximum gain error from this simplification is 3 dB (at ω_c) and the maximum phase error is 5° (at $0{\cdot}1\omega_c$ and $10\omega_c$).

If a first order lag is plotted onto the Nyquist diagram the result is a semicircle as shown in Fig. 2.32. The radius of the circle is half the low frequency gain, and is centred on the point x = half the low frequency gain, y = 0.

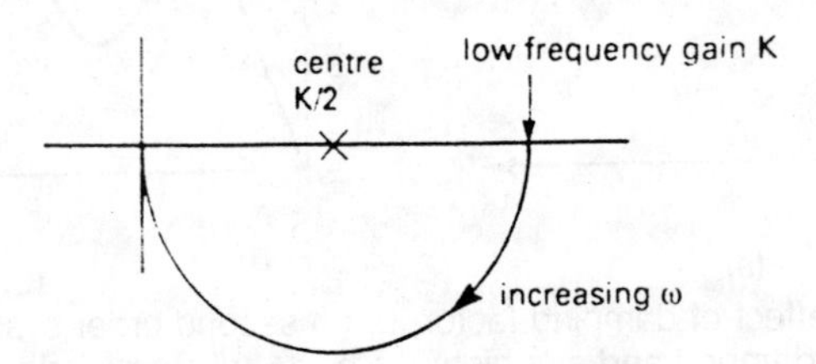

Fig. 2.32 First order lag on a Nyquist diagram.

2.3.9. Second order response

A second order system can be represented by

$$a_2 \frac{d^2y}{dt^2} + a_1 \frac{dy}{dt} + a_0 y = f(t)$$

It is generally more convenient to normalise this to the form

$$\frac{d^2y}{dt^2} + 2b\omega_n \frac{dy}{dt} + \omega_n^2 y = \frac{1}{a_2} f(t)$$

where ω_n is called the natural frequency, and b the damping factor. This apparently random conversion arises from the way the transient response of a second order system is evaluated (see Section 2.2.3.1).

The response of the system is determined by b and ω_n. If b is greater than 1, the system is said to be overdamped, and responds to a step in a sluggish way as Fig. 2.33a. If b is less than 1, the system responds in a damped oscillatory manner as Fig. 2.33b. A value of $b = 1$ represents the transition from overdamped to underdamped behaviour. The lower the value of b, the slower the oscillations decay, until at $b = 0$ continuous oscillations result (of angular frequency ω_n rads/sec).

If follows that there is not a singular frequency response for a second order system but a family of responses depending on the value of b.

Figure 2.34 shows the family drawn on a Bode diagram normalised in terms of ω_n (i.e. a frequency of 1 on the diagram represents ω_n).

It can be seen from the gain curves that at low frequencies (less than $0{\cdot}1\omega_n$) the gain is unity (0 dB). At high frequencies (above $10\omega_n$) the gain falls off at -40 dB/decade. For values of $b < 0{\cdot}7$, a response peak occurs (e.g. $+8$ dB for $b = 0{\cdot}2$).

At low frequencies the phase shift is zero; at high frequencies the

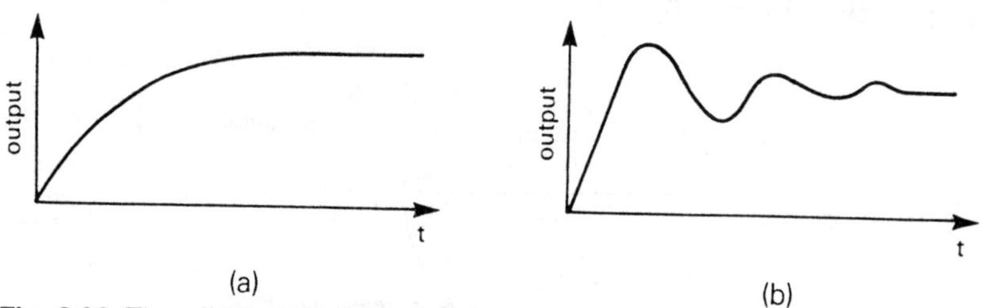

Fig. 2.33 The effect of damping factor, b, on second order systems, (a) $b > 1$, system overdamped and sluggish; (b) $b < 1$, system underdamped and oscillatory.

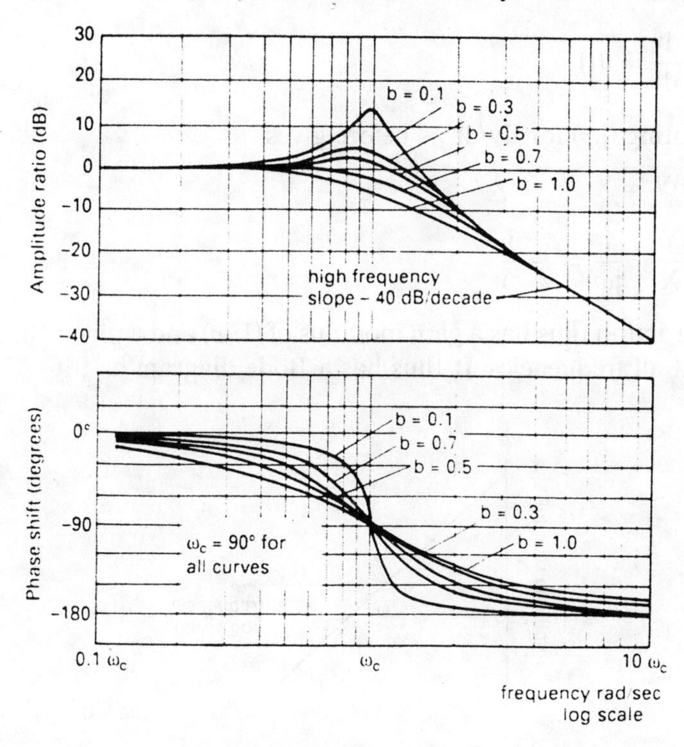

Fig. 2.34 Bode diagram of second order systems.

phase shift tends towards $-180°$. For all values of b, a phase shift of $-90°$ occurs at ω_n. The value of b determines the range over which the phase shift occurs. For most systems a value of $b = 0{\cdot}7$ gives a reasonable compromise between speed of response and minimal overshoot.

The second order system is a 'standard' against which the performance of a closed loop system can be compared. The response of second order systems are discussed further in Section 3.2.

2.3.10. Integral action

Integral action is a characteristic of position and level control systems. It is represented by

$$y = \int f(t)dt$$

$$\text{or } \frac{dy}{dt} = f(t)$$

Substituting ($j\omega$) for dy/dt as before gives

$$(j\omega)Y = X$$

$$\text{or } \frac{Y}{X} = \frac{1}{j\omega} = -\frac{j}{\omega}$$

Integral action thus has a gain modulus of ($1/\omega$) and a phase shift of $-90°$ at all frequencies. It thus has a Bode diagram as Fig. 2.35.

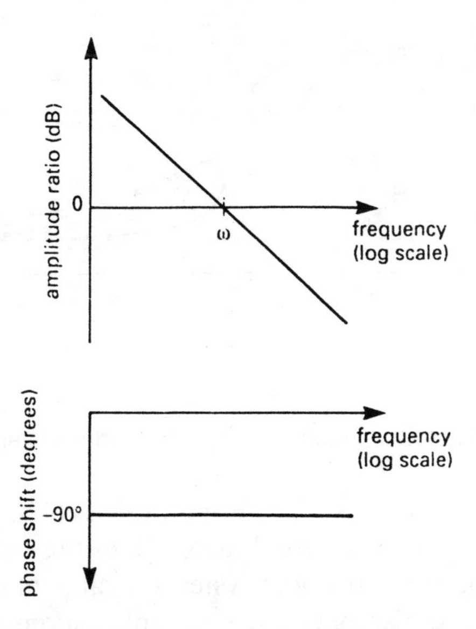

Fig. 2.35 Bode diagram for integral action.

2.3.11. Transit delay

The transit delay is a function of time and distance, and occurs in systems with long pipe runs, sampling or conveyor belts. The output signal is exactly the same as the input signal, but delayed by a fixed time T as shown in Fig. 2.36a.

For a sine wave, the delay represents a phase shift. If the angular frequency is ω rads/sec, the period is $2\pi/\omega$ seconds. A shift of T seconds thus represents a phase shift of

$$\phi = \frac{T}{\text{period}} \times 2\pi \text{ radians}$$

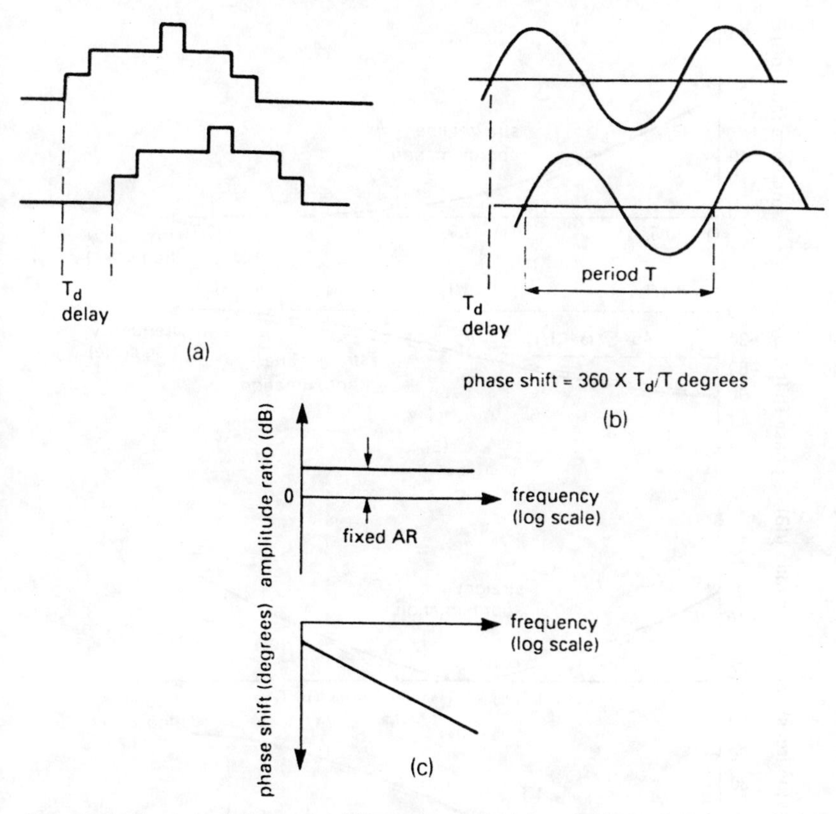

Fig. 2.36 The transit delay, (a) Effect of transit delay; (b) Transit delay and sine wave signal; (c) Bode diagram.

$$= T\omega \text{ radians}$$

$$= \frac{360\ T\omega}{2\pi} \text{ degrees}$$

The modulus is unity regardless of frequency.

A transit delay thus represents constant gain but a phase shift that increases linearly with frequency as summarised in the Bode diagram of Fig. 2.36c. On a Nyquist diagram, a transit delay appears as a unit radius circle centred on the origin.

2.3.12. P + I and P + I + D controllers

The generalised equation for a proportional plus integral (P + I) or a proportional plus integral plus derivative (P + I + D) controller is

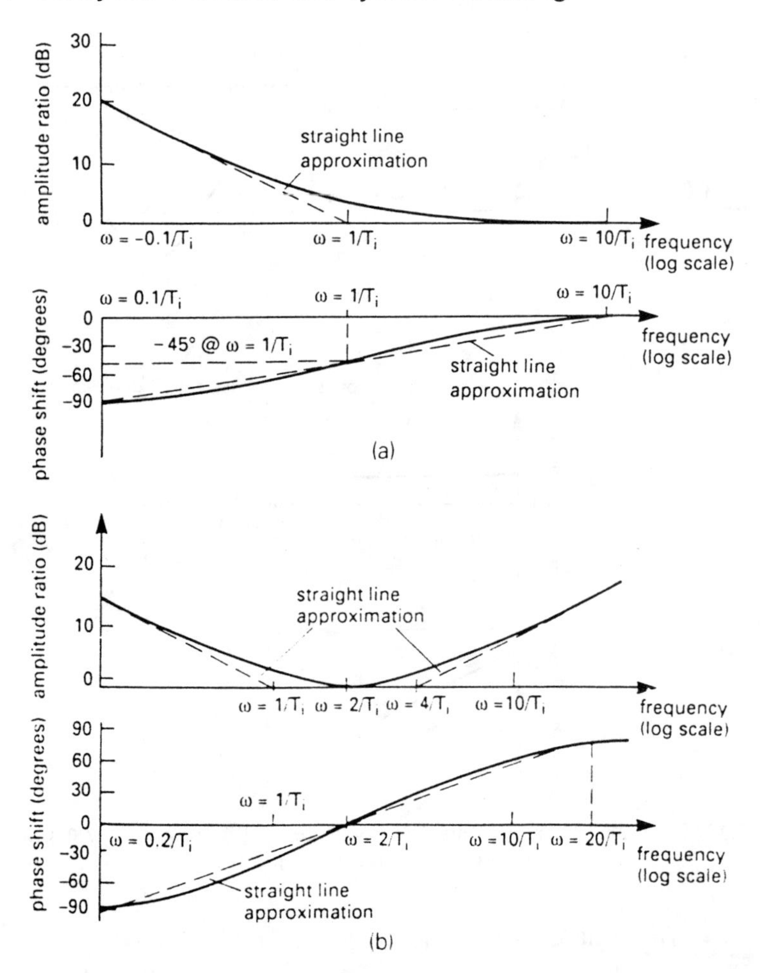

Fig. 2.37 Bode diagrams for P + I and P + I + D controllers, (a) P + I controller; (b) P + I + D controller with $T_i = 4T_d$.

$$V_o = K\left(e + \frac{1}{T_i}\int e\,dt + T_d\frac{de}{dt}\right) \tag{2.52}$$

where V_o is the output signal, K the gain, T_i the integral time and T_d the derivative time. For a P + I controller T_d is zero.

The frequency transfer function is therefore

$$\frac{Y}{X} = K\left(1 + \frac{1}{(j\omega)T_i} + (j\omega)T_d\right)$$

$$= \frac{K((j\omega)T_i + (j\omega)^2T_iT_d + 1)}{(j\omega)T_i} \tag{2.53}$$

for a P + I controller this simplifies to

$$\frac{Y}{X} = K\left(1 - \frac{(j\omega)}{T_i}\right)$$

The Bode diagram of which is shown in Fig. 2.37a for unity gain (K = 1).

There is no standard Bode diagram curve for a P + I + D controller, as the shape depends on the relative values of T_i and T_d. A useful rule of thumb for P + I + D controllers, however, sets $T_i = 4T_d$ (it will be seen later in Chapter 4 that all the standard controller tuning procedures use this ratio). For this specific case, equation 2.53 simplifies to

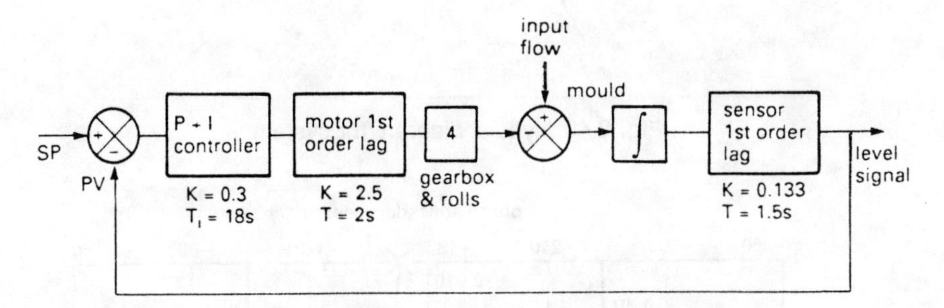

Fig. 2.38 More complete model of caster level control.

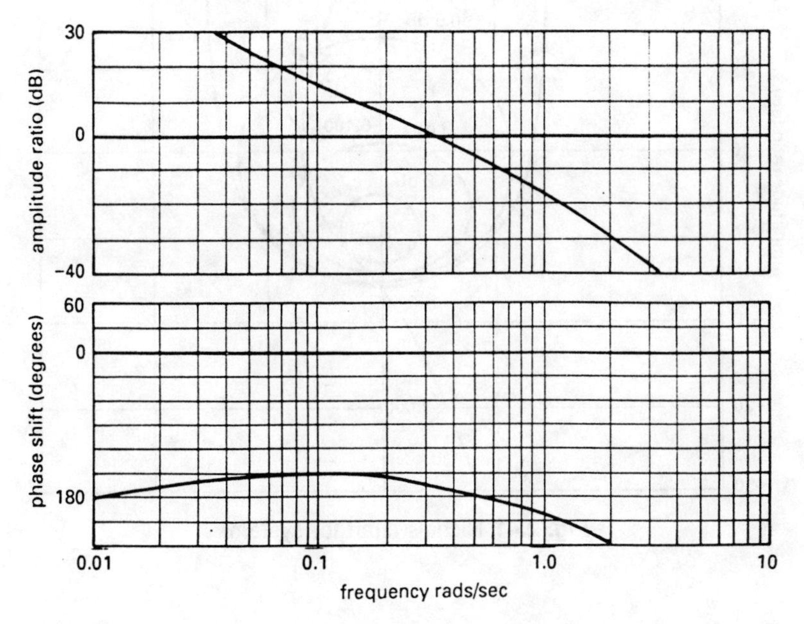

Fig. 2.39 Bode diagram for system.

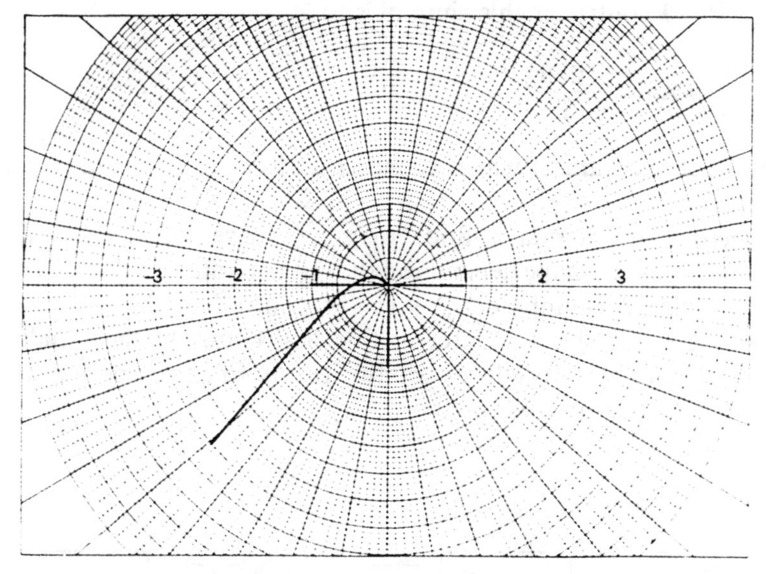

Fig. 2.40 Nyquist diagram for system.

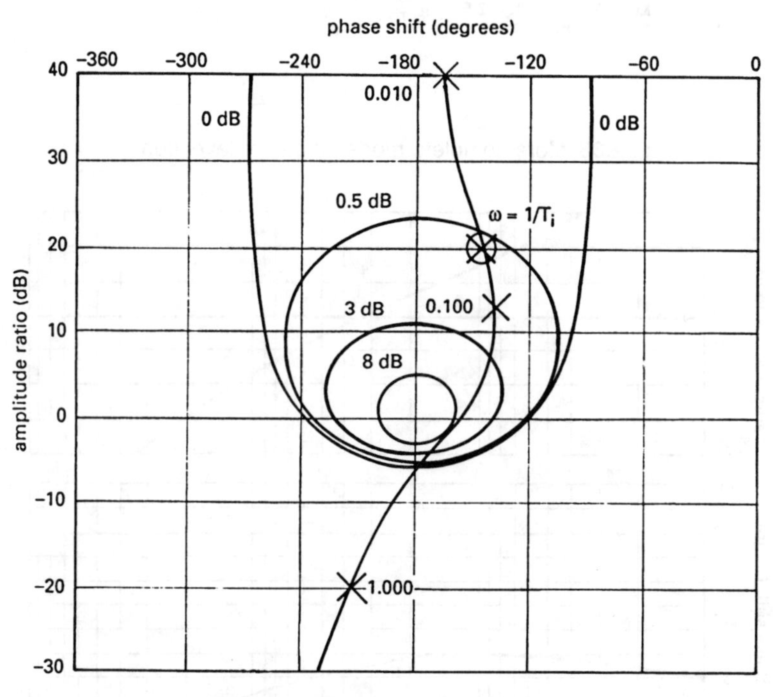

Fig. 2.41 Nichols chart for system.

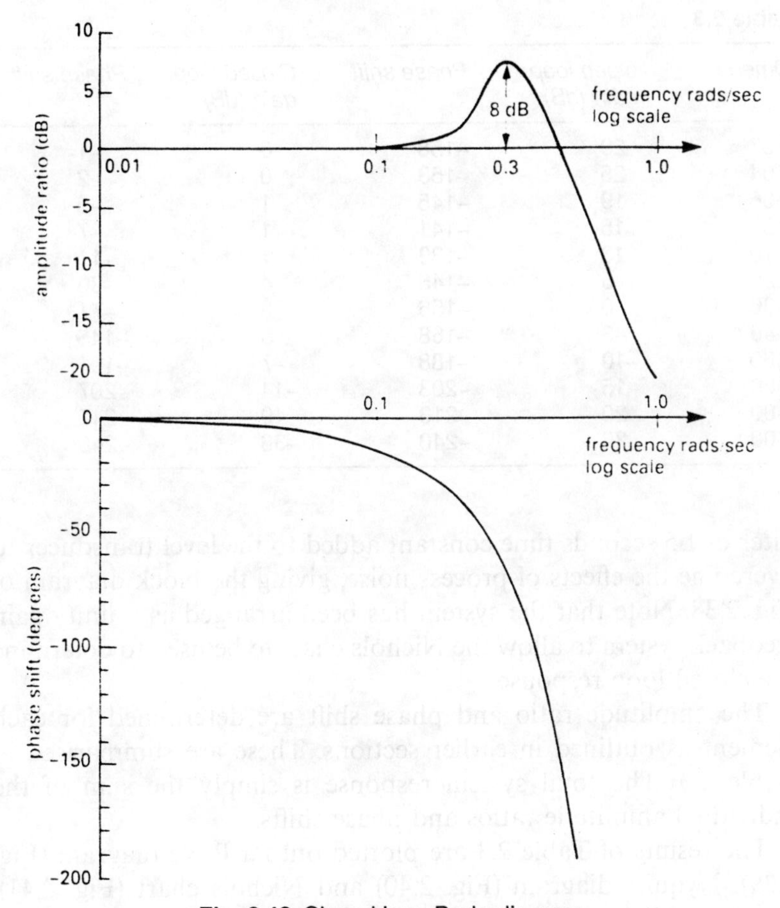

Fig. 2.42 Closed loop Bode diagram.

$$\frac{Y}{X} = \frac{K}{(j\omega)T_i}\left(1 + (j\omega)\frac{T_i}{2}\right)^2$$

This is plotted on a Bode diagram in Fig. 2.37b. Note that the amplitude ratio falls to unity (0 dB) at $\omega = 2/T_i$, and the phase shift goes from $-90°$ to $+90°$, with zero phase shift at $\omega = 2/T_i$.

2.3.13. Analysis of a complete system

To show how a system can be analysed with frequency response methods, we shall analyse a more complete model of the steel casting system introduced in Section 2.2.1. A P + I controller has been added, with a gain of 0·3 and a T_i of 18 seconds, and the first order

Table 2.3

Omega	*Open loop gain (dB)*	*Phase shift*	*Closed loop gain (dB)*	*Phase shift*
0·03	29	–158	0	–1
0.04	25	–153	0	–2
0·06	19	–145	1	–5
0·08	15	–141	1	–7
0·10	13	–139	1	–11
0·20	5	–145	4	–30
0·30	0	–156	8	–71
0·40	–3	–168	5	–145
0·60	–10	–188	–7	–191
0·80	–15	–203	–14	–207
1·00	–20	–213	–19	–217
2·00	–36	–240	–36	–240

filter of 1·5 seconds time constant added to the level transducer to overcome the effects of process noise, giving the block diagram of Fig. 2.38. Note that the system has been arranged as a unity gain feedback system to allow the Nichols chart to be used to determine the closed loop response.

The amplitude ratio and phase shift are determined for each element as outlined in earlier sections. These are summarised in Table 2.3. The total system response is simply the sum of the individual amplitude ratios and phase shifts.

The results of Table 2.3 are plotted onto a Bode diagram (Fig. 2.39), Nyquist diagram (Fig. 2.40) and Nichols chart (Fig. 2.41). From the Nichols chart the closed loop amplitude ratio and phase shift can be read off and compared with the calculated value in Table 2.3.

The closed loop response is transferred to a Bode diagram in Fig. 2.42. We can see a resemblance to a second order system with an 8 dB resonance peak. Rules of thumb can be applied to predict the closed loop response, but these are the subject of the next chapter.

Chapter 3
Stability

3.1. Introduction

In very broad terms, a closed loop system can respond to a disturbance in one of the five ways shown in Fig. 3.1. The first of these is an exponentially increasing output, usually indicative of positive feedback caused by control action in the wrong sense or a plant fault. The increasing amplitude sinusoidal response of Fig. 3.1b is typical of an unstable system, possibly caused by incorrect setting of controller gain.

Figure 3.1d also exhibits a sinusoidal response, but here the amplitude decreases with time. Such a system is oscillatory but stable. Figure 3.1c, with constant amplitude sinusoid response, marks the transition from unstable to stable response. Finally, responses such as Fig. 3.1e exhibit no sinusoidal characteristics at all. Such curves are typical of systems similar to first order lags or overdamped second order systems.

Considerations of stability of a system generally centre around the design of a suitable control algorithm or a study of the plant elements which limit plant performance. Obviously responses of Fig. 3.1a–c are undesirable, and the 'ideal' lies somewhere between Fig. 3.1d and e. A common rule of thumb aim is the quarter damped oscillatory response of Fig. 3.2 where each overshoot is one quarter the amplitude of its predecessor. Such a response could not, however, always be tolerated.

In the roll gap position control system of Fig. 3.3, where the roll gap is required to cover the range 100 mm down to 5 mm any overshoot would lead to the rolls touching on long movements (followed by an expensive stripping of screws and gearboxes and some embarrassing meetings for the control engineer).

This chapter describes various techniques to allow the stability of a design to be predicted.

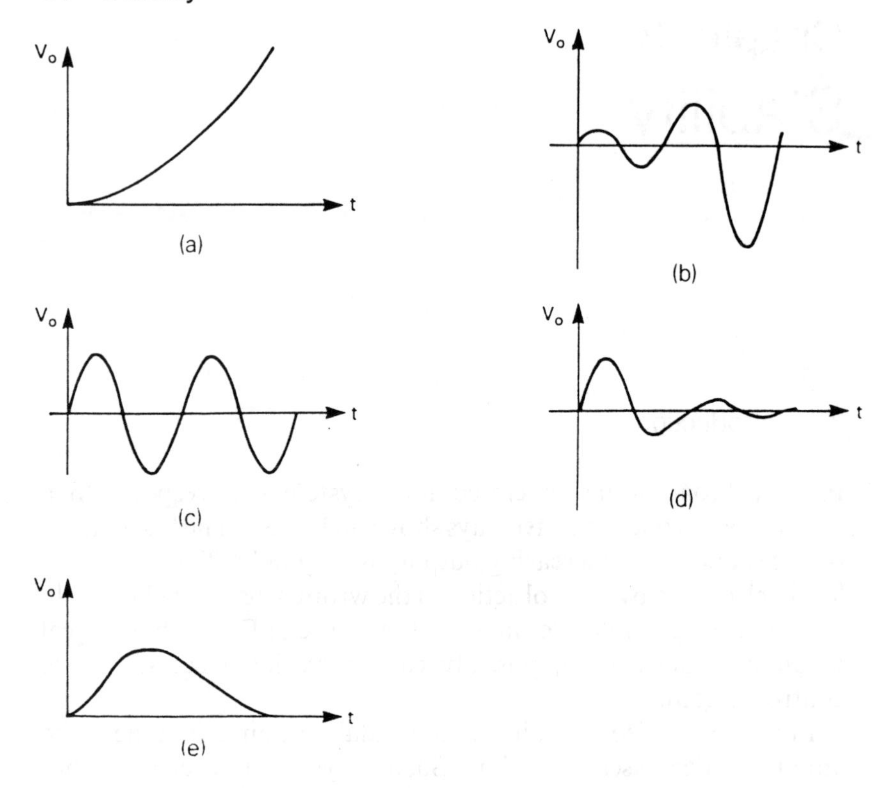

Fig. 3.1 Various forms of response to a disturbance, (a) Unstable runaway; (b) Unstable and oscillatory; (c) Unstable constant amplitude oscillation; (d) Stable and oscillatory; (e) Stable and overdamped.

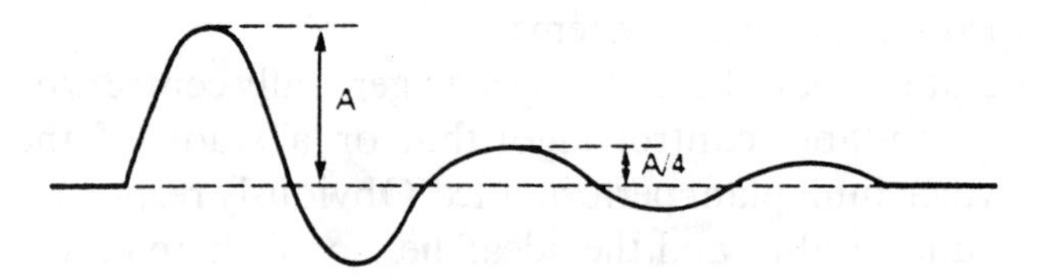

Fig. 3.2 Quarter amplitude response to a disturbance.

3.2. A study of second order systems

A second order system can exhibit all the characteristics of Fig. 3.1 which suggests that a study of a second order system can give an insight into the performance of closed loop behaviour. A second order system driven by a function f(t) can be represented by

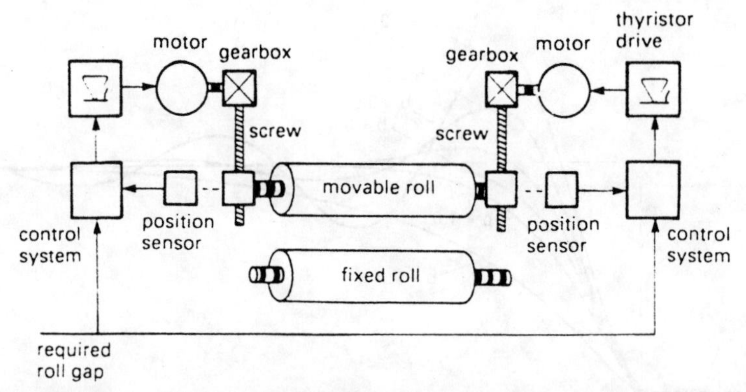

Fig. 3.3 Rolling mill gap position control, a system where no overshoot can be tolerated.

$$A\frac{d^2x}{dt^2} + B\frac{dx}{dt} + Cx = f(t)$$

which can be rearranged

$$\frac{d^2x}{dt^2} + 2b\omega_n\frac{dx}{dt} + \omega_n^2 x = f'(t) \tag{3.1}$$

where ω_n is called the natural frequency, and b the damping factor. This apparently arbitrary substitution arises out of the determination of the transient response via the auxiliary equation (see Section 2.2.3.1).

For example, the system described by

$$\frac{d^2x}{dt^2} + \frac{dx}{dt} + 4x = f(t)$$

has an ω_n of 2 rads/sec and a damping factor of 0·25.

The damping factor determines the form of the response. For negative damping factors, the response increases as Fig. 3.1a or b. For zero damping factor, continuous oscillations occur as Fig. 3.1c, with angular frequency ω_n rads/sec. With a damping factor between zero and unity, damped oscillations occur as Fig. 3.1d. The step response for $0 < b < 1$ has the form

$$x(t) = 1 - \frac{e^{-b\omega_n t}}{\sqrt{1-b^2}}\sin(\omega_d t + \phi) \tag{3.2}$$

where ω_d is the actual decaying oscillation frequency, called the damped natural frequency. It is always lower than ω_n, as shown in Fig. 3.4a, and is related to b by

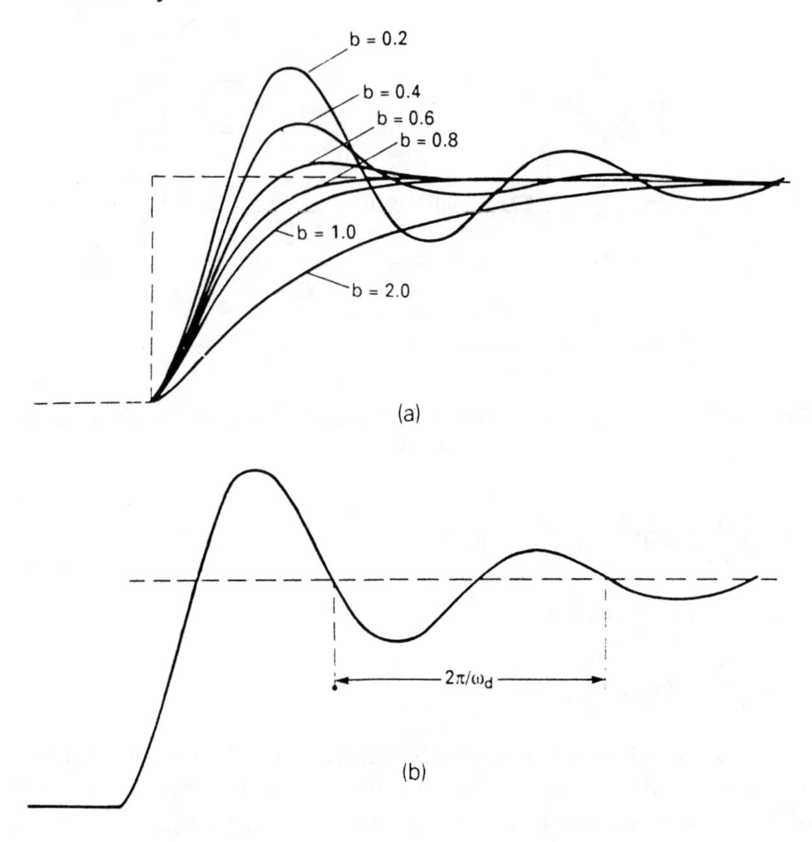

Fig. 3.4 Response of second order systems, (a) Step response for various damping factors; (b) Damped natural frequency, ω_d.

$$\omega_d = \omega_n\sqrt{1 - b^2} \tag{3.3}$$

the value of ϕ is

$$\phi = \tan^{-1}(\sqrt{1 - b^2}/b) \tag{3.4}$$

For a damping factor greater than unity, a response similar to Fig. 3.1e is obtained. The transitional case, $b = 1$, is called critical damping and denotes the point at which there is just no overshoot for a step driving function. For $b = 1$ the step response is

$$x(t) = 1 - e^{-\omega_n t}(1 + \omega_n t) \tag{3.5}$$

for $b > 1$, the step response is

$$x(t) = 1 - \left(\frac{m_2}{m_3}e^{m_1 t} - \frac{m_1}{m_3}e^{m_2 t}\right) \tag{3.6}$$

where $m_1 = -b + \sqrt{b^2 - 1}$
and $m_2 = -b - \sqrt{b^2 - 1}$
and $m_3 = m_1 - m_2 = 2\sqrt{b^2 - 1}$

The case of most interest is the underdamped condition represented by equations 3.2–3.4. The generalised step response of Fig. 3.5a defines some terms of interest.

The first, and largest, overshoot occurs at time T_{max} and has a value x_{max}. These can be found by differentiation of equation 3.2 and setting $dx/dt = 0$. It is found that

$$T_{max} = \frac{\pi}{\omega_n\sqrt{1 - b^2}} = \frac{\pi}{\omega_d} \tag{3.7}$$

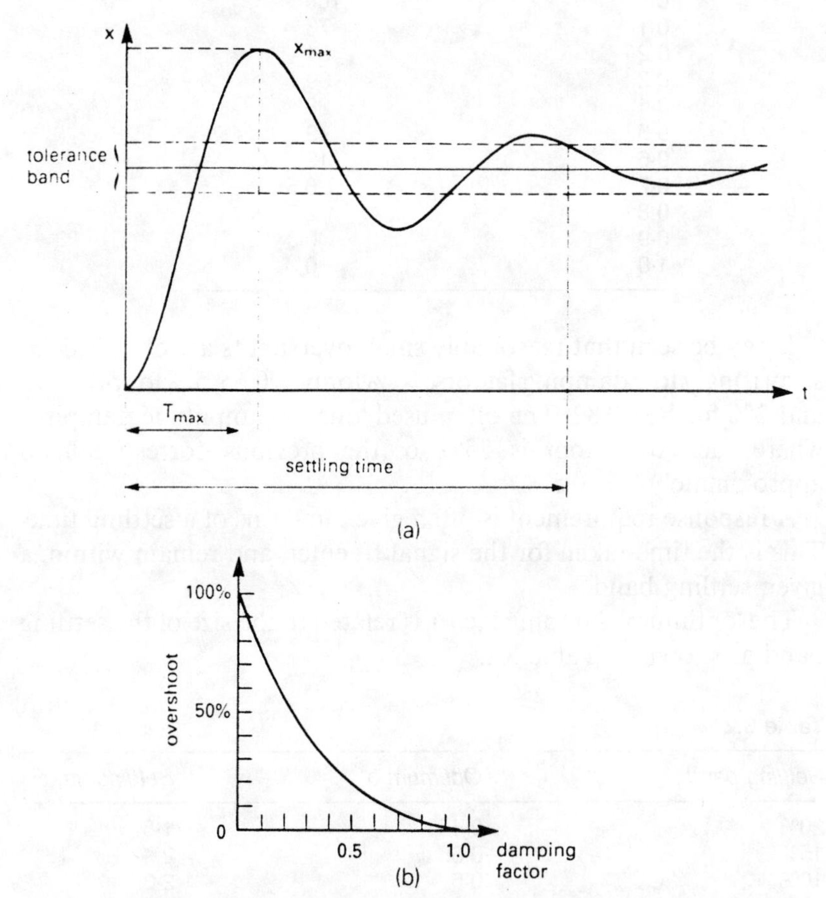

Fig. 3.5 Overshoot on a second order response, (a) Terms of interest on a second order response; (b) Relationship between overshoot and damping factor.

and by substitution

$$x_{max} = 1 + \frac{e^{-\pi b/\sqrt{1-b^2}}}{\sqrt{1-b^2}} \tag{3.8}$$

i.e. the percentage overshoot is purely dependent on the damping factor b. Equation 3.8 has a value of 2 for b = 0 (representing 100% overshoot and continuous oscillation) and unity for b = 1 (representing no overshoot). Figure 3.5(b) shows the relationship between overshoot and damping factor which is also given in Table 3.1.

Table 3.1

Damping factor	*Overshoot (%)*
0	100
0·1	73
0·2	54
0·3	39
0·4	28
0·5	19
0·6	12
0·7	6·5
0·8	2·5
0·9	0·3
1·0	0

It can be seen that reasonably small overshoots are obtained for surprisingly low damping factors; 10% for b = 0·65, 5% for b = 0·75 and 2% for b = 0·82. The often used 'quarter amplitude damping' where each overshoot is 25% of the previous corresponds to approximately b = 0·5.

A response requirement is often given in terms of a settling time. This is the time taken for the signal to enter, and remain within, a given settling band.

The 'optimum' damping factor is related to the size of the settling band as shown in Table 3.2.

Table 3.2

Settling band	*Optimum b*	*Settling time*
20%	0·45	1·8
15%	0·55	2
10%	0·6	2·3
5%	0·7	2·8
2%	0·8	3·5

The settling time is normalised in terms of ω_n (e.g. the settling time corresponding to a settling band of 15% is $2/\omega_n$). In each case, the fastest settling time is obtained with a damping factor which gives an overshoot just less than the specified tolerance based as shown in Fig. 3.6 for a 20% band.

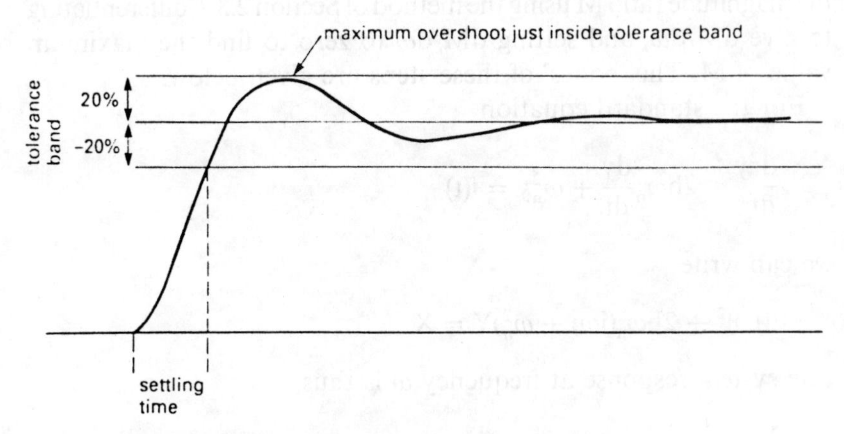

Fig. 3.6 Optimum damping factor for given tolerance band.

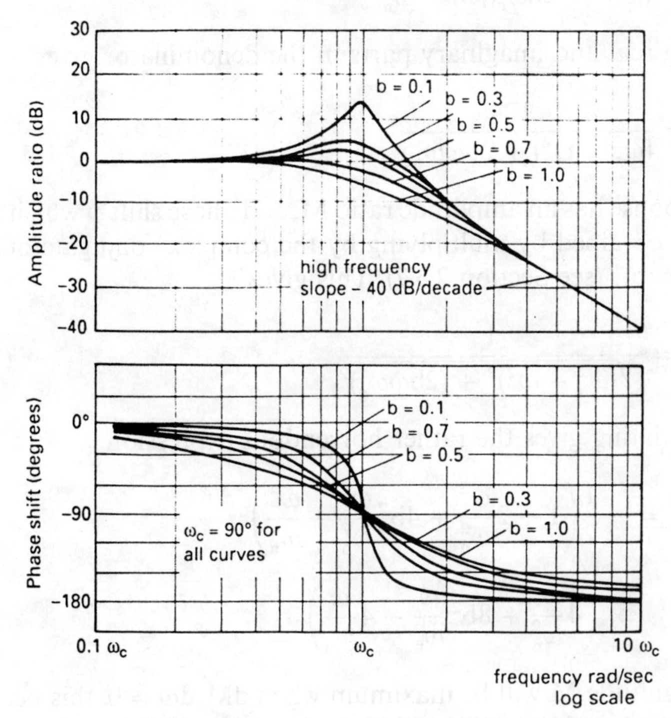

Fig. 3.7 Bode diagram for various values of damping factor.

The frequency response of a second order system can be calculated using the techniques outlined in Section 2.3.7. Bode diagrams for various values of damping factor are shown in Fig. 3.7. These are characterised by a resonance peak which occurs at a frequency lower than the natural frequency ω_n, and a high frequency roll-off of 40 dB/decade. The resonant frequency can be found by evaluating the magnitude ratio M using the method of Section 2.3.7, differentiating to give $dM/d\omega$, and setting $dM/d\omega$ to zero to find the maximum value of M. The 'bones' of these steps are given below.

For the standard equation

$$\frac{d^2y}{dt^2} + 2b\omega_n\frac{dy}{dt} + \omega_n^2 y = f(t)$$

we can write

$$((j\omega)^2 + 2b\omega_n(j\omega) + \omega_n^2)Y = X$$

The system response at frequency ω is thus

$$\frac{Y}{X} = \frac{1}{(j\omega)^2 + 2b\omega_n(j\omega) + \omega_n^2}$$

Splitting real and imaginary parts in the denominator gives

$$\frac{Y}{X} = \frac{1}{(\omega_n^2 - \omega^2) + 2b\omega\omega_n j}$$

The response has an amplitude ratio M, and phase shift ϕ which can both be obtained by multiplying by the complex conjugate of the denominator (see Section 2.3.7). This gives

$$M = \frac{1}{\sqrt{(\omega_n^2 - \omega^2)^2 + (2b\omega\omega_n)^2}} \tag{3.9}$$

Differentiating gives the rather horrendous expression

$$\frac{dM}{d\omega} = -\frac{1}{2}\left(1 - 2\frac{\omega^2}{\omega_n^2} + 4b^2\frac{\omega^2}{\omega_n^2} + \frac{\omega^4}{\omega_n^4}\right)^{-3/2}$$

$$\times\left(4\frac{\omega^3}{\omega_n^2} + 8b^2\frac{\omega}{\omega_n} - 4\frac{\omega}{\omega_n}\right)$$

The magnitude M will be maximum when $dM/d\omega = 0$, this occurs for a resonant frequency ω_r where

$$4\frac{\omega_r^3}{\omega_n^3} + 8b^2\frac{\omega_r}{\omega_n} - 4\frac{\omega_r}{\omega_n} = 0$$

from which

$$\frac{\omega_r^2}{\omega_n^2} + 2b^2 - 1 = 0$$

$$\text{giving } \omega_r = \omega_n\sqrt{1 - 2b^2} \quad (3.10$$

Substituting back into equation 3.9 gives the resonant amplitude ratio

$$M_r = \frac{1}{2b\sqrt{1 - b^2}} \quad (3/.11)$$

Note that from equation 3.10, a maximum value of M is only obtained for $2b^2 < 1$ or $b < 0{\cdot}707$, i.e. there is no resonant peak for $b > 0{\cdot}707$.

To summarise, a second order system with $b < 0{\cdot}707$ has a resonant frequency given by equation 3.10 which is lower than the natural frequency. The magnitude of the resonant peak increases for decreasing b. The maximum amplitude ratios corresponding to various damping factors are shown in Table 3.3.

Table 3.3

Damping factor	*Peak*	*Amplitude ratio (dB)*
0·1	5·02	14·0
0·2	2·55	8·1
0·3	1·75	4·8
0·4	1·36	2·7
0·5	1·15	1·2
0·6	1·04	0·3
0·7	1·00	0

The phase shift at frequency ω is given by

$$\phi = \tan^{-1}\left(\frac{b\omega\omega_n}{(\omega_n^2 - \omega^2)}\right) \quad (3.12)$$

Note that the phase shift tends towards zero for low frequencies and $-\pi$ radians ($-180°$) at high frequencies. At a frequency ω_n, ϕ is $-\pi/2$ ($-90°$) regardless of the value of b. The damping factor determines the range of frequencies over which the phase shift occurs as shown on the Bode diagram of Fig. 3.7.

3.3. Stability prediction from frequency response methods

3.3.1. Introduction

Figure 3.8a is the by now familiar block diagram of a unity gain feedback system, where G represents the model of the controller and plant, SP the set point, PV the process variable (i.e. the thing we are trying to control) and the error (SP − PV). Let us assume the system is unstable and oscillating with constant amplitude. The output PV is thus a constant amplitude sine wave, as Fig. 3.8b.

The error signal is (SP − PV), but as the set point is constant the error signal will be −PV, i.e. a sine wave of the same frequency and amplitude as PV, but shifted in phase by −180° as shown.

The output PV is related to the error signal by the characteristics of the plant G. To maintain constant amplitude oscillations, the plant/controller must have unity gain and contribute −180° phase shift at the oscillation frequency. The error signal then maintains PV oscillating at constant amplitude.

This result should come as little surprise. Any oscillation, mechanical, electronic or process control, requires a loop with unity gain and

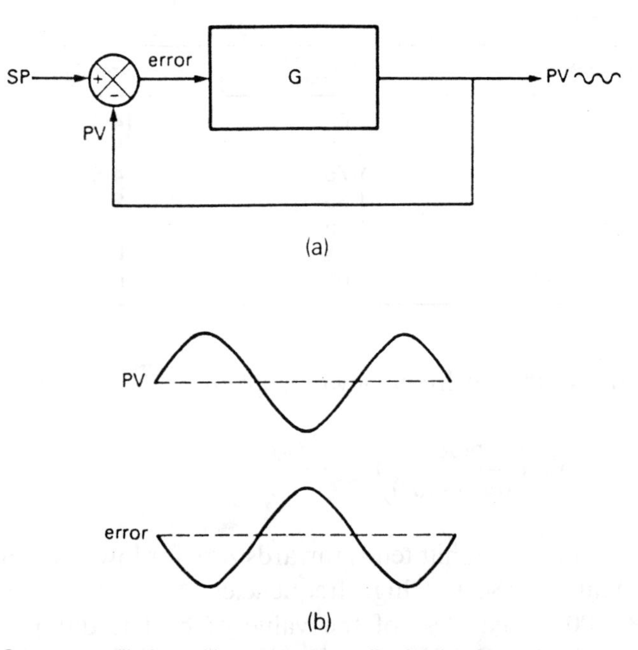

Fig. 3.8 System oscillating with constant amplitude, (a) Closed loop system with unity feedback; (b) Relationship between PV and error.

360° phase shift. In process control, the error subtractor (SP – PV) provides –180° phase shift, leaving –180° to be provided by the plant/controller.

The frequency at which the combined controller and plant exhibits –180° phase shift is thus crucial in determining stability. Somewhat simplisticly if the open loop gain is greater than unity at this frequency the plant will be unstable, and an exponentially increasing sinusoidal output will occur. Practically, if the system is left some component will eventually saturate on the extremes of the sinusoid (e.g. the controller reaches 100% or 0%, or a valve reaches full stroke). This saturation effectively reduces the gain of the loop and constant amplitude oscillations occur with one or more plant items just reaching the limits of their travel or signal range. In reality, a rather panicky manual intervention usually occurs before the situation gets too far out of control or plant damage occurs.

If the loop gain is less than unity at the –180° phase shift frequency, oscillations cannot be maintained. Suppose, for example, the open loop gain is 0·5 when the phase shift is –180°, and PV is forced to follow a constant amplitude sinusoid (possible methods of achieving this are by keeping the controller in manual and driving the actuator sinusoidally, or by introducing a sinusoidal disturbance). At time t = 0 (a zero crossing point) the disturbance ceases and closed loop control is enabled. Each half cycle will now be one half the amplitude of the previous as shown in Fig. 3.9 and each cycle will be one quarter the amplitude of its predecessor. This is known as

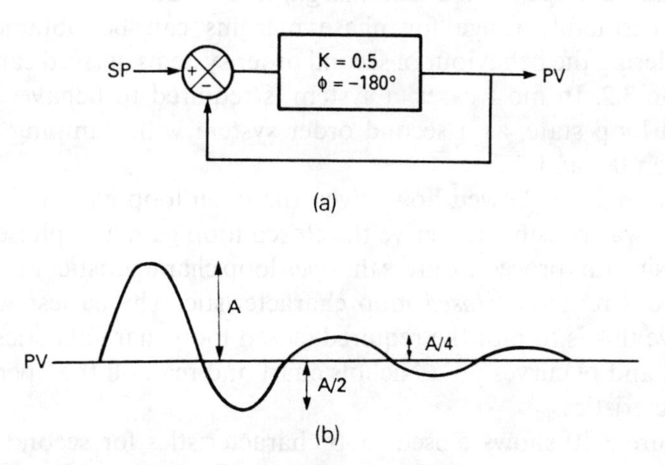

Fig. 3.9 Quarter amplitude damping, (a) System with loop gain of 0·5 and phase shift of –180°; (b) System response.

'quarter amplitude damping' and approximates to 50% overshoot on a step set point change.

Engineers used to dealing with electronic amplifiers should note the relatively low frequencies at which the critical $-180°$ phase shift is reached. For a 'fast' flow loop 0·5 Hz would be typical. Periods of tens of minutes are not unusual on temperature loops and analytical loops on, say, a large blast furnace can have periods of several hours. Patience is thus a virtue for a control engineer, as instability in a slow loop may not be apparent until some considerable time has elapsed.

If the critical point for stability is open loop unity gain and open loop $-180°$ phase shift, it is reasonable to give two figures of 'merit':

(a) The amount by which the gain can be increased at the critical frequency before oscillation can occur. This is known as the 'gain margin', and is simply the inverse of the gain at the critical frequency. A quarter amplitude damped system, described above, has an open loop gain of 0·5 at the critical frequency, so the gain margin is 2. This can also be expressed in decibels as a gain margin of 6 dB.
(b) The amount of additional phase shift that can be tolerated at the frequency at which the open loop gain is unity. This is termed the 'phase margin'. If a system has $-130°$ phase shift when the gain is unity, for example, the phase margin is 50°.

Rules of thumb can be derived for the gain and phase margin. We have already seen that a gain margin of 6 dB corresponds to quarter amplitude damping. Too low a gain leads to a sluggish system, so a reasonable range for the gain margin is 6–12 dB.

A reasonable range for phase margins can be obtained by considering the behaviour of second order systems derived earlier in Section 3.2. In most cases, a system is required to behave, in the closed loop state, as a second order system with damping factor between 0·4 and 0·7.

Section 2.3.3 showed how, given the open loop gain and phase shift, it was possible to derive the closed loop gain and phase shift. Reversing this procedure gives the *open* loop characteristics necessary to give a required *closed* loop characteristic. The easiest way to achieve this is to plot the required closed loop characteristics onto the M and N curves on a Nichols chart and read off the open loop characteristics.

Figure 3.10 shows closed loop characteristics for second order equations with damping factors of 0·4 and 0·7 plotted onto the M and N curves of a Nichols chart. From these we can see directly that

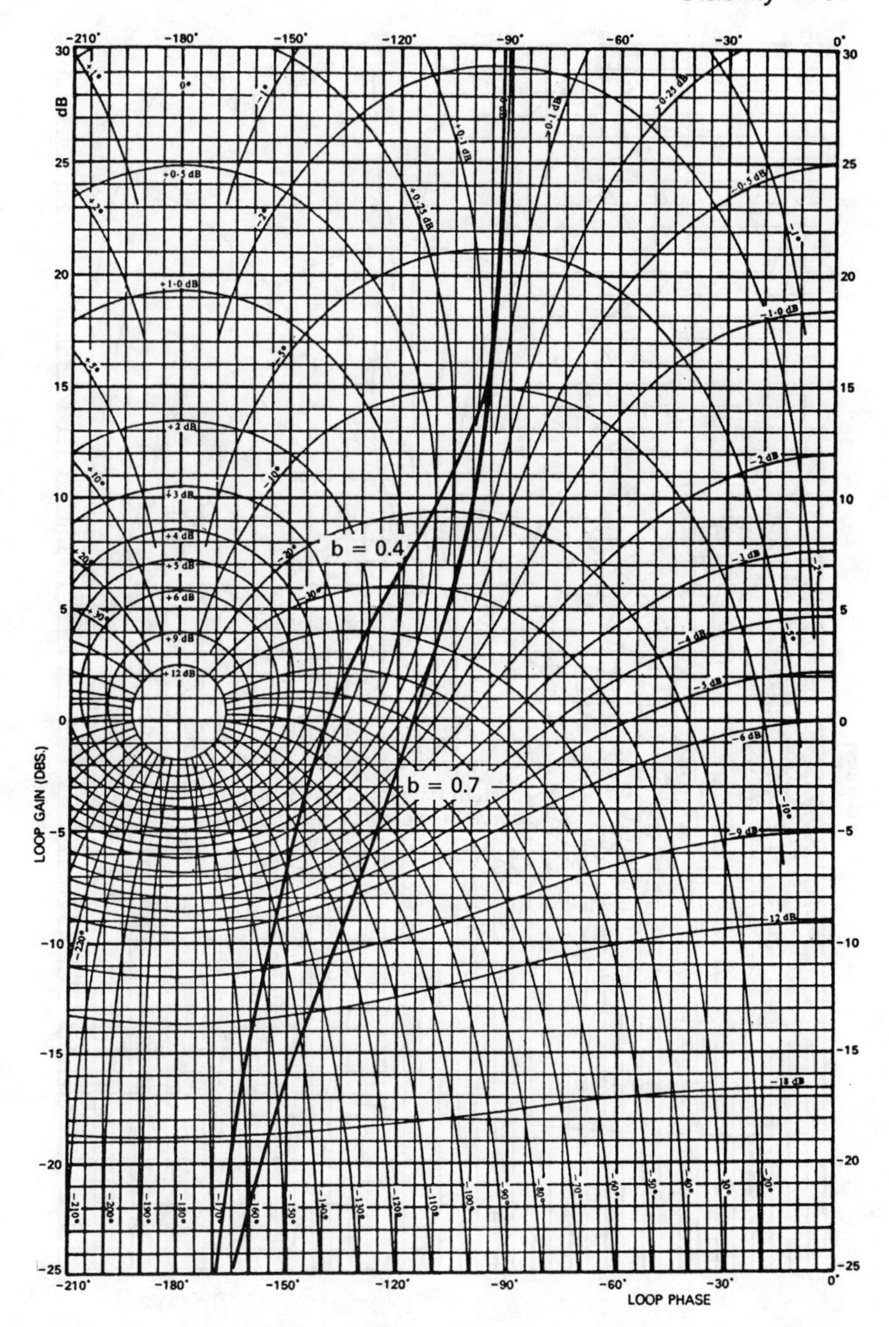

Fig. 3.10 Second order response for damping factors 0·4 and 0·7 drawn onto M and N curves of a Nichols chart.

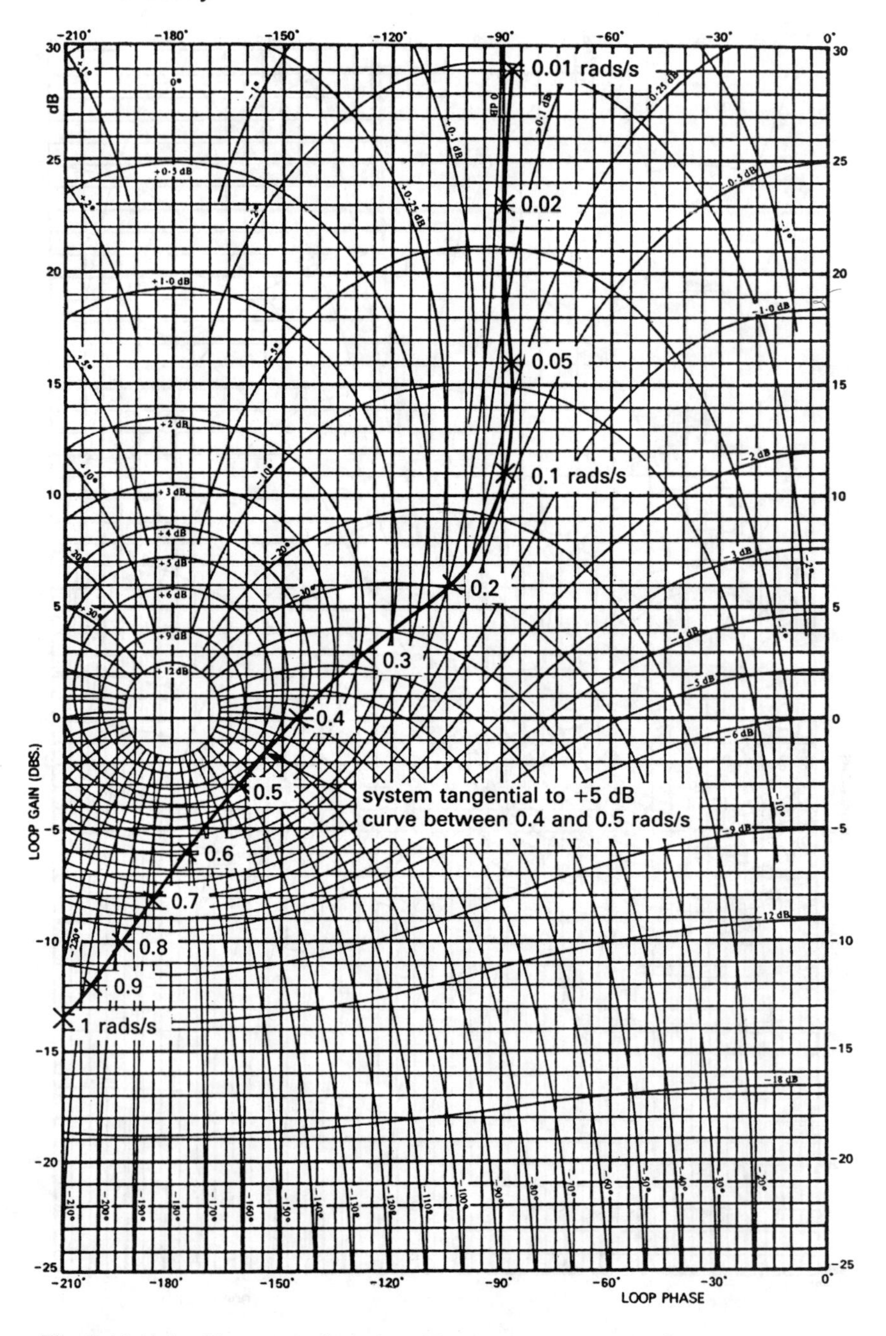

Fig. 3.11 Using M curve to find closed loop resonance peak. System drawn has P + I controller of gain 2·8, T_i 10 s, lags of 3 s and 4 s, time constant of gain 1 and a transit delay of 1 sec.

a damping factor of 0·4 implies a phase margin of about 40° and a damping factor of 0·7 a phase margin of about 65°.

The above analogies suggest a range for gain margins of 6–12 dB and phase margins of 40°–65° according to the degree of damping required. In the majority of systems, these values give perfectly acceptable results.

Figure 3.10 and Table 3.3 suggest an alternative quick 'rule of thumb' method of determining the response of a closed loop system from the M curves on a Nyquist diagram or Nichols chart.

Figure 3.11 is a Nichols chart for some system being studied. It can be seen that the closed loop response just brushes the +5 dB M curve, implying a resonance peak of +5 dB. Comparison with Table 3.3 suggests that a reasonable analogy will be a second order system with damping factor of 0·3. Table 3.1 suggests that this will respond to a set point change with an overshoot of about 40%. This agrees surprisingly well with the actual step response of Fig. 3.12. Applying equations 3.10 to the resonant frequency read from Fig. 3.11 allows the natural frequency to be determined. It is reassuring to note that the system of Fig. 3.11 and Fig. 3.12 has a gain margin of 7 dB and a phase margin of 35°, which both tally with a damping factor of 0·3–0·4.

The reader may feel, with some justification, that we are stretching the analogy between closed loop systems and second order systems a little far. In most systems, however, system models are idealised and inaccurate and the characteristics of the system are dominated by the controller and the elements which introduce phase shift and gain changes at the lowest frequencies. The *dominant* behaviour of a system will be determined by about four items in the majority of

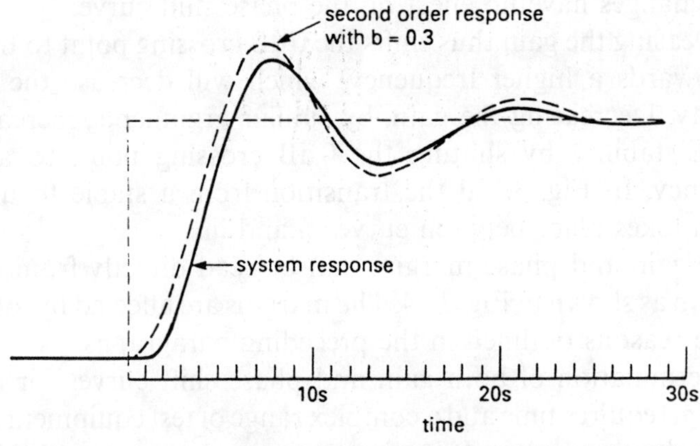

Fig. 3.12 Comparison of step response of system of Fig. 3.11 with second order system with damping factor 0·3.

cases, and for these the gain margin/phase margin/resonant peak analysis gives results that are at least as accurate as most models.

3.3.2. Analysis from Bode diagrams

The Bode diagram introduced in Section 2.3.4 can be used to predict stability and investigate system response (via comparisons with the idealised second order system described above).

It was shown earlier that absolute stability is determined by the open loop gain at the frequency at which the phase shift is $-180°$. If the open loop gain is greater than, or equal to, unity at this frequency the system is unstable.

Unity gain on a Bode diagram corresponds to 0 dB, so the criterion for stability is the gain curve must cross the zero dB axis *before* the phase shift reaches $-180°$. The system represented by Fig. 3.13a is thus stable and Fig. 3.13b is unstable.

In some uncommon, and badly behaved, systems where the phase shift curve crosses the $-180°$ line several times, it can be difficult to determine stability by simple observation. On Fig. 3.13c, for example, curve i represents an unstable system and curve ii a stable system. Such difficult cases are best studied with Nyquist diagrams or Nichols charts described below.

The relationship between loop gain and stability can be seen on Fig. 3.13d. Almost all practical systems have a gain curve with a slope in the area of interest of $-20n$ dB/decade where $n = 1,2,3$ etc. depending on the make up of the system. Changing the loop gain does not alter the shape of the curve, but merely shifts it vertically. Gain changes have no effect on the phase shift curve.

Increasing the gain thus shifts the 0 dB crossing point to the right (i.e. towards a higher frequency) which will decrease the system stability. Decreasing the gain, by similar arguments, increases the system stability by shifting the 0 dB crossing point to a lower frequency. In Fig. 3.13d the transition from a stable to unstable system takes place between curves ii and iii.

The gain and phase margins can be read directly from a Bode diagram as shown in Fig. 3.14. The margins are affected by loop gain for the reasons outlined in the preceding paragraphs.

Determination of both gain and phase shift curves for a Bode diagram requires time and a complex range of test equipment such as UV or chart recorders. It is often easier to determine only the gain

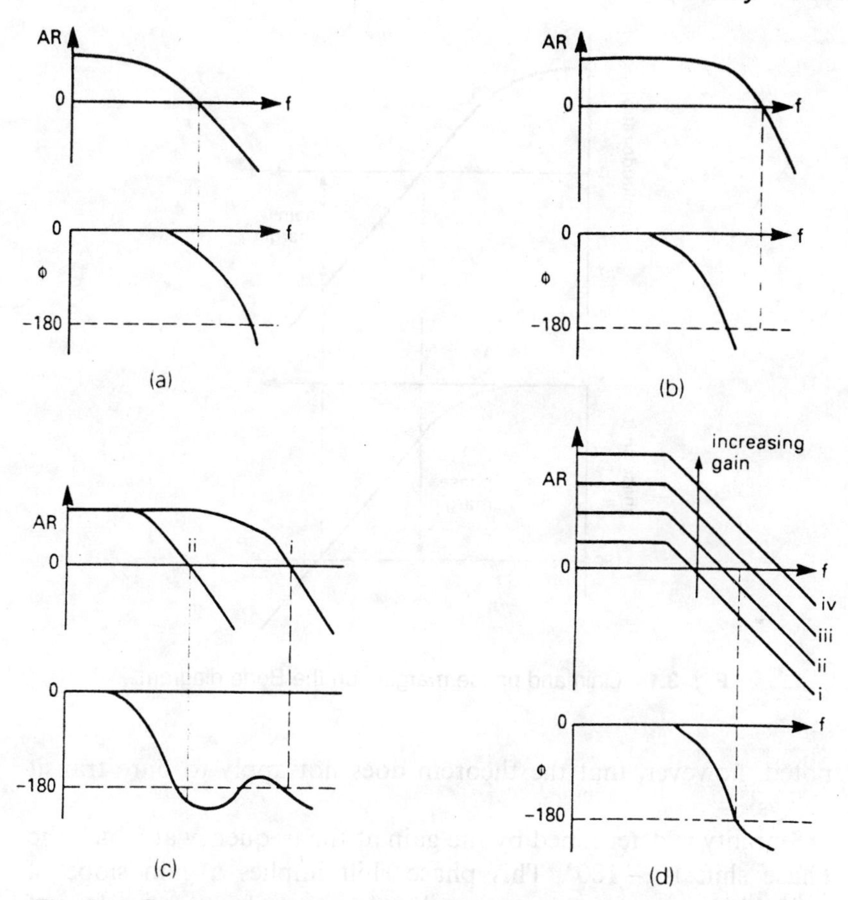

Fig. 3.13 Stability predictions from the Bode diagram, (a) A stable system; (b) An unstable system; (c) System with multiple crossings of the $-180°$ phase shift line; (d) The effect of gain on stability.

curve from which, surprisingly, information about system performance can be obtained.

Bode's theorem, somewhat simplified, relates phase shift to the *slope* of the gain curve in dB/decade. If a system at some frequency has a gain curve slope of m dB/decade the phase shift at that frequency is approximately 4·5 m degrees. An integrator, for example, has a constant slope of -20 dB/decade and a constant phase shift of $-90°$. A first order lag has asymptotes with slopes 0 dB/decade at low frequencies (phase shift 0°) and -20 dB/decade at high frequencies (phase shift $-90°$). Both of these agree with Bode's theorem; a similar examination of other functions such as second order systems and differentiators will similarly confirm the theorem. It should be

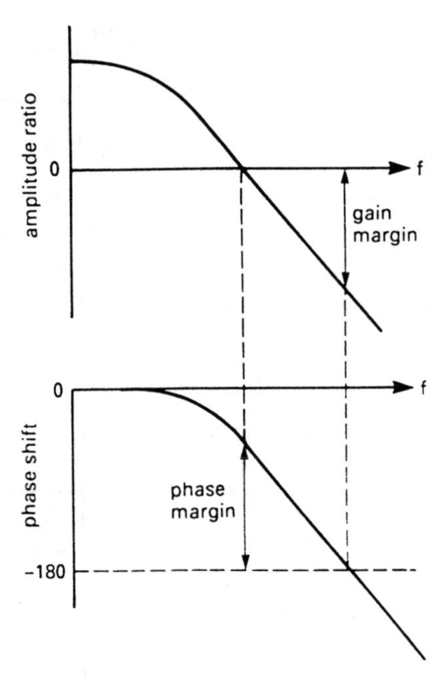

Fig. 3.14 Gain and phase margins on the Bode diagram.

noted, however, that the theorem does not apply to pure transit delays.

Stability is determined by the gain at the frequency at which the phase shift is $-180°$. This phase shift implies a gain slope of -40 dB/decade. A stable system, therefore, must have a gain slope of less than -40 dB/decade at the 0 dB crossing frequency.

System response can also be expressed in terms of gain curve slope. A phase margin of 50° requires a phase shift of $-130°$ at the 0 dB crossing point. This implies a gain curve slope of -29 dB/decade at this point.

For a system which is well behaved and reasonably set up, it is possible to deduce an approximate response from a Bode diagram. The damping factor, and hence the shape, can be inferred from the gain and phase margin. The closed loop resonant frequency ω_r is approximately the 0 dB crossing frequency. Applying equations 3.10, 3.3 and 3.7 gives the decay frequency ω_d (approximately $0{\cdot}85\omega_r$ for reasonable systems) and the time to first maximum T_{max} of the step response (approximately $2{\cdot}8/\omega_r$ for systems with acceptable damping). Given the damping factor T_{max} and ω_d the predicted step response can be drawn.

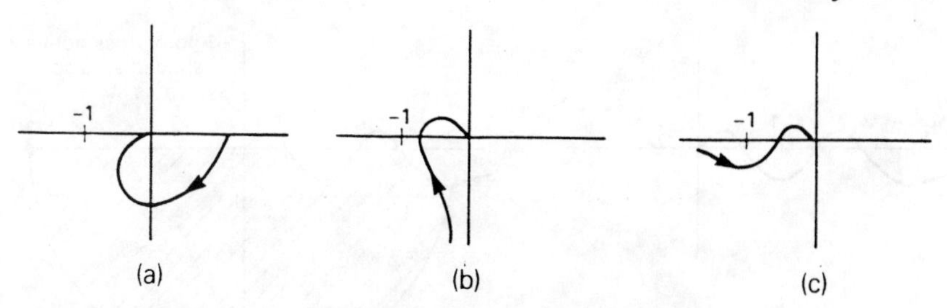

Fig. 3.15 Various forms of Nyquist diagram, (a) Simple system; (b) System with single integral action originates at infinity at −90°; (c) System with P + I controller and integral action originates at infinity at −180°.

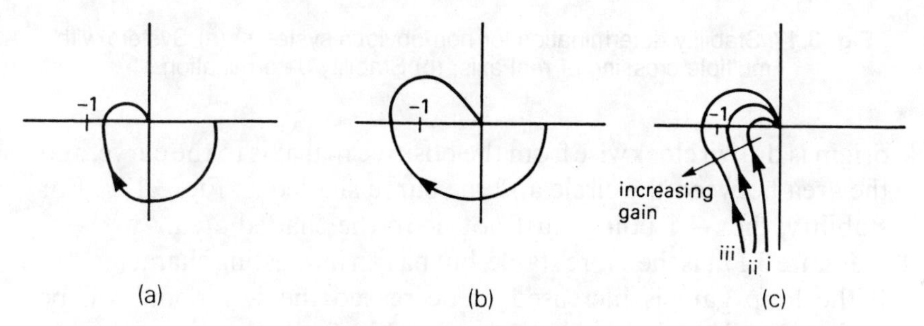

Fig. 3.16 Stability prediction on the Nyquist diagram, (a) Stable system; (b) Unstable system; (c) Effect of gain.

3.3.3. Analysis using Nyquist diagrams

The critical unity gain/−180° phase shift point on a Nyquist diagram is represented by the −1 point on the real axis. All Nyquist diagrams of real systems have zero gain at infinite frequency, and as such originate on the real axis (as Fig. 3.15a) or at infinity on one axis (when one or more integral terms are present as Fig. 3.15b, c) and approach the origin.

For stability, the curve must not enclose the −1 point. Figure 3.16a thus represents a stable system and Fig. 3.16b an unstable system. The effect of increasing the loop gain is shown in Fig. 3.16c. Increased gain moves the crossing point of the real axis to the left. Curves i and ii are stable, the transition to instability coming between curves ii and iii.

'Enclosure' is easy to see for Fig. 3.16, but is less obvious for curves such as Fig. 3.17a which crosses the negative real axis several times. In such cases, an imaginary circle of infinite radius centred on the

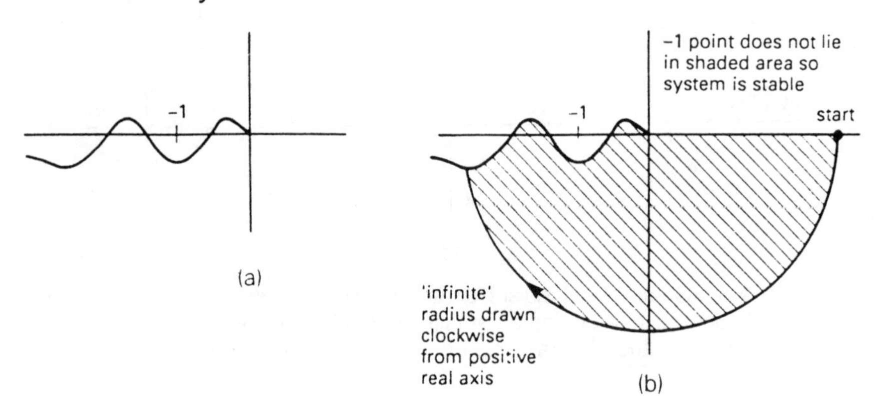

Fig. 3.17 Stability determination for non-obvious systems, (a) System with multiple crossing of real axis; (b) Stability determination.

origin is drawn clockwise from the positive real axis to the curve, and the area between the circle and the curve shaded as Fig. 3.17b. For stability, the -1 point must not lie in the shaded area.

Figure 3.17a is therefore stable, but has an interesting characteristic. If the loop gain is increased *or* decreased the -1 point will be enclosed and the system become unstable. Such characteristics are not unknown in real processes, but are not desirable as the system becomes difficult to set up and maintain.

Gain and phase margins can be read directly from a Nyquist diagram with the aid of a unity radius circle constructed on the origin as Fig. 3.18.

Closed loop response can be inferred from the gain and phase margins, or by using the closed loop M circles (described in Section 2.3.5) which represent the loci of constant closed loop gain. These are shown on Fig. 3.19, along with the Nyquist curve for a system.

This curve is just tangential to the $M = 1{\cdot}7$ curve, so the closed loop system corresponding to Fig. 3.19 has a resonance peak with gain 1·7, which from Table 3.3 corresponds to a damping factor of about 0·3. The resonant frequency ω_r can be read directly from the Nyquist diagram, allowing the decay frequency ω_d of the step response to be predicted from equations 3.10, 3.3 and 3.7.

It has been shown previously that the damping factor of a closed loop system can be adjusted by varying the open loop gain. A graphical construction technique, called Brown's construction, allows the gain to be determined for a given resonant peak gain (and hence a given damping factor).

Figure 3.20 shows the mathematical background to the technique.

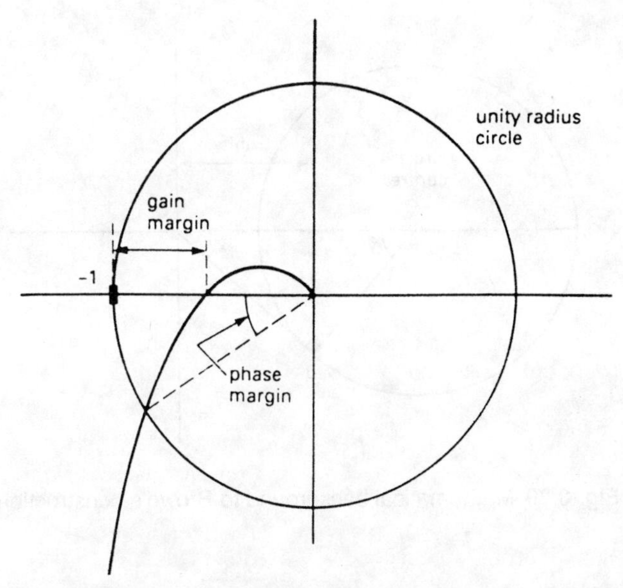

Fig. 3.18 Gain and phase margins on a Nyquist diagram.

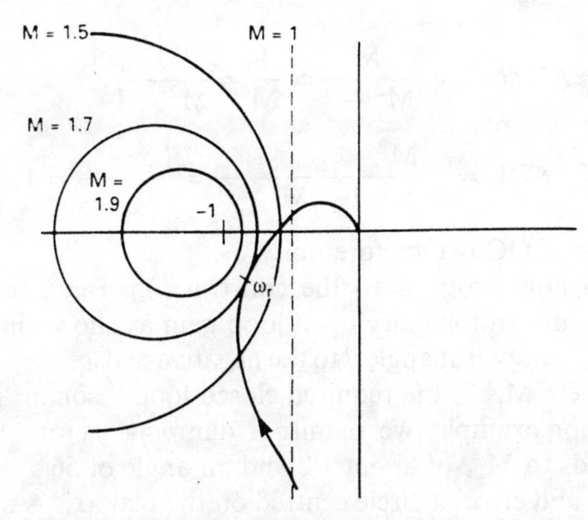

Fig. 3.19 Determination of maximum amplitude ratio from M curves on Nyquist diagram.

The circle, on centre A, is an M circle, and OB is a tangent from the origin to the circle. BC is a perpendicular from B to the axis. From equations 2.43 and 2.44 we have

$$\text{OA} = M^2(M^2 - 1) \quad \text{(origin)}$$
$$\text{and AB} = M/(M^2 - 1) \quad \text{(radius)}$$
$$\text{hence } x = \sin^{-1}(1/M)$$

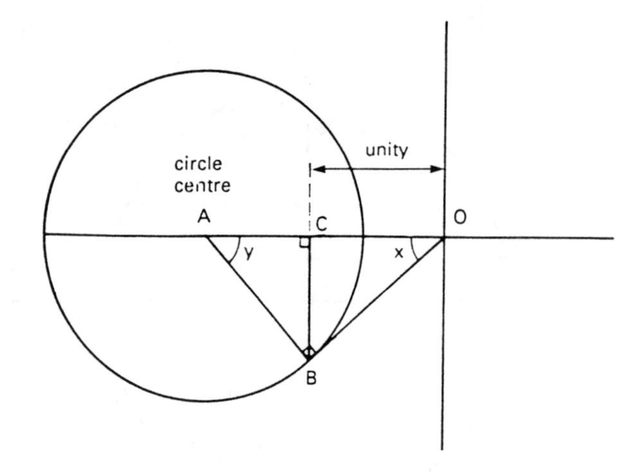

Fig. 3.20 Mathematical background to Brown's construction.

Since ABO is a right angle triangle, cos y = sin x, hence cos y = 1/M and

$$AC = AB \cos y = \frac{M}{M^2 - 1} \times \frac{1}{M} = \frac{1}{M^2 - 1}$$

$$\text{therefore } OC = \frac{M^2}{M^2 - 1} - \frac{1}{M^2 - 1} = \frac{M^2 - 1}{M^2 - 1} = 1$$

The length of OC is therefore unity.

We can now progress to the construction. First, the Nyquist diagram is drawn for unity open loop gain as shown in Fig. 3.21. Next a line is drawn at angle θ to the negative real axis where θ = sin $(1/M_{pk})$ where M_{pk} is the required closed loop resonant peak gain. Suppose, for example, we require a damping factor of 0·5. This corresponds to M_{pk} of about 1·2 and an angle of 56°.

By trial and error, a circle centred on the real axis is now drawn which is tangential to both the Nyquist curve and the line drawn above. A perpendicular BC is next drawn from the tangential point shown to the real axis. The required gain is now the reciprocal of the length OC.

In effect, we are determining the required M circle by calculating the gain required to make OC unity whilst maintaining the required tangential angle x in Fig. 3.20. The closed loop resonant frequency ω_r can be read off the Nyquist diagrams and the system response predicted as described above. It should be noted that Brown's

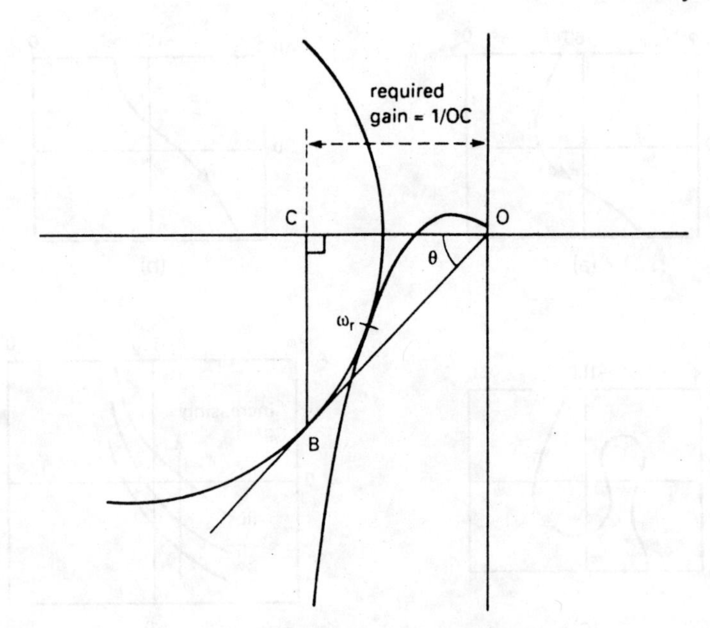

Fig. 3.21 Brown's construction.

construction is limited to reasonably well-behaved systems. It could not, for example, be applied to Fig. 3.17.

3.3.4. Analysis using Nichols charts

On a Nichols chart, the critical unity gain/−180° phase shift point is simply the intercept of the 0 dB and −180° axis. For stability, the critical point must be to the right of the curve when traversed in the direction of increasing frequency. Figure 3.22a is thus stable and Fig. 3.22b unstable. It is usually simple to establish stability even for complex curves such as Fig. 3.22c (which is, incidentally, stable).

Gain changes correspond to a vertical shift of the curve, as shown on Fig. 3.22d. Curves i and ii are stable, with the transition to instability coming between curves ii and iii.

Gain and phase margins can be read directly from the Nichols chart as shown in Fig. 3.23.

Nichols charts are normally obtained preprinted with closed loop M and N curves. This allows the closed loop resonance peak (if any) to be determined. The curve on Fig. 3.24, for example, just brushes the +3 dB M curve which from Table 3.3 corresponds to a damping factor of approximately 0·4. The closed loop resonant frequency ω_r

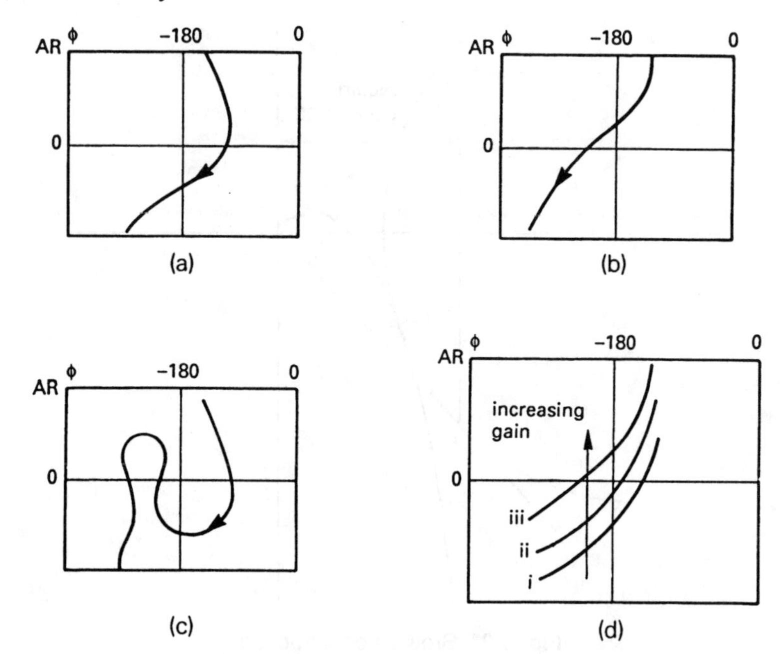

Fig. 3.22 Stability prediction on Nichols chart, (a) Stable system; (b) Unstable system; (c) Complex stable system; (d) Effect of gain on stability.

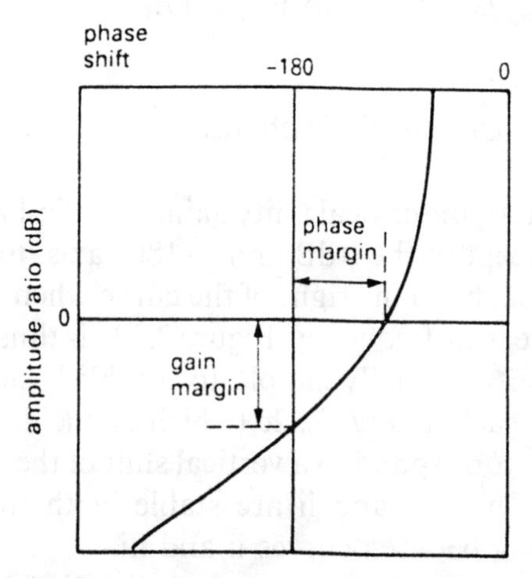

Fig. 3.23 Gain and phase margins on Nichols chart.

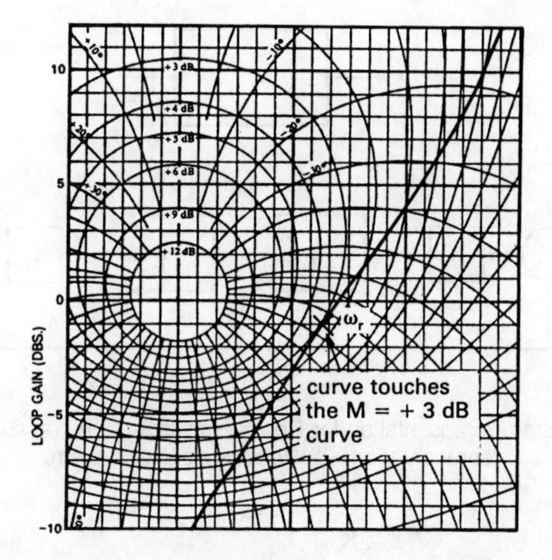

Fig. 3.24 Determination of maximum amplitude ratio from M curves on Nichols chart.

can be read directly off the curve, allowing system response to be determined as described above.

3.4. Routh–Hurwitz criteria

Any system, closed or open loop, can be represented by Fig. 3.25, where

$$\frac{Y}{X} = \frac{A^0 + A_1 D + A_2 D^2 + \ldots + A_m D^m}{B_0 + B_1 D + B_2 D^2 + \ldots + B_n D^n} \tag{3.13}$$

where $m < n$ in practical systems. Figure 3.25, for example, represents a position control system. This has open loop gain.

$$G = \frac{2K}{D(1 + (T_1 + T_2)D + T_1 T_2 D^2)}$$

The relationship between Y and X is

$$\frac{Y}{X} = \frac{G}{1+G}$$

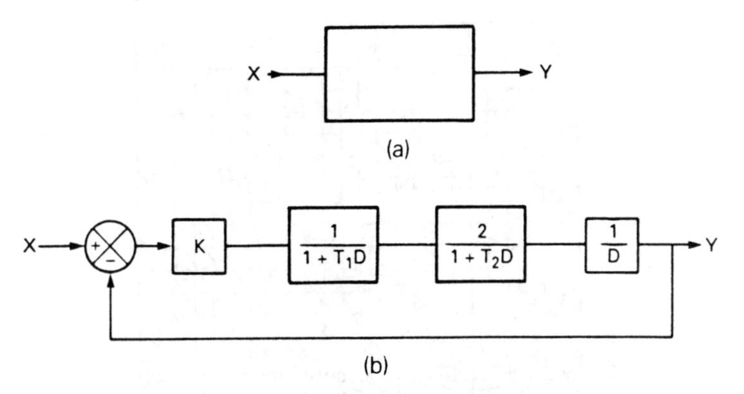

Fig. 3.25 System representation for Routh–Hurwitz criteria, (a) General system representation; (b) Position control system.

$$= \frac{\dfrac{2K}{D(1 + (T_1 + T_2)D + T_1T_2D^2)}}{1 + \dfrac{2K}{D(1 + (T_1 + T_2)D + T_1T_2D^2)}}$$

$$= \frac{2K}{2K + D + (T_1 + T_2)D^2 + T_1T_2D^3}$$

which has the form of equation 3.13.

The transient response and stability of the system is determined by the denominator, called the characteristic equation. (The reader should compare this with the formation of the auxiliary equation in Section 2.2.3.)

The Routh–Hurwitz criteria apply two tests to the characteristic equation

$$B_0 + B_1D + B_2D^2 + \ldots + B_nD^n$$

The first test is simple, all terms $B_0, B_1, B_2 \ldots B_n$ must have the same sign or the system is unstable. Note that this is a test for instability, not a test for stability. A system can pass this test but still be unstable.

To apply the second test, an array is formed as below from the coefficients of the characteristics equation.

First the coefficients are arranged

$$\begin{matrix} B_n & B_{n-2} & B_{n-4} \\ B_{n-1} & B_{n-3} & B_{n-5} \end{matrix} \quad \text{etc.}$$

A new row is now generated

$C_1 \quad C_2 \qquad$ etc.

$$\text{where } C_1 = B_{n-2} - \frac{B_n}{B_{n-1}} \times B_{n-3}$$

$$\text{and } C_2 = B_{n-4} - \frac{B_n}{B_{n-1}} \times B_{n-5} \qquad \text{etc.}$$

The terms in the next row are similarly calculated

$$D_1 = B_{n-3} - \frac{B_{n-1}}{C_1} \times C_2$$

and so on until the array is reduced to a single term.

When complete, the left-hand elements of each row are examined. For stability these must all have the same sign.

An alternative (but mathematically identical) approach to the second criteria is to form (n − 1) determinants as below.

$$R_1 = B_1$$

$$R_2 = \begin{vmatrix} B_1 & B_0 \\ B_3 & B_2 \end{vmatrix}$$

$$R_3 = \begin{vmatrix} B_1 & B_0 & 0 \\ B_3 & B_2 & B_1 \\ B_5 & B_4 & B_3 \end{vmatrix} \qquad \text{etc.}$$

For stability the values of the determinants $R_1, R_2, R_3 \ldots R_{n-1}$ must all have the same sign.

Like most mathematical operations, the procedures are best demonstrated with examples.

First consider the characteristic equation

$$D^5 + 2D^4 + 2D^3 + 2D^2 + D + 1$$

The first two rows are formed by writing down the coefficients

1 2 1
2 2 1

For the next row

$$C_1 = 2 - \frac{1}{2} \times 2 = 1$$

and

$$C_2 = 1 - \frac{1}{2} \times 1 = 0{\cdot}5$$

The next row has one element.

$$D_1 = 2 - \frac{2}{1} \times 0{\cdot}5 = 1$$

The final row is simply

$$E_1 = 0{\cdot}5 - \frac{1}{1} \times 0 = 0$$

Giving the complete array

$$\begin{array}{lll} 1 & 2 & 1 \\ 2 & 2 & 1 \\ 1 & 0{\cdot}5 & \\ 1 & & \\ 0{\cdot}5 & & \end{array}$$

All the left-hand terms are positive, so the system is stable.

Similarly the system

$$D^5 + 2D^4 + D^3 + 4D^2 + D$$

gives the array

$$\begin{array}{rrr} 1 & 1 & 1 \\ 2 & 4 & 0 \\ -1 & 1 & \\ 6 & & \\ 1 & & \end{array}$$

This has a sign reversal in the third row, so the system is unstable.

The Routh–Hurwitz criteria, whilst simple to apply, can be prone to simple routine slips in positioning and calculation. Figure 3.26 gives a useful 'aide mémoire' for forming the elements of a row. Note that the division A/B appears in each element.

The Routh–Hurwitz criteria can also be used to determine the

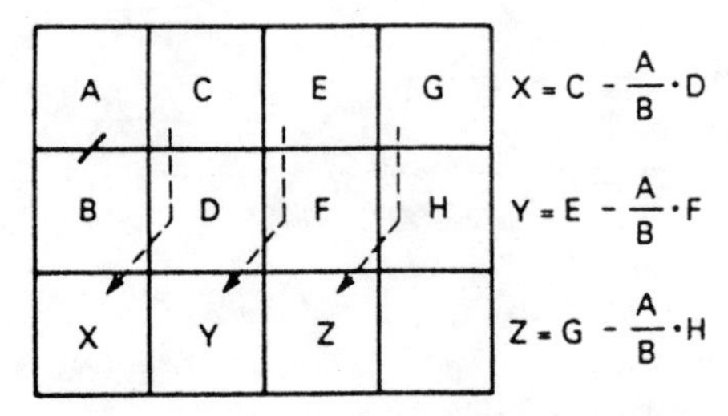

Fig. 3.26 Formation of row elements in Routh–Hurwitz array.

critical gain. Consider the position control system of Fig. 3.25b. This has the characteristic equation

$$T_1T_2D^3 + (T_1 + T_2)D^2 + D + 2K$$

Assigning time constants $T_1 = 1$ sec and $T_2 = 3$ sec yields

$$3D^3 + 4D^2 + D + 2K$$

The first test simply requires K to be positive. This is hardly a surprising result, so we proceed to the second test and directly write down the first two rows of the array

$$\begin{matrix} 3 & 1 \\ 4 & 2K \end{matrix}$$

There is one element in the third row which is given by

$$1 - \frac{3}{4} \times 2K = 1 - 1{\cdot}5K$$

There is similarly one element in the fourth row, simply 2K.

The complete array is therefore

$$\begin{matrix} 3 & 1 \\ 4 & 2K \\ (1 - 1{\cdot}5K) & \\ 2K & \end{matrix}$$

To keep all the elements of the left-hand column the same sign, K must be positive and $(1 - 1{\cdot}5K)$ must also be positive, i.e.

$$1 - 1{\cdot}5K > 0$$

therefore $K < 1/1{\cdot}5$

$K < 0{\cdot}67$

The Routh–Hurwitz criteria has thus given us the maximum gain for stability for the position control system. It cannot give us the predicted response (although a gain setting of half the critical gain is always a good starting point) and is unable to deal with transit delays.

3.5. Root locus

3.5.1. Introduction

In Sections 2.2.3.1 and 3.4, it was shown that the transient response of a system can be determined from the characteristic equation, and is in general the sum of

(a) Terms of the form $Ae^{\alpha t}$
(b) Terms of the form $(A_1 + A_2t + \ldots + A_nt^{n-1})e^{\alpha t}$
(c) Terms representing a sine wave with an exponential envelope: i.e.

$$e^{\alpha t}(A_1 \sin \omega t + A_2 \cos \omega t)$$

In each case, α determines the rate of exponential decay (α negative) or increase (α positive) and the A terms are constants of integration determined by the driving function and the initial conditions.

It is possible to represent these solutions graphically. In Fig. 3.27, the so-called 's-plane' (for reasons which will become apparent) has two axes. One axis represents the frequency of the transient response (denoted by ω) and the other the exponential component (denoted conventionally by σ, and directly equivalent to α in Section 2.2.3.1). It will be noted that both positive and negative frequencies are shown

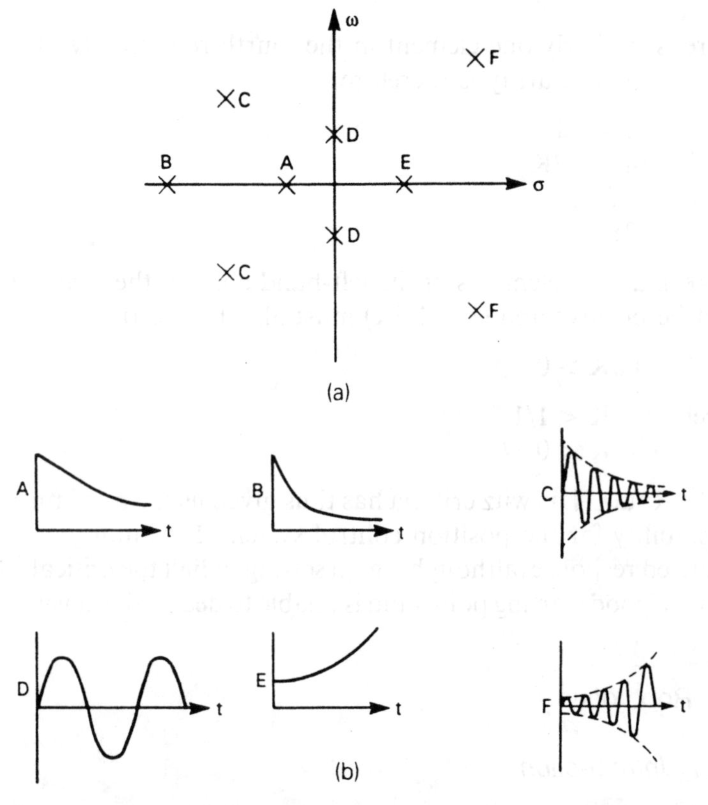

Fig. 3.27 Interpretation of points on the s-plane, (a) Points on the s-plane; (b) Interpretation of points on (a). Note that B has a faster decay than A, and frequency increases from D to C to F.

in Fig. 3.27. Negative frequencies have no practical significance, but arise out of the underlying mathematics.

If we denote a solution of the characteristic equation by X (called 'poles' for reasons explained later) they can be plotted directly onto the s-plane; several examples being shown in Fig. 3.27. Note that pure exponential terms lie on the σ axis, and complex conjugates (i.e. terms representing a sine wave with exponential envelope) occur in pairs (examples C, D and F). Example D represents a continuous sine wave.

Acceptable control requires a transient response which dies away, hence the exponential part, represented by σ, must be negative. The condition for stability, therefore, is that the poles must lie in the left-hand part of the s-plane.

For a given closed loop system, the position of the poles will be determined by gain of the controller. Figure 3.28a represents two first order lags with time constants one and five seconds, controlled by a proportional controller of gain K. We shall determine the position of the poles (i.e. the solution of the characteristic equation) for various values of K.

We have $G = \dfrac{K}{(1 + D)(1 + 5D)}$

$$\text{Giving } \frac{Y}{X} = \frac{G}{1+G} = \frac{\dfrac{K}{(1 + D)(1 + 5D)}}{1 + \dfrac{K}{(1+ D)(1 + 5D)}}$$

$$= \frac{K}{5D^2 + 6D + (1 + K)}$$

The characteristic equation is therefore

$$5D^2 + 6D + (1 + K)$$

By simple algebra, this has a solution

$$\frac{-6 \pm \sqrt{36 - 20(1 + K)}}{10} \tag{3.14}$$

The pole positions for various values of K can be found by substitutions into equation 3.14. Some values are shown in Table 3.4 and plotted on the s-plane on Fig. 3.28b.

It can be seen that there is a pattern to these results. For low values of K the poles appear as separate pairs on the σ axis (denoting a

Table 3.4

	K	*Poles*	
A	0·1	−0·23	−0·97
B	0·3	−0·28	−0·91
C	0·5	−0·36	−0·84
D	0·7	−0·46	−0·74
E	1	−0·6	j0·20
F	1·5	−0·6	j0·37
G	2	−0·6	j0·49

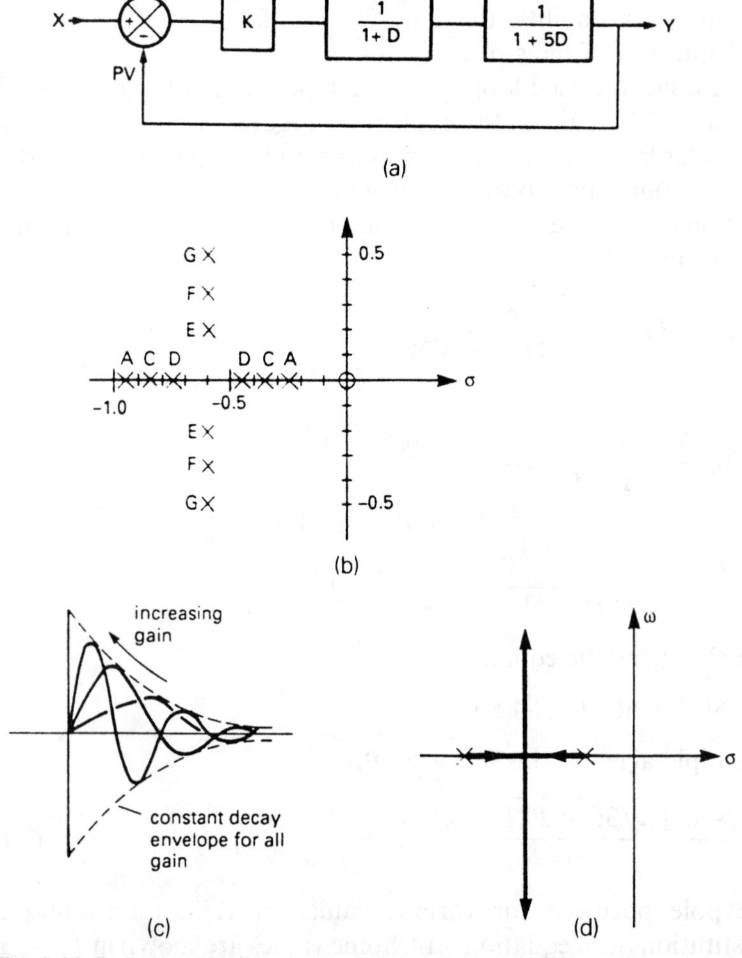

Fig. 3.28 Root locus plot for a simple system, (a) Simple system; (b) Pole positions for various values of gain as calculated in text; (c) Gain affects the frequency but not the envelope for points E–G; (d) The form of the root locus.

purely exponential response). As the gain increases, the poles move towards each other, meet (where $36 = 20(1 + K)$ from equation 3.14), then break away into complex conjugate pairs.

As the gain increases further the poles appear to move vertically implying a transient response with increasing frequency with increasing gain but a constant decaying exponential envelope as Fig. 3.28c. We can plot the path of the poles as the gain is changed, and the above results would suggest that the path has the form of Fig. 3.28d. Such a sketch is called a root locus. The root locus of Fig. 3.28d tells us the effect of changing gain on our simple system. For low gain the system behaves like two cascaded first order lags. For high gain, the system has a damped oscillatory response with the gain determining the frequency, but not the decay envelope. Finally we can see that the system is stable for all values of gain.

Plotting a root locus from the characteristic equation, as we have done, is tedious and does not show anything that we could not have seen from the solutions of the characteristic equation. As the complexity of systems increases, it becomes difficult to find the solution of the characteristic equation. It would be convenient, therefore, to find some method of drawing the root locus of a system given the components of the system. Simple rules for drawing root loci are given in Section 3.5.3, but first we must examine some of the background mathematics.

3.5.2. Background mathematics

The position of the closed loop poles along the root loci can be found by solving the characteristic equation. Unfortunately for all bar the simplest and most trivial systems this involves equations with high powers of D.

Plotting the position of the open loop poles, however, is usually quite simple, as the open loop transfer function is composed of blocks representing the system components, and is inherently already factorised. Plotting the position of the open loop poles rarely involves solving equations higher than a quadratic. Figure 3.29, for example, represents a position control system with two first order lags (representing the drive and the system inertia) and an under-damped non-rigid structure. From this block diagram we can immediately plot the position of the open loop poles as shown.

The s-plane can also represent the response of an open loop system to an exponential enveloped sine wave driving function. In

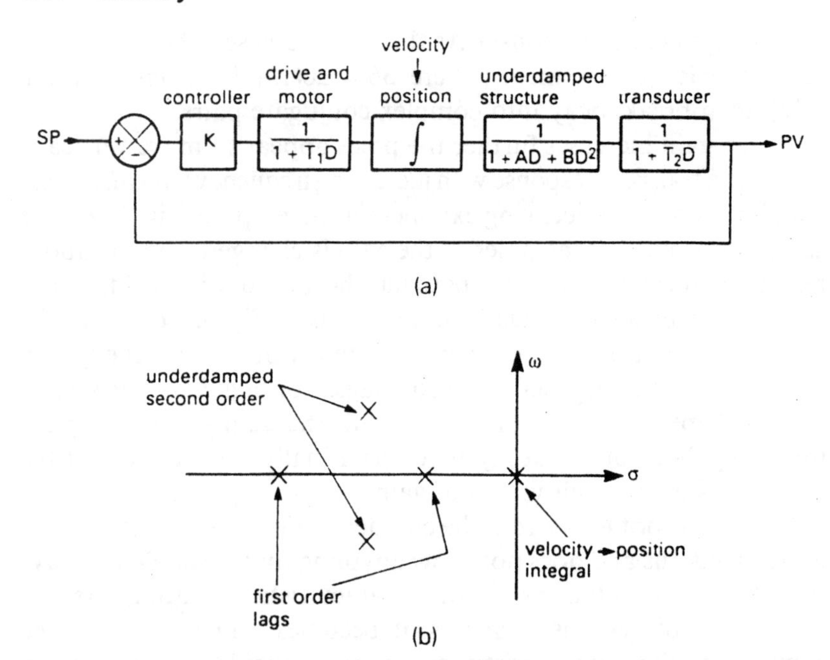

Fig. 3.29 Plotting open loop poles on the s-plane, (a) Position control system; (b) Position of open loop poles.

Section 2.2.3.2, methods for calculating the response of a system to a steady state sine wave were defined. The response to an exponentially changing sine wave (admittedly a mathematical abstraction rather than a practical experimental technique) can be calculated in a similar manner. We could envisage a 'thought experiment' similar to Fig. 3.30 where the system is driven by an exponential sine wave and the output observed. As observed for steady state sine waves, for each set of driving conditions, we observe a system gain (magnitude of output/magnitude of input) and a phase shift. These can be plotted, but because we have two variables under our control (σ and ω) we need a three-dimensional Bode diagram to plot the results as shown in Fig. 3.31a, b. The result is drawn as Fig. 3.31c which shows the gain plot for a simple first order lag. The phase shift curve is more difficult to represent, but can be envisaged as a spiral centred on the pole. The response to a steady sine wave (i.e. the usual frequency response curve) is simply the curve for $\sigma = 0$, shown bold in Fig. 3.31c.

Calculation of the system response to exponentially varying sine waves is performed in a similar manner to the steady state sine wave response calculated in Section 2.2.3.2. There, terms in $d^n x/dt^n$ were replaced by $(j\omega)^n$; as shown below we substitute $(\sigma + j\omega)$ to give the enveloped sine wave response.

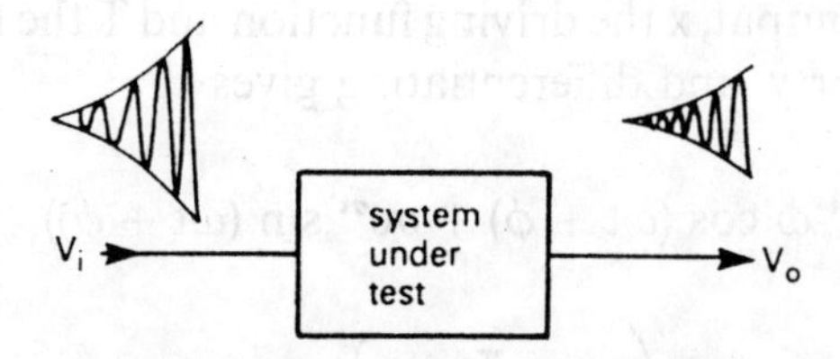

Fig. 3.30 A thought experiment plotting the response of a system to an exponentially varying sine wave.

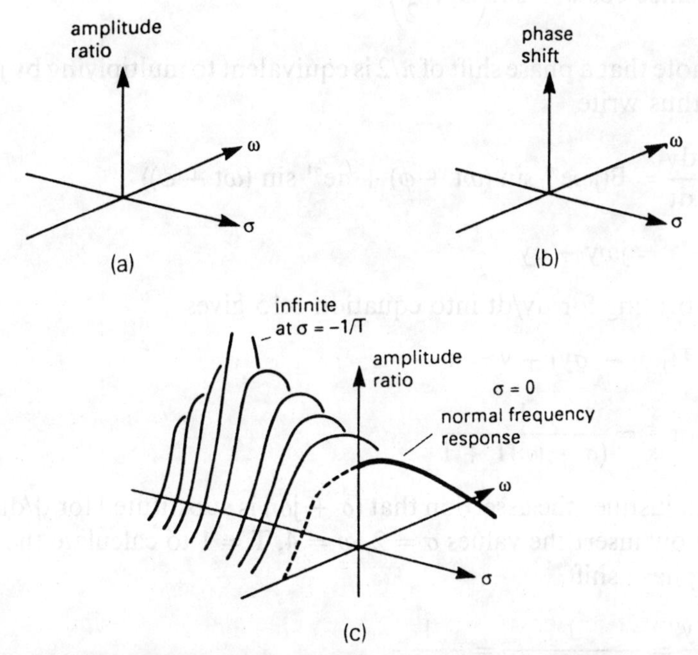

Fig. 3.31 Plotting the response to a sine wave with an exponential envelope, (a) Amplitude ratio; (b) Phase shift; (c) Response for first order lag.

As an example, we shall calculate the response to a first order lag of time constant 1 second to a sine wave of angular frequency 4 rads/sec and an envelope with decay factor (σ) of 2/sec.

The driving function has the form $Ae^{\sigma t} \sin \omega t$ (where $\sigma = 2$ and $\omega = 4$). The output will have the form $Be^{\sigma t} \sin (\omega t + \phi)$; i.e. different amplitude, phase shift ϕ, but the same decay factor and angular frequency.

For a first order system we have

$$T\frac{dy}{dt} + y = x \tag{3.15}$$

where y is the output, x the driving function and T the time constant. Substituting for y, and differentiating gives

$$\frac{dy}{dt} = B(e^{\sigma t}\,\omega\cos(\omega t + \phi) + \sigma e^{\sigma t}\sin(\omega t + \phi))$$

$$= B\left(e^{\sigma t}\,\omega\sin\left(\omega t + \frac{\pi}{2} + \phi\right) + \sigma e^{\sigma t}(\sin\omega t + \phi)\right)$$

since $\cos\theta = \sin\left(\theta + \frac{\pi}{2}\right)$

We note that a phase shift of $\pi/2$ is equivalent to multiplying by j. We can thus write

$$\frac{dy}{dt} = B(j\omega e^{\sigma t}\sin(\omega t + \phi) + \sigma e^{\sigma t}\sin(\omega t + \phi))$$

$$= j\omega y + \sigma y$$

Substituting for dy/dt into equation 3.15 gives

$$T(j\omega y + \sigma y) + y = x$$

$$\text{or } \frac{y}{x} = \frac{1}{(\sigma + j\omega)T + 1}$$

which justifies the assertion that $(\sigma + j\omega)$ is substituted for d/dt. We can now insert the values $\sigma = 2$, $\omega = 4$, $T = 1$ to calculate the gain and phase shift.

$$\frac{y}{x} = \frac{1}{(2 + 4j) + 1} = \frac{1}{3 + 4j}$$

$$= \frac{1}{3 + 4j} \times \frac{3 - 4j}{3 - 4j}$$

$$= \frac{3}{25} - \frac{4j}{25}$$

The amplitude ratio is therefore

$$\sqrt{\left(\frac{3}{25}\right)^2 + \left(-\frac{4}{25}\right)^2} = 0{\cdot}2$$

and the phase shift is $\tan^{-1}(-4/3)$.

The above lengthy, but straightforward, calculation positions one

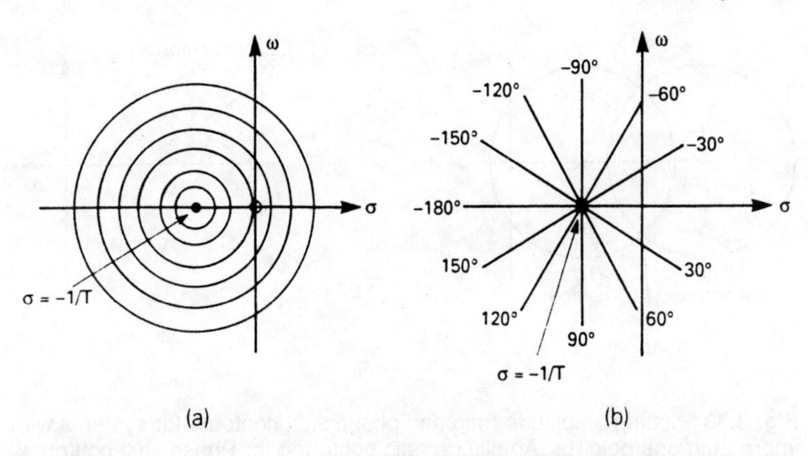

Fig. 3.32 Amplitude ratio and phase shift contours, (a) Amplitude ratio contours; (b) Phase shift contours.

point on the surface of Fig. 3.31. Many such calculations allow the full surface to be drawn.

An alternative representation is to use contour lines. It can be seen from Fig. 3.31c that the gain of the first order system becomes infinite at $\sigma = -1/T$. This corresponds to the open loop pole position calculated earlier. Represented by contour lines, the gain surface appears as concentric circles as Fig. 3.32a, and the phase shift as radiating lines as Fig. 3.32b, both centred on the pole at $\sigma = -1/T$. The gain and phase surfaces for systems with more than one pole can be found by adding the surfaces for each individual open loop pole. Figure 3.33a, b shows how the surface for a system with two poles at $\sigma = -a$, $\sigma = -b$ is formed. Note that the original pole positions are unaffected. Composite surfaces for any open loop system comprising first or second order blocks could be constructed, albeit laboriously, in a similar manner.

Given the position of the open loop poles, we now require to find the loci of the closed loop poles as the controller gain K is changed. Figure 3.34 shows the by now familiar closed loop block diagram. This has a response

$$\frac{y}{x} = \frac{KG}{1 + KG}$$

where K is the gain and G the open loop transfer function. The closed loop poles will be located where $(1 + KG) = 0$, i.e. the open loop phase shift is an odd multiple of $-180°$ (i.e. $-180°$, $-540°$ etc.)

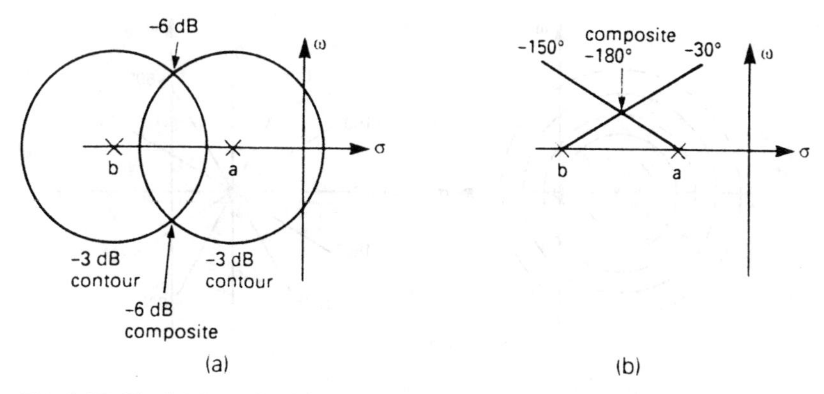

Fig. 3.33 Plotting amplitude ratio and phase shift contours for systems with more than one pole, (a) Amplitude ratio contours; (b) Phase shift contours.

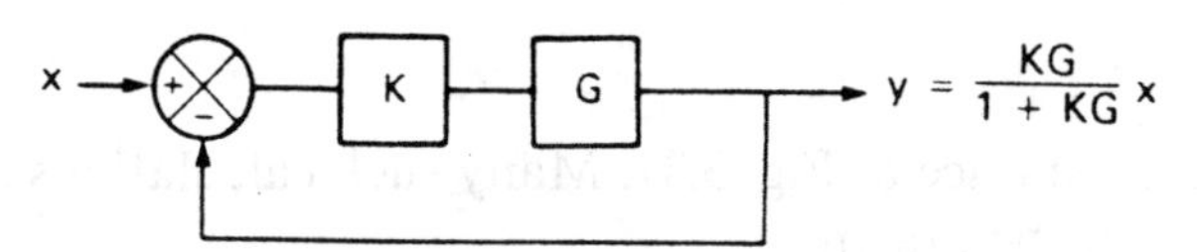

Fig. 3.34 Closed loop block diagram.

and the open loop gain is 1/K. These can be read directly off the open loop contours.

Figure 3.35 shows a composite gain/phase shift contour map for a third order position control system, with a pole from the velocity to position integrator at $\sigma = 0$, $\omega = 0$, and two first order lags each contributing one pole. The contours are calculated as above.

Let us assume a controller gain of 15 dB. The closed loop poles will lie where the $-180°$ or $-540°$ phase shift contours cross the -15 dB contour, since at these points $KG = -1$. There are three such points, each denoted by the pole symbol X.

Closed loop poles, therefore, lie along the phase shift contours for odd multiples of $-180°$. These contours form the root loci. The actual pole positions are determined by the controller gain K and lie where the gain contour 1/K crosses the phase shift contours forming the root loci.

The root loci in Fig. 3.35 is emphasised. It can be seen that the system is non-oscillatory for low gain, but becomes oscillatory as the controller gain rises above 3 dB and the right-hand poles leave the σ axis. At the controller gain of 26 dB the root loci pass into the right-hand half of the s-plane; giving positive σ and an unstable system. A controller gain of 26 dB will thus drive the system into instability.

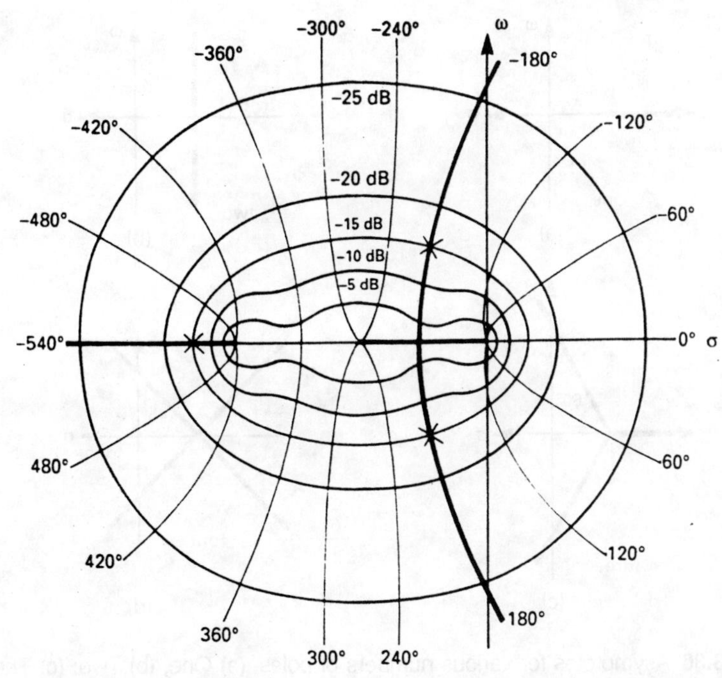

Fig. 3.35 Combined amplitude ratio and phase shift contours for a third order system. X denotes the positions corresponding to a controller gain of +15 dB. The system becomes unstable for a controller gain of +26 dB where the loci pass into the right-hand portion of the plot (σ positive).

The above procedure seems, at first sight, to require a lot of calculation to draw the composite contours. Fortunately it is possible to draw root loci by simple rules of thumb, given in the next section.

The root locus method of investigating stability substitutes $(\sigma + j\omega)$ for d/dt. The letter 's' is conveniently used to denote $(\sigma + j\omega)$ (hence the description of the $\sigma/j\omega$ diagram as the s-plane) and provides the link to Laplace transforms described later.

It is also usually more convenient to express blocks in the form

$$\frac{A}{(s + a)} \quad \text{or} \quad \frac{B}{(s^2 + as + b)}$$

as this allows roots to be directly observed; $s = -a$ for the first example. The first order delay, time constant T, low frequency gain A, i.e.

$$\frac{A}{(1 + TD)} \quad \text{becomes} \quad \frac{A/T}{(s + 1/T)}$$

in s notation showing directly the open loop pole at $-1/T$.

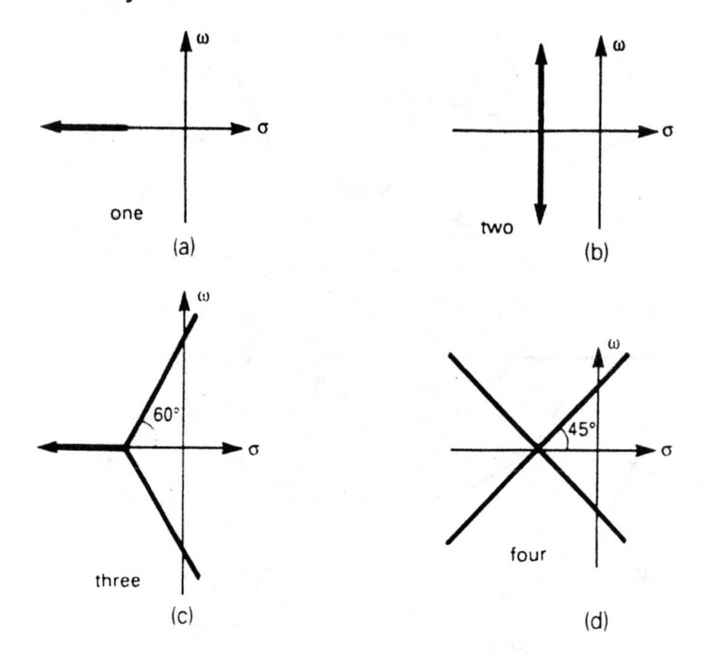

Fig. 3.36 Asymptotes for various numbers of poles, (a) One; (b) Two; (c) Three; (d) Four.

3.5.3. Simple rules for drawing root loci

Root loci start at the open loop poles (corresponding to very low values of K) and proceed to the edges of the s-plane at $\sigma = \pm\infty$ and $\omega = \pm\infty$. There are as many loci as there are poles, and these tend towards asymptotes as the value of K increases. These asymptotes all originate from one centre, and are symmetrical about the σ axis, as shown in Fig. 3.36.

The origin of the asymptotes is the mean position of the poles, i.e. at

$$\sigma = \frac{\Sigma \text{ pole positions}}{\text{number of poles}}$$

The angle between the asymptotes is 360°/(number of poles) arranged to be symmetrical about the σ axis as Fig. 3.36.

A simple position control system is shown in Fig. 3.37, with time constant of 1 sec and 5 sec. This has poles at s = 0, s = − 0·2 and s = − 1. The mean pole position is −0·4, so this is the origin of the asymptotes. The angle between the asymptotes is 360°/3 = 120°, arranged as Fig. 3.36c for symmetry. We can thus sketch the root

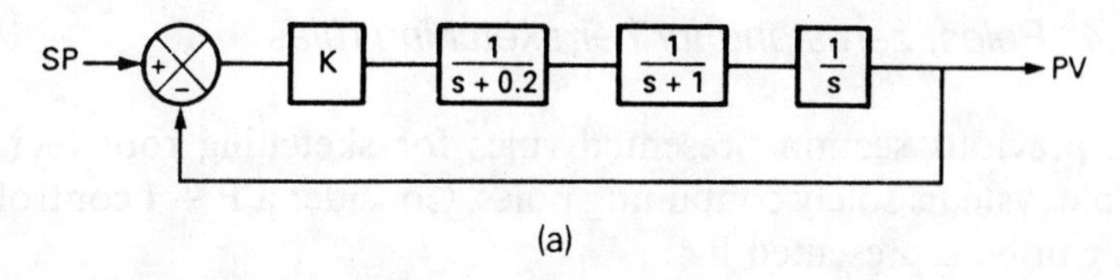

(a)

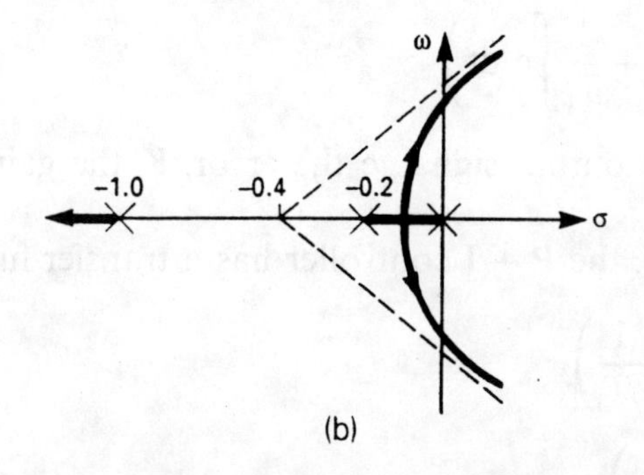

(b)

Fig. 3.37 Drawing the root locus for a simple system, (a) A simple position control system; (b) Root locus for system.

locus as Fig. 3.37b. This shows that the system can become unstable for large values of K. The critical value of controller gain can be found by forming the characteristic equation and applying the Routh–Hurwitz criteria.

The point on the σ axis where the poles split (and the response becomes oscillatory) can be found by solving the equation

$$\frac{1}{(\sigma - P_1)} + \frac{1}{(\sigma - P_2)} + \ldots + \frac{1}{(\sigma - P_n)} = 0$$

where P_n is the position of pole n, and the solution σ is the negative of the critical damping point. For the above example

$$\frac{1}{\sigma} + \frac{1}{\sigma - 0{\cdot}2} + \frac{1}{\sigma - 1} = 0$$

$$(\sigma - 0{\cdot}2)(\sigma - 1) + \sigma^2 - \sigma + \sigma^2 - 0{\cdot}2\sigma = 0$$

$$3\sigma^2 - 2{\cdot}4\sigma + 0{\cdot}2 = 0$$

Giving $\sigma = 0{\cdot}094$ or $\sigma = 0{\cdot}706$

The solution $\sigma = 0{\cdot}094$ is correct by inspection, so the root locus splits at $\sigma = -0{\cdot}094$.

3.5.4. Poles, zeros and further sketching rules

The previous section presented rules for sketching root loci for simple systems solely containing poles. Consider a P + I controller; this can be represented by

$$V_o = K\left(e + \frac{1}{T_i}\int e\,dt\right)$$

where V_o is the output signal, e the error, K the gain and T_i the integral time.

In s notation, the P + I controller has a transfer function

$$K\left(1 + \frac{1}{sT_i}\right)$$

$$= K\left(\frac{s + a}{s}\right)$$

where $a = 1/T_i$.

This transfer function has a pole at the origin (from the denominator) but it also has zero value at $s = -a$. Positions on the s-plane where a transfer function goes to zero are known, not surprisingly, as 'zeros' and are conventionally represented by an 'O'. A P + I controller on its own therefore appears as Fig. 3.38.

The inclusion of zeros serves to stabilise and improve the transient performance of a closed loop system, but requires some minor modifications of the root locus sketching rules given in the previous section.

(a) The root loci start at the open loop poles (corresponding to low gain) and proceed to the edge of the s-plane or to a zero.
(b) The root locus lies on the σ axis to the left of odd numbered poles

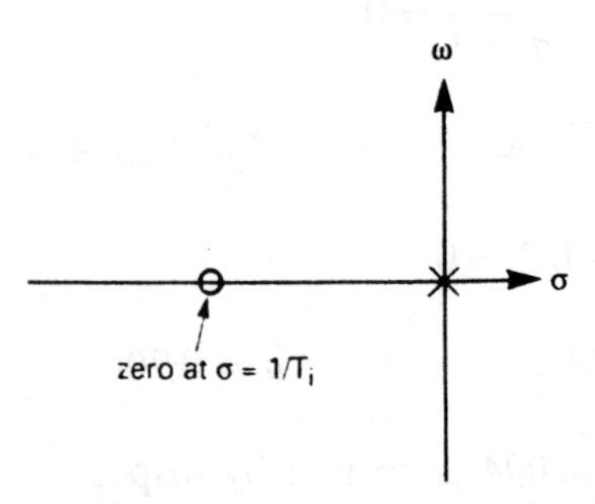

Fig. 3.38 A P+I controller open loop pole and zero on s-plane.

and zeros (counted from the pole with most positive σ).

(c) There are N asymptotes, where

$$N = (\text{number of poles} - \text{number of zeros})$$

(d) The asymptotes originate from

$$\sigma = \frac{\Sigma \text{ pole positions} - \Sigma \text{ zero positions}}{N}$$

(e) The angle between asymptotes is 360°/N.

(f) The root locus is symmetrical about the σ axis.

(g) Splits off the σ axis occur at $(-\sigma)$ where σ is the solution of

$$\frac{1}{\sigma - P_1} + \frac{1}{\sigma - P_2} + \ldots + \frac{1}{\sigma - P_n} - \frac{1}{\sigma - Z_1} - \ldots - \frac{1}{\sigma - Z_m} = 0$$

and P_n and Z_m are the pole and zero positions respectively.

Figure 3.39 represents a level control system controlled by a P + I controller. It will be noted that the P + I controller, in conjunction with the integral action of the tank, has introduced a double pole at the origin (s^2). Double poles (or double zeros) can be considered for analytical purposes as two poles (or zeros) separated by a small distance as Fig. 3.39b.

There are two possible conditions for the root locus, $1/T_i$ greater than $1/T$ (as Fig. 3.39c) or $1/T_i$ less than $1/T$ (as Fig. 3.39d).

In both cases, the locus will lie between the left-hand pole and the zero (from rule b above) and there will be two asymptotes (number of poles = 3, number of zeros = 1). The basic shape of the root locus will therefore be similar.

Let us assume T = 5 seconds, and find the origin of the asymptote for T_i = 10 sec, corresponding respectively to cases in Fig. 3.39c and d.

For case c, the origin is at $(0 + (-0{\cdot}2) - (-0{\cdot}5))/2$, i.e. $\sigma = 0{\cdot}15$. The asymptote is therefore to the right of the origin, giving a root locus as shown on Fig. 3.39e. This system is unstable for all values of gain. (This, incidentally, illustrates one of the rules of thumb for controller settings; the integral time must be longer than the dominant time constant in systems with integral action.)

For case d, the origin is at $(0 + (-0{\cdot}2) - (0{\cdot}1))/2$, i.e. $\sigma = -0{\cdot}05$. The asymptote is to the left of the origin with a root locus as Fig. 3.39f. Such a system is stable, but oscillatory, for all gains.

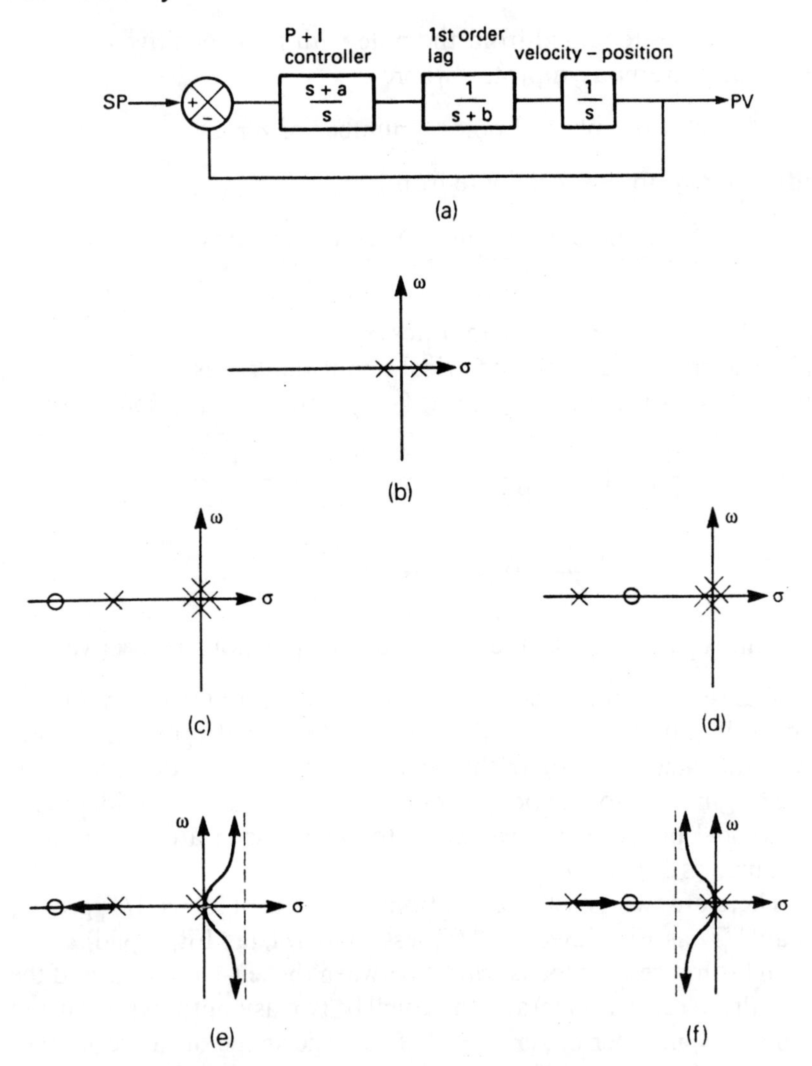

Fig. 3.39 Stability of position control system, (a) Position control system; (b) Representation of double pole; (c) Open loop poles/zero with $1/T_i < 1/T$; (d) Open loop poles/zero with $1/T < 1/T_i$; (e) Root locus with $1/T_i < 1/T$; (f) Root locus with $1/T < 1/T_i$.

Figure 3.40 illustrates a non-obvious aspect of drawing root loci, the system representing a single first order lag controlled by a P + I controller. From rule b, the root loci must lie as emphasised, but how do these link up? There is one asymptote, along the σ axis, from rules e and f. The root loci follows the shape of Fig. 3.40b, splitting briefly rejoining then one pole going to the zero and one to $\sigma = -\infty$. Such a system is non-oscillatory at low gain, becoming oscillatory but

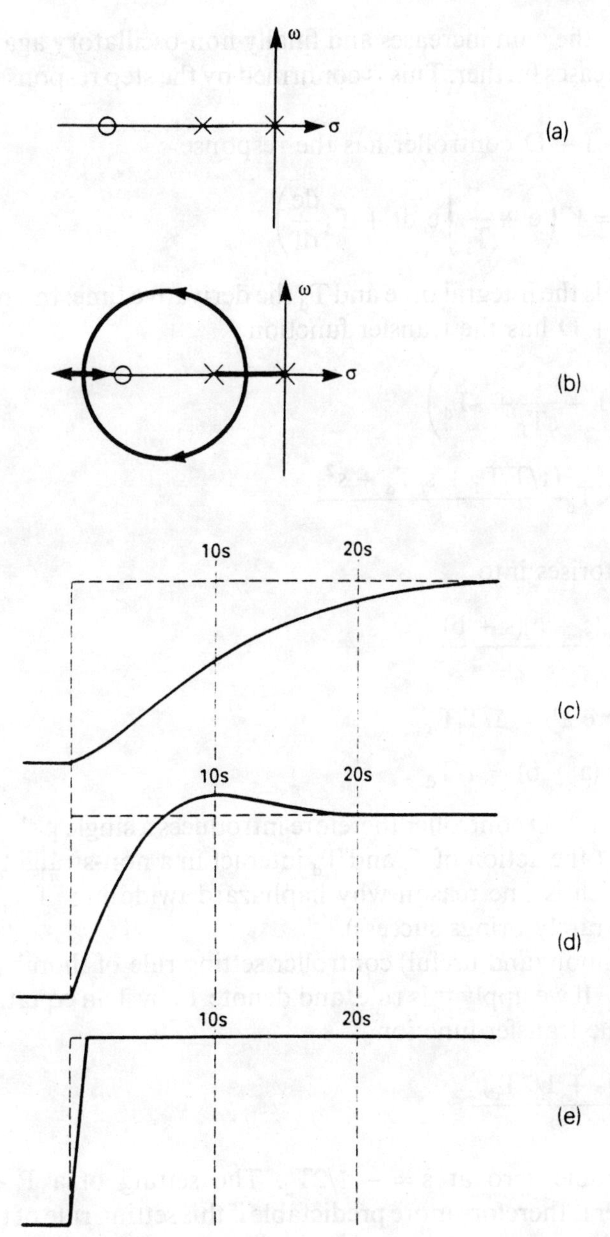

Fig. 3.40 System with non-obvious root locus. Step responses are based on first order lag of time constant 5 s controlled by a P + I controller with T_i of 2 s and gain as shown, (a) Open loop poles/zeros for first order lag controlled by P + I controller; (b) Root locus for system; (c) Low gain (K = 0·2); (d) Medium gain (K = 1); (e) High gain (K = 100).

stable, as the gain increases and finally non-oscillatory again as the gain increases further. This is confirmed by the step responses of Fig. 3.40c–e.

A P + I + D controller has the response

$$V_o = K\left(e + \frac{1}{T_i}\int e\,dt + T_d\frac{de}{dt}\right)$$

where T_i is the integral time and T_d the derivative time. In s notation, a P + I + D has the transfer function

$$K\left(1 + \frac{1}{sT_i} + sT_d\right)$$

$$\text{or } KT_d\frac{(1/T_iT_d + s/T_d + s^2)}{s} \tag{3.16}$$

This factorises into

$$\frac{KT_d(s + a)(s + b)}{s}$$

$$\text{where } ab = 1/T_iT_d$$

$$\text{and } (a + b) = 1/T_d$$

A P + I + D controller therefore introduces a single pole and two zeros, but the action of T_i and T_d interact in a non-straightforward way (which is one reason why haphazard twiddling of T_i and T_d controls rarely brings success).

A common (and useful) controller setting rule of thumb is to set $T_i = 4T_d$. If we apply this rule, and denote T_d by T in equation 3.16, we get the transfer function

$$\frac{KT(s + 1/2T_d)^2}{s}$$

i.e. a double zero at $s = -1/2T_d$. The setting of a P + I + D controller is therefore more predictable if this setting rule of thumb is followed.

Where there are many poles and zeros, it can sometimes be difficult to work out what path the root locus follows. In these cases it is often useful to start off with a simple locus based on the dominant poles and zeros, and refine it by introducing further poles/zeros. Consider, for example, the system of Fig. 3.41a. This has four poles and one zero, so there are three asymptotes centred on

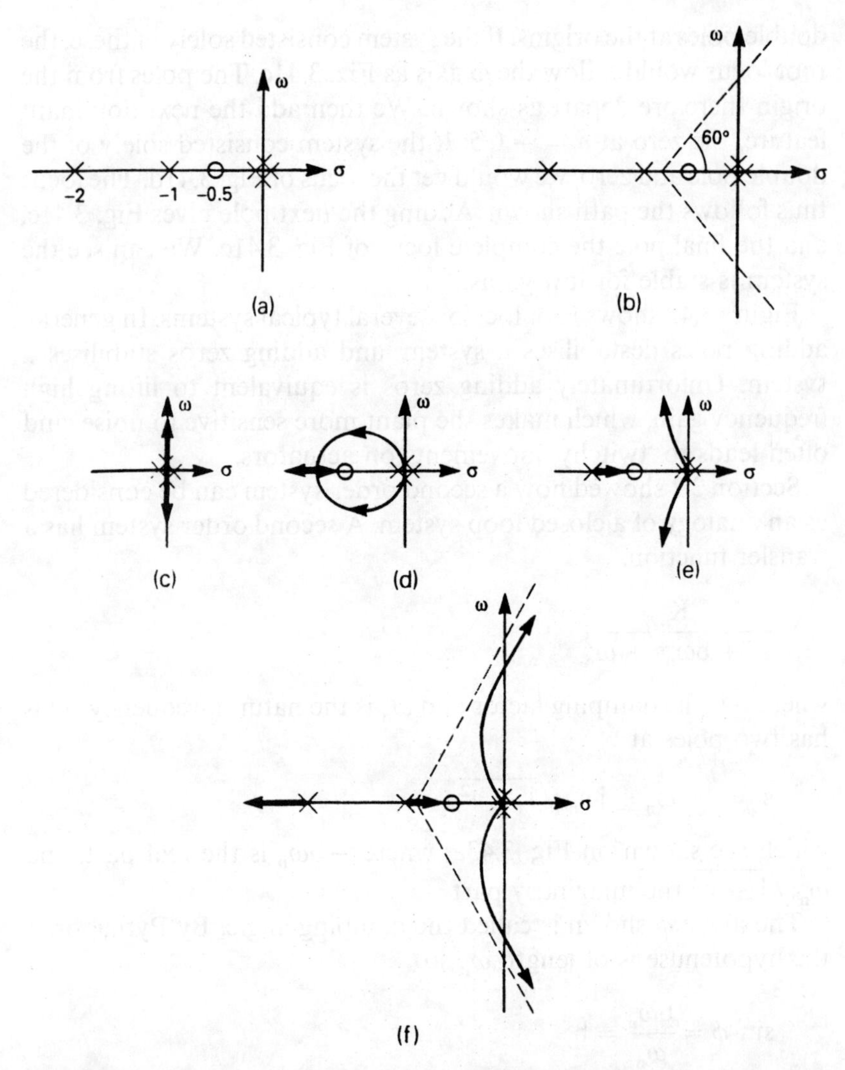

Fig. 3.41 Construction of root locus in stages, (a) Open loop poles and zero; (b) Asymptotes; (c) Locus near origin; (d) Addition of dominant zero; (e) Addition of dominant zero and pole; (f) Complete locus.

$((-3 + 0{\cdot}5)/3)$, i.e. $-0{\cdot}833$. Applying the rules above, we obtain the general shape of Fig. 3.41b, with the two right-hand arms being the path of the poles originating from the origin. This tells us the system is unstable for large gains, but we cannot deduce immediately if it is stable for all gains.

To resolve what happens in the region of low gains, we build the root locus up in stages. The most dominant poles/zeros are the double poles at the origins. If the system consisted solely of these, the root locus would follow the ω axis as Fig. 3.41c. The poles from the origin therefore depart as shown. We then add the next dominant feature, the zero at $\sigma = -0{\cdot}5$. If the system consisted solely of the double pole and zero, we would get the locus of Fig. 3.41d. The locus thus follows the path shown. Adding the next pole gives Fig. 3.41e, and the final pole the complete locus of Fig. 3.41c. We can see the system is stable for low gains.

Figure 3.42 shows root loci for several typical systems. In general, adding poles destabilises a system, and adding zeros stabilises a system. Unfortunately adding zeros is equivalent to lifting high frequency gain, which makes the plant more sensitive to noise, and often leads to 'twitchy' movements on actuators.

Section 3.2 showed how a second order system can be considered as an analogy of a closed loop system. A second order system has a transfer function.

$$\frac{K}{s^2 + b\omega_n s + \omega_n^2}$$

where b is the damping factor and ω_n is the natural frequency. This has two poles at

$$s = -b\omega_n \pm j\omega_n\sqrt{1 - b^2}$$

which are shown on Fig. 3.43a, where $-b\omega_n$ is the real part, and $\omega_n\sqrt{1 - b^2}$ the imaginary part

The angle ϕ shown is called the damping angle. By Pythagoras, the hypotenuse is of length ω_n so

$$\sin \phi = \frac{b\omega_n}{\omega_n} = b$$

This gives us a way to position the closed loop poles on the root locus. Section 3.21 related design criteria such as overshoot and settling time to the damping factor. In Fig. 3.43b a line (called the damping line) is drawn at the damping angle corresponding to a required damping factor (an overshoot of 20%, for example, would correspond to a damping factor of 0·5, and a damping angle of 30°). The required pole positions occur where the damping line crosses the root locus.

The values of σ and ω can be read off the sketch and the required gain can, in theory, be found by substituting $s = (\sigma + j\omega)$ with values for σ and $j\omega$, into the characteristic equations and solving for K by

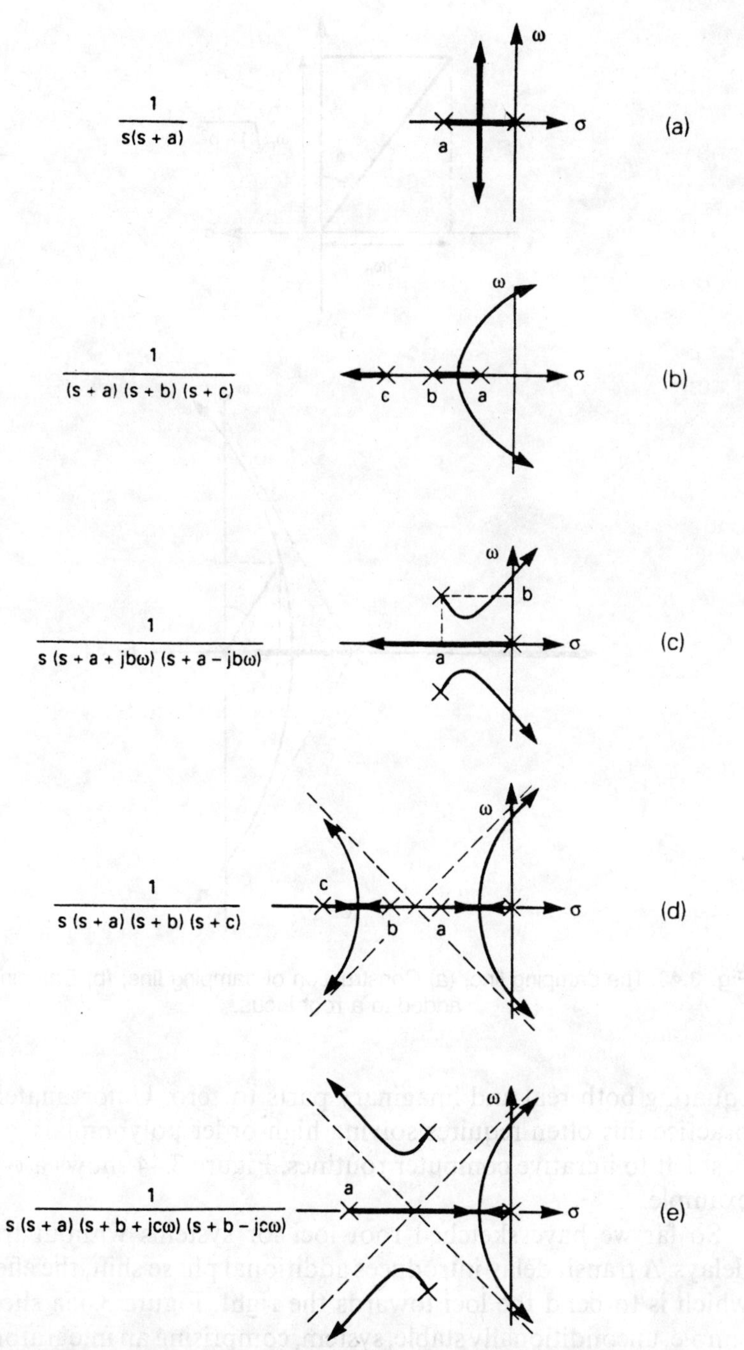

Fig 3.42 Various common root loci, (a) Simple position or level control; (b) Three cascaded lags (c) Position control with non-rigid structure or level control with resonance; (d) Position control with three lags; (e) Position control with lag and non-rigid structure.

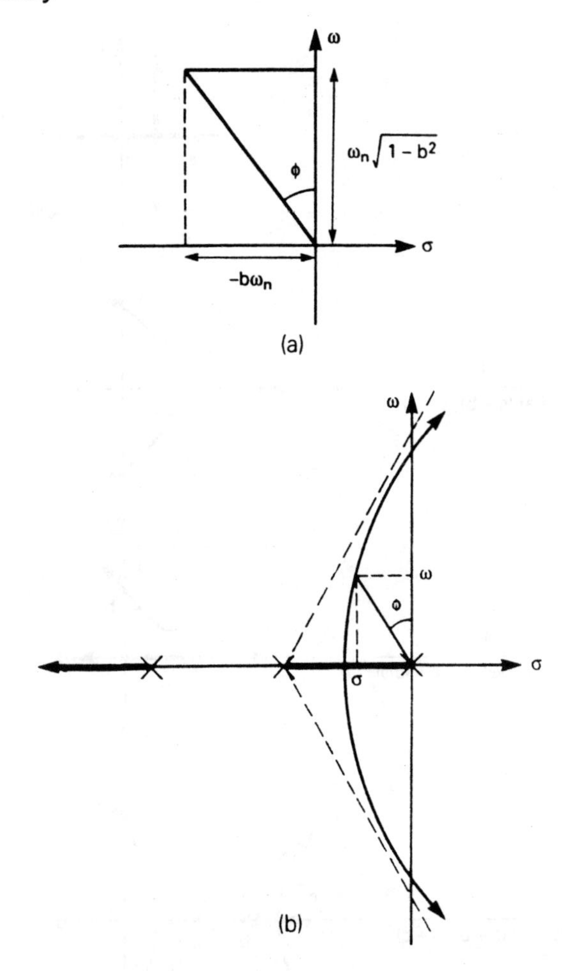

Fig. 3.43 The damping line, (a) Construction of damping line; (b) Damping line added to a root locus.

equating both real and imaginary parts to zero. Unfortunately, in practice this often requires solving high order polynomials, a task best left to iterative computer routines. Figure 3.44 shows a typical example.

So far we have sketched root loci for systems without transit delays. A transit delay introduces additional phase shift, the effect of which is to bend the loci towards the right. Figure 3.45a shows a simple, unconditionally stable, system, comprising an integrator and a single first order lag. If a ten second transit delay is added, the additional phase shift will be added as Fig. 3.45b. This additional phase shift is added to the phase shift from the poles and bends the

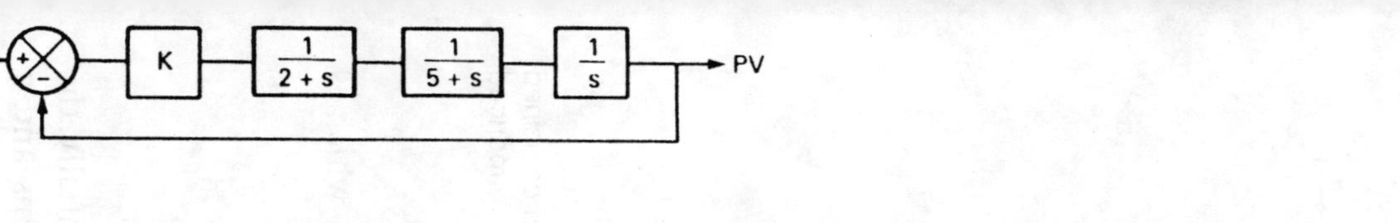

System

Pole positions for various values of K

K	Poles		
1	−5·06	−1·83	−0·11
4	−5·24	−1·0	−0·76
5	−5·29	−0·86 ± 0·46j	
8	−5·43	−0·78 ± 0·92j	
10	−5·51	−0·74 ± 1·12j	
20	−5·88	−0·56 ± 1·76j	
40	−6·41	−0·29 ± 2·48j	
60	−6·82	−0·09 ± 2·96j	
80	−7·16	+0·08 ± 3·34j	
100	−7·46	+0·23 ± 3·65j	

From Plot:
system becomes oscillatory for $K \approx 4{\cdot}5$
unstable for $K \approx 70$
damping line drawn for $b = 0{\cdot}5$
required gain $= K \approx 12$
predicted $\sigma = 0{\cdot}7$, $\omega \approx 1{\cdot}25$

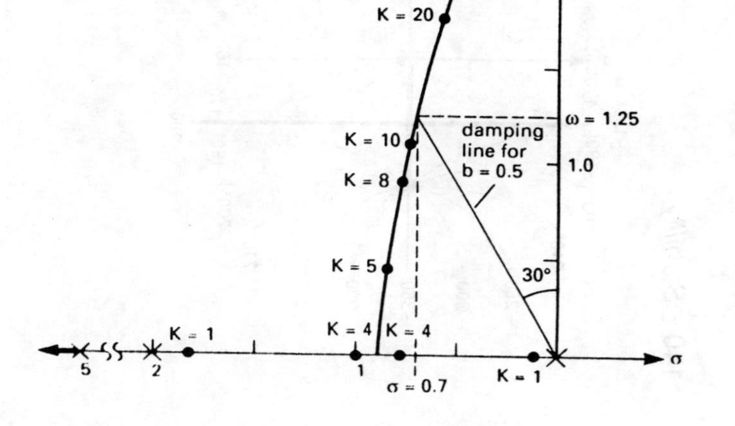

Fig. 3.44 Root locus calculation for sample third order system.

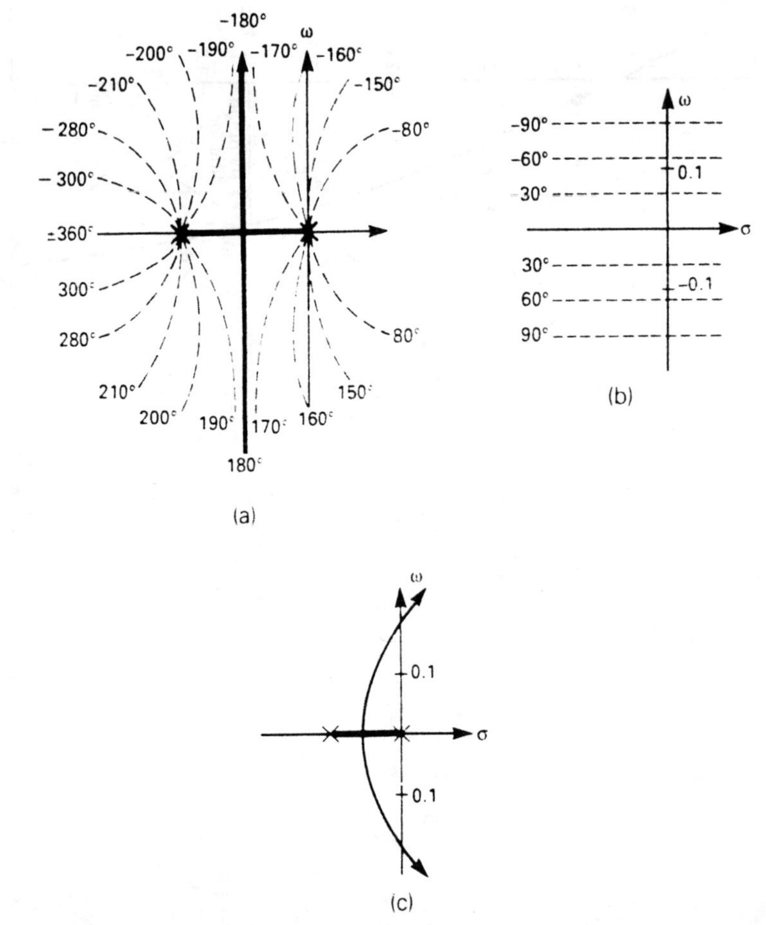

Fig. 3.45 Transit delay plotted on root locus, (a) Simple system, with combined phase shifts, which is stable for all gains; (b) Additional phase shift introduced by 10 s transit delay; (c) Modified root locus.

root locus, giving the shape of Fig. 3.45c and a system which is unstable at high gains.

3.6. Laplace transforms

Determination of the response of a control system is, essentially, the solution of a differential equation given initial conditions and a driving function (e.g. a step, ramp, sine wave, etc.) The Laplace transform converts a differential equation, initial conditions and driving function to a functional expression which can be manipulated as a straightforward mathematical expression. Inverse transformations allow conversion back to give the time related response.

A study of Laplace transforms could (and does) easily fill a book, and it is beyond the introductory nature of this book to justify the underlying mathematics. The treatment given below therefore treats Laplace transforms as a tool to be used; the reader should consult more advanced works for mathematical derivations and the link between Laplace transforms and the root locus technique described above.

The Laplace transform of a function x(t) is defined as

$$\mathscr{L}(x) = \int_0^\infty e^{-st}\, x(t)\, dt \tag{3.17}$$

Often $\bar{x}(s)$ is used to denote the Laplace transform. Transforms of function can be obtained by performing the integration defined in equation 3.17 but it is generally more convenient to use reference tables such as Table 3.5. This shows, for example, that the Laplace transform of e^{at} is $1/(s - a)$.

The Laplace transform of each term in a differential equation is given by

$$\mathscr{L}\left(\frac{d^n x(t)}{dt}\right) = s^n \bar{x}(s) - s^{n-1} x(0) - s^{n-2} x'(0) \tag{3.18}$$

where x(0) is the value of x at t = 0, x′(0) the value of dx/dt at t = 0 and so on.

The derivation of the response of a system to a given driving function essentially falls into three stages:

(a) The equations representing the system and the driving function are transformed to 's' notation using Table 3.5 and equation 3.18 (including the initial conditions required by equation 3.18).
(b) The resulting equation in 's' is manipulated by standard rules of algebra into standard forms from the right-hand side of Table 3.5 (often via partial fractions).
(c) The inverse of the standard forms are read off the table. The sum of these is the solution.

To see how this works in practice, we shall derive the response of a second order system to a step input of height 2 units. The equation describing the system is

$$\frac{d^2x}{dt^2} + 3\frac{dx}{dt} + 2x = 2u(t) \tag{3.19}$$

Table 3.5 *Common Laplace Transforms*

Signal	*Description*	*Representation*	*Transform*
	unit step	$u(t)$	$\frac{1}{s}$
	ramp	t	$\frac{1}{s^2}$
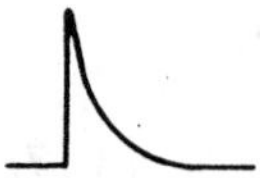	exponential	e^{-at}	$\frac{1}{s+a}$
	exponential	te^{-at}	$\frac{1}{(s+a)^2}$
	sine wave	$\sin \omega t$	$\frac{\omega}{s^2+\omega^2}$
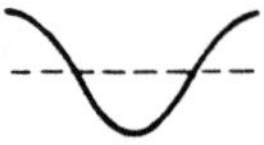	cosine wave	$\cos \omega t$	$\frac{s}{s^2+\omega^2}$
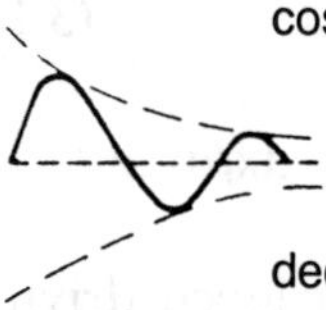	decaying sinusoid	$e^{-at} \sin \omega t$	$\frac{\omega}{s^2+2as+a^2+\omega^2}$
	impulse	$\delta(t)$	1

We will add the conditions that at $t = 0$, $x = 2$ and $dx/dt = 1$. Using equation 3.18 we have

$$\mathscr{L}\left(\frac{d^2x}{dt^2}\right) = s^2\bar{x}(s) - 2s - 1$$

$$\mathscr{L}\left(3\frac{dx}{dt}\right)3(s\bar{x}(s) - 2)$$

$$\mathscr{L}(2x) = 2\bar{x}(s)$$

The Laplace transform of 2u(t) is, from Table 3.5, 2/s. Substituting into equation 3.19 gives

$$(s^2\bar{x}(s) - 2s - 1) + 3(s\bar{x}(s) - 2) + 2\bar{x}(s) = \frac{2}{s}$$

$$\bar{x}(s)(s^2 + 3s + 2) - 2s + 6 = \frac{2}{s}$$

$$\text{or } \bar{x}(s) = \frac{2s^2 + 6s + 2}{s(s^2 + 3s + 2)} \qquad (3.20)$$

To obtain the inverse of equation 3.20, we need to convert equation 3.20 into one of the standard forms of Table 3.5. The denominator of 3.20 factorises into s(s + 1)(s + 2), so we can use partial fractions and write

$$\bar{x}(s) = \frac{A}{s} + \frac{B}{s+1} + \frac{C}{s+2} \qquad (3.21)$$

where A, B, C are to be determined. Putting equation 3.21 into the form of equation 3.20 yields

$$\bar{x}(s) = \frac{(A + B + C)s^2 + (3A + 2B + C)s + 2A}{s(s^2 + 3s + 2)}$$

$$\begin{aligned} \text{Hence}\quad & A + B + C = 2 \\ & 3A + 2B + C = 6 \\ & 2A = 2 \end{aligned}$$

Solving these simultaneous equations gives A = 1, B = 2, C = − 1

$$\text{Hence } \bar{x}(s) = \frac{1}{s} + \frac{2}{s+1} - \frac{1}{s+2}$$

The inverse transform for each function can now be read directly off Table 3.5 to give

$$x(t) = u(t) + 2e^{-t} - e^{-2t}$$

which is the system response to a two unit step.

3.7. Sampled systems and the Z transform

3.7.1. Introduction

So far the systems discussed have been based on analog signals, i.e. the value of the plant variables are represented by continuous voltages, currents or pneumatic pressures. These variables can have any value in the signal range, and are constantly monitored.

The increasing use of computers and digital techniques requires a different view of how a process is controlled. A computer deals with integer numbers, not continuous signals. A common control standard uses a 12 bit binary number to represent signals. Unscaled this can only represent from 0 to 4095 in steps of one. If a computer uses scaling to represent a flow from 0 to 15 000 litres per minute as a 12 bit binary number, the best resolution is 12 000/4095 or about three litres per minute. Some values (e.g. 5 lpm) will never appear as the computer will go directly from 3 to 6 in one jump. The digital nature of a computer inherently introduces additional resolution error, about 0·025% for a 12 bit representation.

A more subtle problem, however, occurs because the signal is not constantly monitored, but is sampled at regular intervals. A computer can only do one operation at a time. In a control system it will have to read inputs, calculate the control algorithm, update the outputs and drive the all important operator interface. At any instant the computer can only be doing one task so inputs, for example, are read at regular intervals (say once per second). Having read an input the computer assumes it stays at this value until the next sample one second later. Outputs are similarly updated at regular time intervals, and not driven continuously.

Figure 3.46a shows a typical computer control system. The switches represent the sampling process. Analog inputs are read at regular time intervals by a device called an analog to digital converter (or ADC). This provides a digital representation of an analog input signal. The computer performs its control algorithm on the value provided by the ADC. There is no point in performing the control algorithm more than once per ADC value, as the plant signal from the ADC is not going to change until the next ADC read.

The output of the control algorithm is passed to a digital to analog converter (or DAC) which converts the number from the control algorithm to an analog signal, an 8 bit signal with a range of 0 to 255, say, being converted to 0 to 10 v. Note that although the output is analog at this stage, it can only assume certain voltages, and consequently changes in steps. For our 8 bit output, each step would be about 50 mV. The operation can be visualised as the scan of Fig. 3.46b, with the scan being performed at regular time intervals.

Intuitively one can see that this scan introduces a time delay into the control loop, and from previous discussions it should be expected that this time delay will be destabilising.

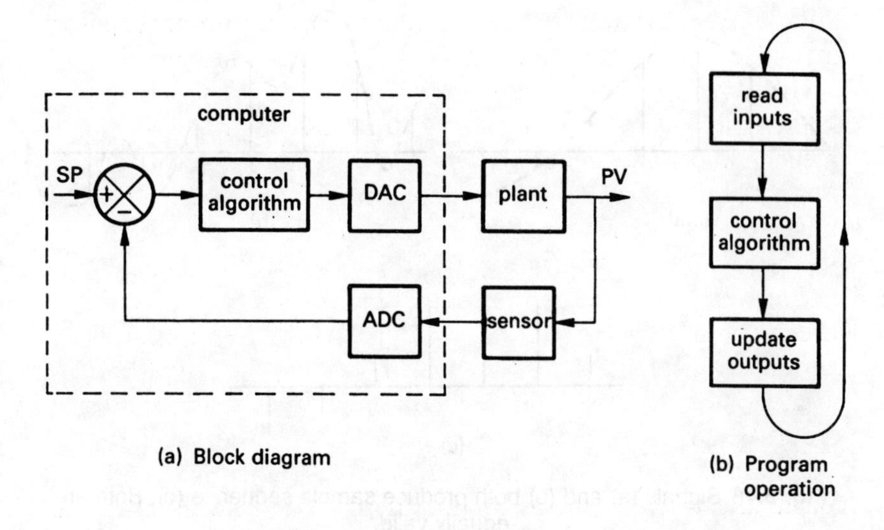

Fig. 3.46 A computer control system.

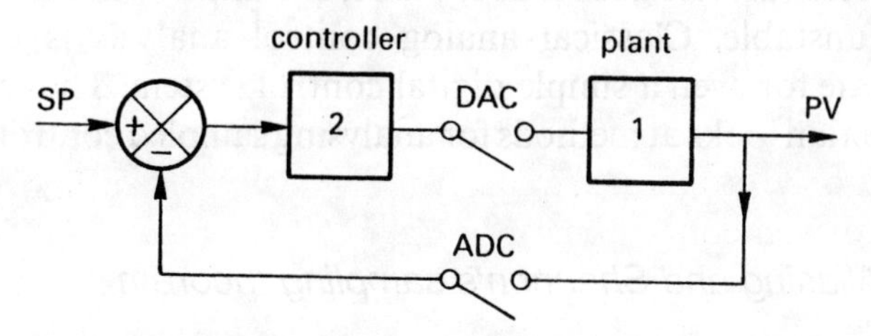

Fig. 3.47 A simple sampled control system. The switches denote the sampling process.

Consider the simple system of Fig. 3.47. Here a simple proportional only digital controller of gain two is driving a simple plant of unity gain. There are no dynamic effects. Suppose we introduce a set point change from 0 to 100. Continuous control theory would say the plant would be stable and settle at an output value of 67.

Our digital control system, though, behaves in a most undesirable manner.

Scan 1: Set point = 100, Process variable = 0, Error = 100
Controller output = 200, Plant output = 200
Scan 2: Set point = 100, Process variable = 200, Error = − 100
Controller output = − 200, Plant output −200
Scan 3: Set point = 100, Process variable = − 200, Error = 300
Controller output = 600, Plant output = 600

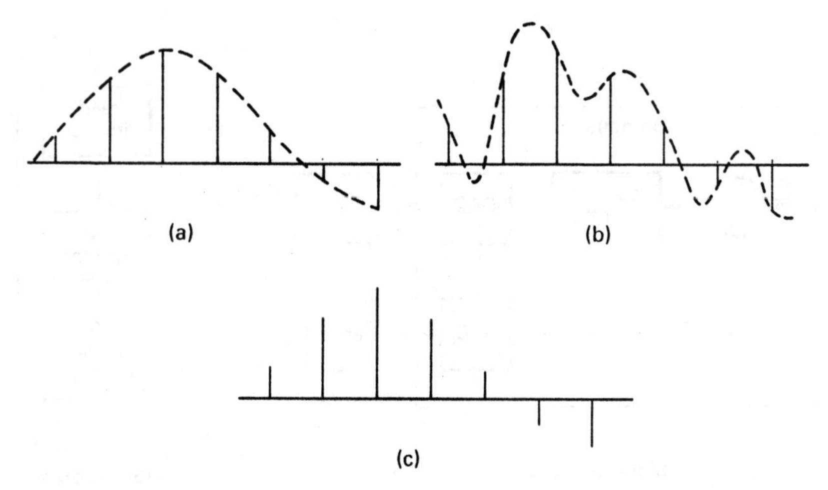

Fig. 3.48 Signals (a) and (b) both produce sample sequence (c). Both are equally valid.

The process variable goes 0, 200, −200, 600 and so on. The system is plainly unstable. Classical analog control analysis is obviously inadequate for even a simple digital control system. The remainder of this section looks at methods for analysing sampled control systems.

3.7.2. Aliasing and Shannon's sampling theorem

A sampled control system only knows about the values of its samples. It cannot infer any other information about the signals it is dealing with. Both of the continuous signals in Fig. 3.48a and b (and an infinite number of others) would produce the identical samples of Fig. 3.48c. An obvious question, therefore, is what sample rate we need to choose if our control performance is to be satisfactory.

In Fig. 3.49a, a sine wave is being sampled at a relatively fast rate. Intuitively one would assume the sample rate is adequate. In Fig. 3.49b the sample rate and the sine wave frequency are the same and the samples are delivering a constant value. Obviously this sample rate is too slow.

In Fig. 3.49c the sample rate is slower than the frequency, and the sample values are implying a sinusoidal frequency of much lower frequency than the signal. This latter case is called aliasing. A visual effect can often be seen on a cinema screen where moving wheels appear stationary or move backward. This effect occurs because the camera samples the world about 50 times per second.

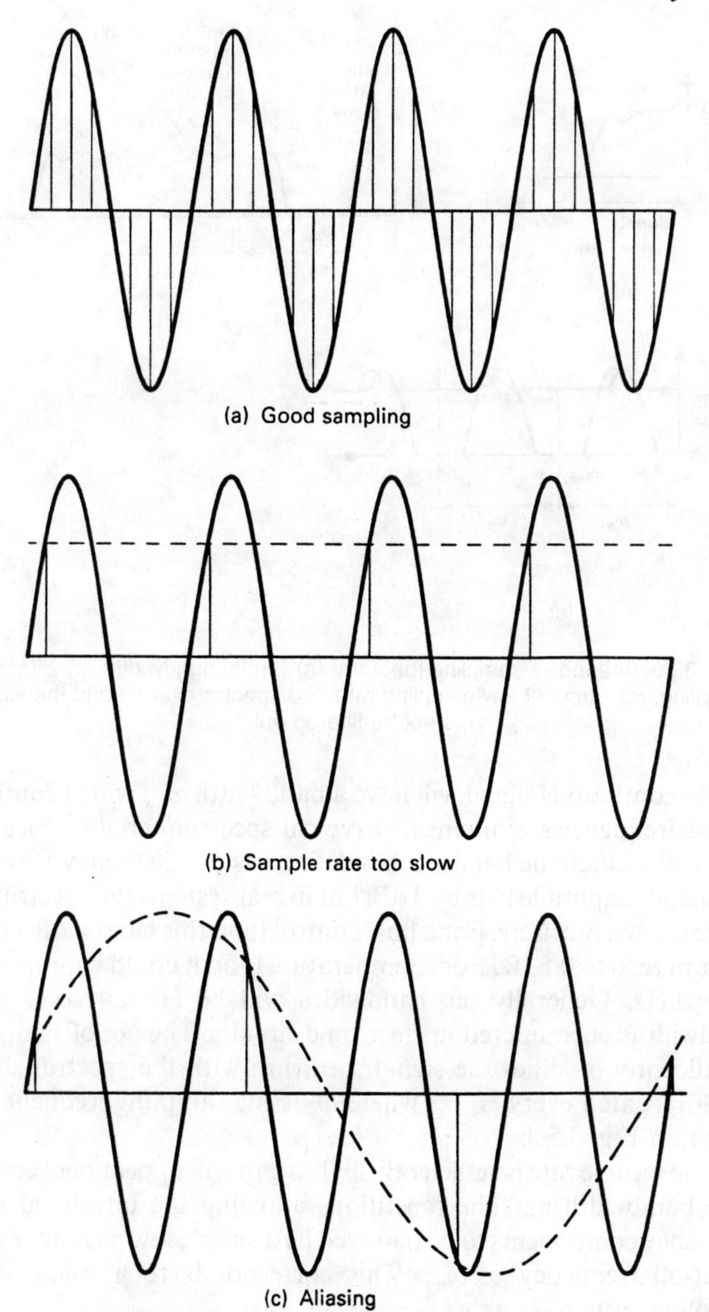

Fig. 3.49 The effect of the sampling rate.

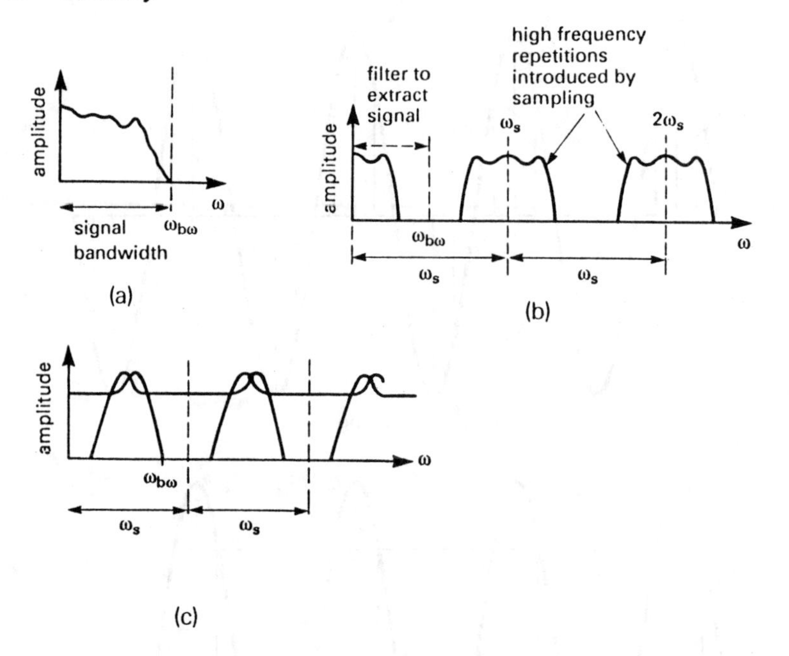

Fig. 3.50 Shannon's sampling theorem, (a) Signal bandwidth; (b) Effect of sampling; (c) Effect of low sampling rate, the spectra overlap and the signal cannot be filtered out.

Any continuous signal will have a bandwidth ω_{bw} which contains all the frequencies of interest. A typical spectrum could appear as Fig. 3.50a. Often the bandwidth is defined as the frequency at which the signal amplitude falls by 3 dB, but in real systems the spectrum is never known precisely. For a flow control loop this bandwidth could be from zero to 0·5 Hz. For a temperature loop it could be from zero to 0·01 Hz. Generally the bandwidth will be far narrower than bandwidths encountered in electronic circuits. The act of sampling significantly modifies the signal spectrum with the spectral shape being repeated every ω_s Hz where ω_s is the sampling frequency as shown in Fig. 3.50b.

If the sample rate is sufficiently high, a gap will appear between the base bandwidth and the repetitions allowing the unwanted high frequency components to be removed by a simple low pass filter with a cut-off frequency of ω_{bw}. This corresponds to a well-defined sampling system.

If, however, the sampling frequency ω_s is less than twice ω_{bw} the frequency spectra will overlap as shown in Fig. 3.50c and it will not be possible to extract the original signal by any means.

It follows that the sample frequency must be at least twice the bandwidth of interest. This is known as Shannon's sampling theorem.

Any system, however, will probably not have a spectrum as clearly defined as Fig. 3.50a, and the signal will probably also contain high frequency noise. The unwanted part of the spectrum and the noise may cause aliasing, so a low pass filter with sharp cut-off is normally placed before the ADC. This clearly defines the bandwidth and is known as an anti-aliasing filter.

Many sampled data systems use very high sampling frequencies (by very high we mean about 10–50 Hz), and can be analysed by normal continuous theory. Shannon's sampling theorem states the minimum sampling rate is twice the bandwidth. An obvious question is what happens in practical systems.

The lower sampling limit is usually 5–10 times the bandwidth; a temperature loop with a bandwidth of 0·01 Hz could be sampled at once every ten seconds, but at this sample rate the techniques described later in this section would have to be used. At sample frequencies greater than 50 times the bandwidth the sampling can normally be ignored and normal classical continuous analysis performed.

3.7.3. Sampling and the Z transform

A sampling system produces a string of numbers at regular time intervals, for example

1, 1·6, 1·96, 2·176, etc.

where 1 occurred at sample 0, 1·6 at sample 1 and so on. One way of representing this is as a polynomial series:

$$a_0 + a_1Z^{-1} + a_2Z^{-2} + a_3Z^{-3} + \ldots$$

where a_0, a_1, a_2 etc. are the amplitude of the samples and Z^{-n} can be considered as a marker saying this is sample n. A more accurate interpretation is to view Z^{-n} as a delay operator, shifting the time by n samples. The reasons for the rather arbitrary choice of Z^{-n} instead of, say, Z^n are buried in the history of the topic. For the earlier sequence we would therefore have

$$1 + 1{\cdot}6Z^{-1} + 1{\cdot}96Z^{-2} + 2{\cdot}176Z^{-3} + \ldots$$

If we are sampling a continuous signal which is a function of time f(t), the sample sequence will be

$$f(0) + f(1)Z^{-1} + f(2)Z^{-2} + f(3)Z^{-3} + \ldots \qquad (3.22)$$

where f(n) is the value of the function at time nΔt with Δt being the sample period. Equation 3.22 is known as the Z transform of the original function f(t). Note that the sample time Δt appears in the transform, albeit hidden in each sample as f(n).

The Z transform is the sum of all the terms in equation 3.22 so it can be more conveniently written as

$$Z[f(n)] = F(Z) = \sum_{n=0}^{n=\infty} f(n)Z^{-n} \tag{3.23}$$

It can be shown that the series in Z^{-1} can be dealt with as simple geometric converging series. From normal school mathematics the geometric series

$$S = 1 + r + r^2 + r^3 + r^4 + \ldots$$

simplifies to

$$S = \frac{1}{1 - r} \tag{3.24}$$

First consider a simple ramp starting at zero and increasing by one per sample. This will have the Z transform:

$$0 + Z^{-1} + 2Z^{-2} + 3Z^{-3} + \ldots$$

Applying equation 3.23 gives

$$Z[\text{ramp}] = \frac{Z^{-1}}{(1 - Z^{-1})^2}$$

$$= \frac{Z}{(Z - 1)^2} \tag{3.25}$$

The idea of summing infinite series is central to an understanding of the Z transform, so three more examples of common signals will be given.

A step is a standard test signal for process control. A sampled step will have the form of Fig. 3.51 where f(n) = 1 for n > = 0, and hence the series

$$Z[\text{step}] = 1 + Z^{-1} + Z^{-2} + Z^{-3} + \ldots \tag{3.26}$$

which (using the earlier formula for a geometric series with r = Z − 1) converges to

$$Z[\text{step}] = \frac{1}{1 - Z^{-1}}$$

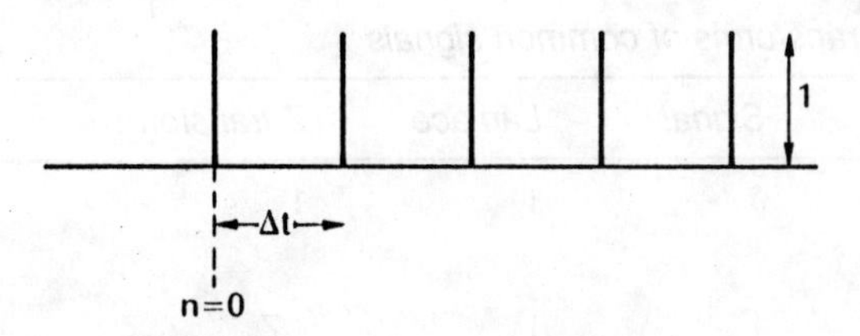

Fig. 3.51 The sampled step.

$$= \frac{Z}{Z-1} \tag{3.27}$$

Next consider a decaying exponential

$$f(t) = e^{-at}$$

This will have value one at time t = 0, and when sampled at time intervals of Δt will produce the sample series:

$$1, e^{-a\Delta t}, e^{-2a\Delta t}, e^{-3a\Delta t}, \ldots$$

The Z transform is thus:

$$Z[\exp] = 1 + e^{-a\Delta t}Z^{-1} + e^{-2a\Delta t}Z^{-2} + \ldots$$

which as before with $r = e^{-a\Delta t}Z^{-1}$ sums to

$$Z[\exp] = \frac{1}{1 - e^{-a\Delta t}Z^{-1}}$$

$$= \frac{Z}{Z - e^{-a\Delta t}} \tag{3.28}$$

One final, and very important, transform is the unit pulse of Fig. 3.52. This has the sampled sequence

$$1, 0, 0, 0, 0, \ldots$$

and hence

$$Z[\text{pulse}] = 1 \tag{3.29}$$

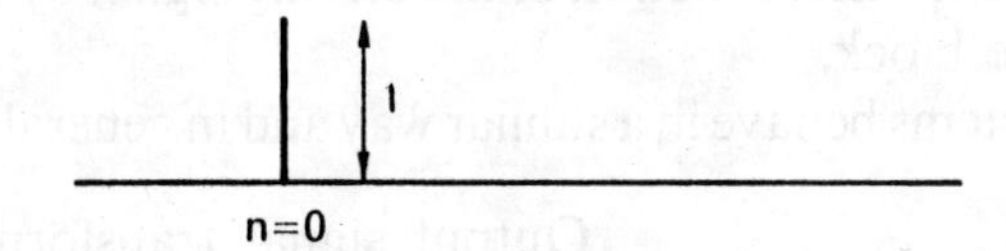

Fig. 3.52 The unit pulse.

Table 3.6 *Z transforms of common signals*

	Signal	*Laplace*	*Z transform*
Impulse	δ	1	1
Step	1	$\frac{1}{s}$	$\frac{Z}{Z-1}$
Ramp	t	$\frac{1}{s^2}$	$\frac{\Delta t Z}{(Z-1)^2}$
Exponential	e^{-at}	$\frac{1}{(s+a)}$	$\frac{Z}{Z-e^{-a\Delta t}}$
Exponential rise	$1-e^{-at}$	$\frac{a}{s(s+a)}$	$\frac{Z(1-e^{-a\Delta t})}{(Z-1)(Z-e^{-a\Delta t})}$
Sine wave	$\sin(\omega t)$	$\frac{\omega}{s^2+\omega^2}$	$\frac{Z\ \sin(\omega\Delta t)}{Z^2-2(\cos(\omega\Delta t))Z+1}$
Enveloped sine	$e^{-a\Delta t}\sin(\omega t)$	$\frac{\omega}{(s+a)^2+\omega^2}$	$\frac{Ze^{-a\Delta t}\ \sin(\omega\Delta t)}{Z^2-2(e^{-a\Delta 2t}\ \cos(\omega\Delta t))Z+e^{-2a\Delta t}}$

Equations 3.25 to 3.29 show the Z transforms of four common driving signals. Others can be derived in a similar way, but it is much easier to look up the transforms from tables such as Table 3.6.

3.7.4. Block transfer functions

Blocks in a continuous system have a defined transfer function which can be used to analyse their response. In the s-plane the transfer function of the output signal from the block can be found by multiplying the transfer function of the driving signal by the transfer function of the block.

Sampled systems behave in a similar way and in general we can say:

$$\text{Block transfer function} = \frac{\text{Output signal transform}}{\text{Input signal transform}}$$

This relationship holds regardless of whether we are working with continuous or sampled systems.

If we denote the output signal transform by Y(Z), the input signal transform by X(Z) and the block transfer function by G(Z) we obviously can write

$$G(Z) = \frac{Y(Z)}{X(Z)}$$

from which

$$Y(Z) = X(Z).G(Z) \tag{3.30}$$

We saw earlier (in equation 3.29) that the Z transform of a unit pulse is 1, so if X(Z) is 1, Y(Z) = G(Z). The transfer function of a block is thus the response of the block to a unit pulse.

We will look at ways of deriving block transfer functions later in Sections 3.7.6 and 3.7.7, but for now accept that the transfer function of an integrator (flow to level, velocity to position, etc.) with unity Δt is

$$G(Z) = \frac{1}{Z - 1}$$

The transform of a step signal was given earlier in equation 3.27 as

$$X(Z) = \frac{Z}{Z - 1}$$

If a sampled step is applied to the integrator we have an output signal Y(Z) given by

$$Y(Z) = \frac{Z}{(Z - 1)} \times \frac{1}{(Z - 1)}$$

$$= \frac{Z}{(Z - 1)^2}$$

Examination of Table 3.6 shows this to be the transform of a ramp, so (not surprisingly) applying a step to an integrator has produced a ramp.

There are, in general, two methods that can be used to find the response of a block to a signal. In general, the output signal Y(Z) will end up of the form

$$Y(Z) = \frac{A(Z)}{B(Z)} \tag{3.31}$$

where A(Z) and B(Z) are polynomials in Z. In the example above, A(Z) = 1 and B(Z) = $(Z - 1)^2$. In the first method, long division can be used to give a series in $Z - 1$ from which the sample values (and hence the response) can be seen.

In the second method, equation 3.31 is broken down into partial fractions to allow the solution to be looked up from the standard tables.

To show an example of each method we will calculate the response of a first order lag block represented by

$$G(Z) = \frac{Z}{(Z - 0{\cdot}6)}$$

to a unit step

$$X(Z) = \frac{Z}{(Z - 1)}$$

From equation 3.30,

$$Y(Z) = \frac{Z^2}{Z^2 - 1{\cdot}6Z + 0{\cdot}6} \tag{3.32}$$

For the long division method we simply divide the numerator by the denominator as Fig. 3.53 to give

$$Y(Z) = 1 + 1{\cdot}6Z^{-1} + 1{\cdot}96Z^{-2} + 2{\cdot}176Z^{-3} + \ldots$$

which is the sample sequence for the output from the block. It seems to have the form of a first order response which is encouraging.

Next we will find the response using partial fractions. We start with

$$Y(Z) = \frac{Z}{(Z - 0{\cdot}6)} \cdot \frac{Z}{(Z - 1)}$$

$$1 + 1{\cdot}6Z^{-1} + 1{\cdot}96Z^{-2} + 2{\cdot}176Z^{-3} \text{ etc}$$

$$Z^2 - 1{\cdot}6Z + 0{\cdot}6 \,\overline{)\, Z^2 + 0.Z^1 + 0.Z^0}$$

$$Z^2 - 1{\cdot}6Z + 0{\cdot}6$$

$$1{\cdot}6Z - 0{\cdot}6$$

$$1{\cdot}6Z - 2{\cdot}56 + 0{\cdot}96Z^{-1}$$

$$1{\cdot}96 - 0{\cdot}96Z^{-1}$$

$$1{\cdot}96 - 3{\cdot}136Z^{-1} + 1{\cdot}176Z^{-2}$$

$$2{\cdot}176Z^{-1} - 1{\cdot}176Z^{-2}$$

Fig. 3.53 Step response of first order lag computed by long division.

which can be rearranged to

$$Y(Z) = Z\left(\frac{Z}{(Z - 0{\cdot}6)(Z - 17)}\right) \tag{3.33}$$

To express this via partial fractions we know the solution will have the form

$$\begin{aligned} Y(Z) &= Z\left(\frac{A}{Z - 0{\cdot}6} + \frac{B}{Z - 1}\right) \\ &= Z\left(\frac{A(Z - 1) + B(Z - 0{\cdot}6)}{(Z - 0{\cdot}6)(Z - 1)}\right) \\ &= Z\left(\frac{(A + B)Z - (A + 0{\cdot}6B)}{(Z - 0{\cdot}6)(Z - 1)}\right) \end{aligned} \tag{3.34}$$

where A and B are unknowns. Comparison of equations 3.33 and 3.34 gives the simultaneous equations

$$A + B = 1$$
$$A + 0{\cdot}6B = 0$$

With two equations and two unknowns, we find

$$A = -1{\cdot}5 \text{ and } B = 2{\cdot}5$$

giving the predicted response

$$\begin{aligned} Y(Z) &= \frac{2{\cdot}5Z}{Z - 1} - \frac{1{\cdot}5Z}{Z - 0{\cdot}6} \\ &= 2{\cdot}5\left(\frac{1}{1 - Z^{-1}}\right) - 1{\cdot}5\left(\frac{1}{1 - 0{\cdot}6Z^{-1}}\right) \end{aligned} \tag{3.35}$$

This has two terms. From the standard transforms in Table 3.6 the first term is a step of height 2·5, and the second an exponential decay of the form $-1.5(0.6)^n$ where n is the sample number. The predicted response is thus

$$Y(n) = 2{\cdot}5 - 1{\cdot}5 \times (0{\cdot}6)^n$$

giving the sample sequence

1, 1·6, 1·96, 2·176,...

and the Z transform

$$Y(Z) = 1 + 1{\cdot}6Z^{-1} + 1{\cdot}96Z^{-2} + 2{\cdot}176Z^{-3} + \ldots$$

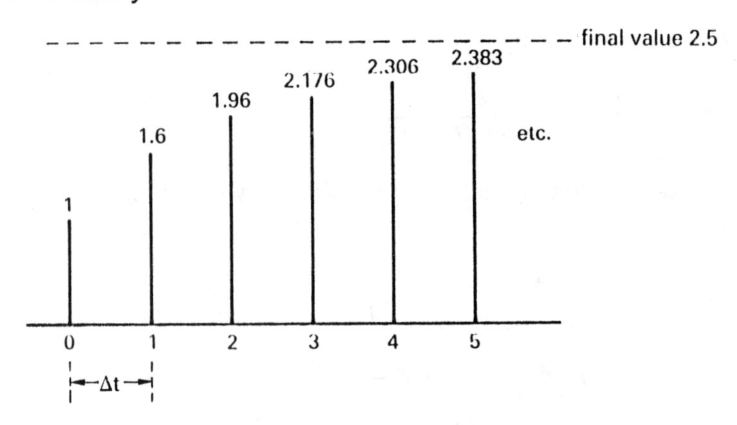

Fig. 3.54 Output sample sequence.

It is reassuring that this is the same as the result obtained from the long division method!

The partial fractions method gives a little more insight into what is going on with the block. Figure 3.54 shows the two signals represented in equation 3.35 and demonstrates that the response is, indeed, a sampled first order lag heading for a final value of 2·5.

3.7.5. Cascaded blocks

Sampled blocks in series behave in a similar manner to continuous blocks in series. In Fig. 3.55a two sampled blocks with transforms G(Z) and H(Z) are connected in series. As might be expected, the transform of the output signal C(Z) is related to the input signal A(Z) by

$$\begin{aligned} C(Z) &= B(Z).H(Z) \\ &= A(Z).G(Z).H(Z) \end{aligned} \tag{3.36}$$

One word of warning is needed, however. Equation 3.36 assumes both blocks are sampled. If we have a system similar to Fig. 3.55b, where two continuous blocks are driven by a sampled signal, we cannot take the Z transform of block G(s) and the Z transform of

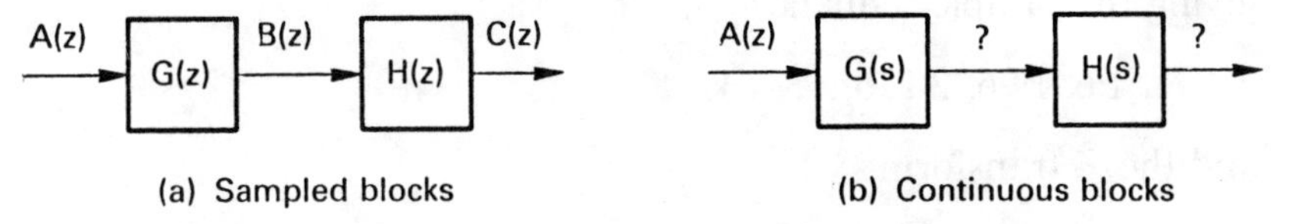

Fig. 3.55 Transfer function of blocks in series.

Table 3.7 *Pulse transform functions*

Laplace	*Pulse transfer function*
$\frac{1}{s}$	$\frac{\Delta t}{Z-1}$
$\frac{1}{s^2}$	$\frac{(\Delta t)^2(Z+1)}{2(Z-1)^2}$
$\frac{1}{(s+a)}$	$\frac{1-e^{-a\Delta t}}{a(Z-e^{-a\Delta t})}$

block H(s) and multiply them together. Here we must take the Z transform of the two blocks in series, i.e. GH(Z) not G(Z).H(Z). We will return to this in Sections 3.7.6 and 3.7.8.

3.7.6. Deriving block transfer functions

A plant can be considered to be constructed of standard blocks such as first and second order lags, integrators and so on. These can most simply be found from standard tables such as Table 3.7, but it is useful to describe the method by which these tables have been derived.

From Section 3.7.4 we know that the transfer function of a block is defined as the response to a unit pulse (which has the Z transform of 1). A unit pulse in the sampled domain has height one (obviously) and width of the sample period.

Suppose we have a first order block. The unit pulse can be considered as a positive step as Fig. 3.56a followed by a negative step as Fig. 3.56b giving the (continuous) output signal as Fig. 3.56c. The Laplace transform of a unit step is 1/s, so the step response of the plant in the s-plane is G(s)/s. The response to the unit step will be G(s)/s followed by −G(s)/s one sample period later, i.e.

$$\frac{G(s)}{s} - \frac{e^{-s\Delta t}G(s)}{s} \tag{3.37}$$

where Δt is the sample period and $e^{-s\Delta t}$ represents a delay of one sample period.

G(s)/s represents a signal; we can look up its equivalent Z transform from Table 3.6. Let us call this G(Z), remembering that G(Z) is the Z transform corresponding to G(s)/s.

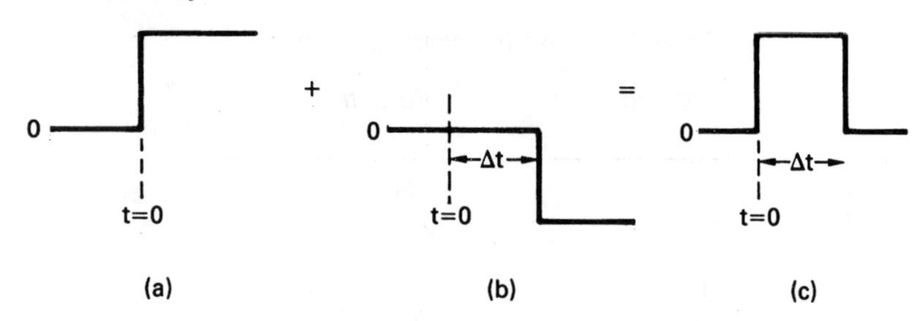

Fig. 3.56 Formation of a unit pulse.

The delay $e^{-s\Delta t}$ in equation 3.37 corresponds to Z^{-1}, so we can now write the Z transform of the block as

$$G(Z) - Z^{-1}G(Z)$$

which with a simple bit of reorganisation becomes

$$\frac{Z-1}{Z}G(Z) \tag{3.38}$$

where G(Z) is the Z transform corresponding to the signal G(s)/s. Some transforms obtained in this manner are shown in Table 3.7.

3.7.7. *Digital algorithms*

First and second order lags, plus functions such as a P + I + D controller can be performed on a sampled basis. The transfer functions of these sampled algorithms are simpler to obtain than the transfer functions of the continuous blocks in a plant. Note that the transfer functions are *not* the same as the Z transforms of the equivalent continuous blocks. It is also important to note that the sample time Δt appears (sometimes hidden) in the transfer functions.

Consider first the very simple (but unknown) block of Fig. 3.57. If a unit pulse is applied we get the sampled output 1, 0·5, 0·25 as shown. By definition, this has the transfer function

$$G(Z) = 1 + 0{\cdot}5Z^{-1} + 0{\cdot}25Z^{-2}$$

The transfer functions of blocks obtained from digital algorithms are obtained in a similar way by considering what the output sample sequence will be when driven by a unit pulse.

A first order filter with time constant T seconds can be represented in continuous form by

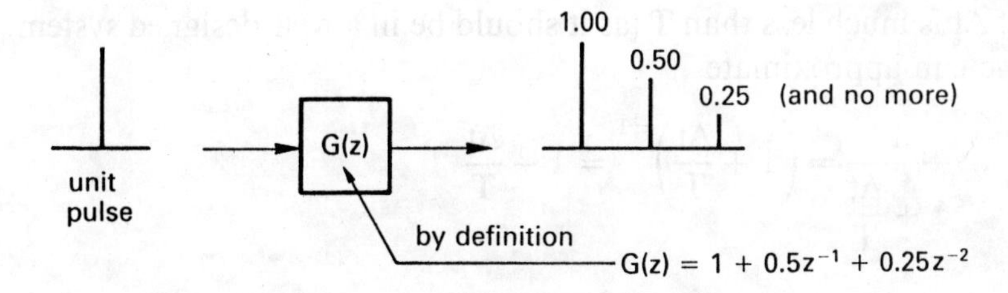

Fig. 3.57 Transfer function of block.

$$T\frac{dy}{dt} + y = x \tag{3.39}$$

In a sampled system, all that are available are the sampled input values x_n, x_{n-1} etc. and the sampled output values y_n, y_{n-1} etc. The rate of change of the output signal approximates to

$$\frac{dy}{dt} \simeq \frac{\Delta y}{\Delta t} = \frac{y_n - y_{n-1}}{\Delta t}$$

Substituting into equation 3.39 gives

$$Ty_n - Ty_{n-1} + y_n\Delta t = x_n\Delta t$$

where Δt is the sample period.

Solving for y_n gives

$$y_n = \frac{y_{n-1}}{1 + \Delta t/T} + \frac{x_n\Delta t/T}{1 + \Delta t/T}$$

This can be constructed in the form of Fig. 3.58 where

$$A = \frac{\Delta t/T}{1 + \Delta t/T} \text{ and } B = \frac{1}{1 + \Delta t/T}$$

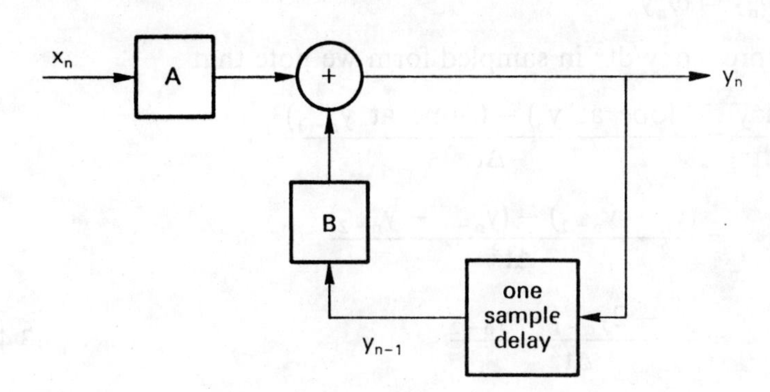

Fig. 3.58 Simulation of first order lag.

If Δt is much less than T (as it should be in a well-designed system), we can approximate

$$\frac{1}{1+\frac{\Delta t}{T}} = \left(1+\frac{\Delta t}{T}\right)^{-1} = 1-\frac{\Delta t}{T}$$

by ignoring terms in $(\Delta t/T)^2$. This leads to $A = \Delta t/T$ and $B = 1 - A$ from which

$$y_n = Ax_n + (1 - A)y_{n-1}$$

In terms of the Z transform

$$y(Z) = Ax(Z) + BZ^{-1}y(Z)$$

which with reorganisation gives

$$\text{transfer function} = \frac{y(z)}{x(z)} = \frac{aZ}{(Z-b)} \tag{3.40}$$

A second order filter can be approximated in a similar manner. In continuous form, a second order filter has the form

$$\frac{d^2y}{dt^2} + 2b\omega_n\frac{dy}{dt} + \omega_n^2 y = \omega_n^2 x \tag{3.41}$$

where ω_n is the natural frequency and b the damping factor.

From the study of the first order filter above we can write

$$2b\omega_n\frac{dy}{dt} \simeq 2b\omega_n\frac{(y_n - y_{n-1})}{\Delta t} \tag{3.42}$$

and obviously

$$\omega_n^2 y = \omega_n^2 y_n$$

To express d^2y/dt^2 in sampled form we note that

$$\frac{d^2y}{dt^2} = \frac{(\text{slope at } y_n) - (\text{slope at } y_{n-1})}{\Delta t}$$

$$= \frac{(y_n - y_{n-1}) - (y_{n-1} - y_{n-2})}{\Delta t^2}$$

$$= \frac{y_n - 2y_{n-1} + y_{n-2}}{\Delta t^2} \tag{3.43}$$

Combining equations 3.41 to 3.43 gives

$$\frac{y_n - 2y_{n-1} + y_{n-2}}{\Delta t^2} + \frac{2b\omega_n(y_n - y_{n-1})}{\Delta t} + \omega_n^2 y_n = \omega_n^2 x_n$$

Some laborious, but straightforward, reorganisation yields an equation for y_n in terms of x_n and previous values of y_{n-1} and y_{n-2}

$$y_n = \frac{2(1 + b\omega_n\Delta t)y_{n-1} - y_{n-2} + \omega_n^2\Delta t^2 x_n}{(1 + 2b\omega_n\Delta t + \omega_n^2\Delta t)} \tag{3.44}$$

which can be more simply represented as

$$y_n = b_0 y_{n-1} + b_1 y_{n-2} + a_0 x_n \tag{3.45}$$

where b_0, b_1 and a_0 are constants. Note that b_1 is negative from equation 3.44.

Equation 3.45 can be represented by Fig. 3.59 which leads to the Z transform

$$\text{transfer function} = \frac{y(z)}{x(z)} = \frac{a_0}{1 + b_0 Z^{-1} + b_1 Z^{-2}} \tag{3.46}$$

Finally we will obtain the transfer function of a three term (P + I + D) controller. To achieve this we have to obtain algorithms for the integral and derivative actions.

The integral term can be approximated by a suitably scaled sum of all the previous error values. At sample (n – 1) let us assume the error sum is M_{n-1}. At sample n, the new error sum M_n will be the old error sum M_{n-1} plus the error E_n for the current sample multiplied by the sample period, i.e.

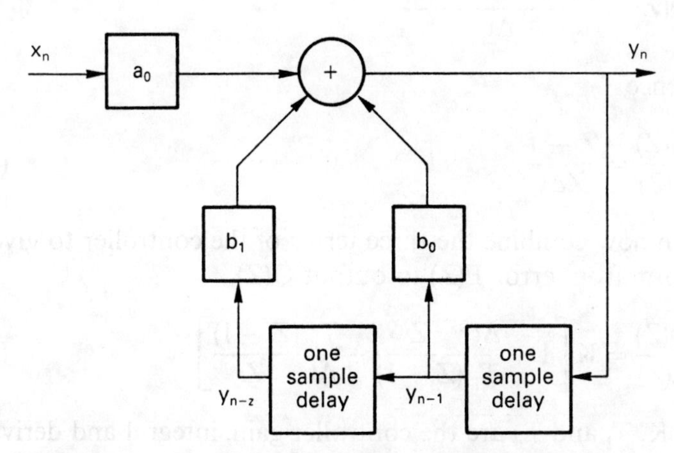

Fig. 3.59 Simulation of second order lag. The Pascal program SIMULATE in Chapter 7 uses the methodology of Figs. 3.58 and 3.59.

$$M_n = M_{n-1} + \Delta t E_n \tag{3.47}$$

where Δt is the sample period as usual.

This can be rewritten using Z transforms as

$$M(Z) = Z^{-1}M(Z) + \Delta t E(Z)$$

and rearranged to

$$\frac{M(Z)}{E(Z)} = \frac{\Delta t Z}{Z - 1} \tag{3.48}$$

which is the Z transform of a simple integral action. A better approximation is the trapezoid algorithm described in Section 4.4.2 and Fig. 4.25c. This has the Z transform

$$\frac{M(Z)}{E(Z)} = \frac{\Delta t(Z + 1)}{2(Z - 1)} \tag{3.49}$$

Differential action can be approximated by

$$\frac{de}{dt} = \frac{\text{change in error}}{\text{change in time}}$$

$$= \frac{E_n - E_{n-1}}{\Delta t}$$

which gives the Z transform

$$D(Z) = \frac{E(Z) - Z^{-1}E(Z)}{\Delta t}$$

and hence

$$\frac{D(Z)}{E(Z)} = \frac{Z - 1}{Z\Delta t} \tag{3.50}$$

We can now combine the three terms of the controller to give the transform from error E(Z) to output Q(Z)

$$\frac{Q(Z)}{E(Z)} = K\left[1 + \frac{\Delta t}{T_i}\frac{Z}{(Z - 1)} + \frac{T_d}{\Delta t}\frac{(Z - 1)}{Z}\right] \tag{3.51}$$

where K, T_i and T_d are the controller gain, integral and derivative times. T_i and T_d must be in the same units (seconds or minutes) as the sample time Δt.

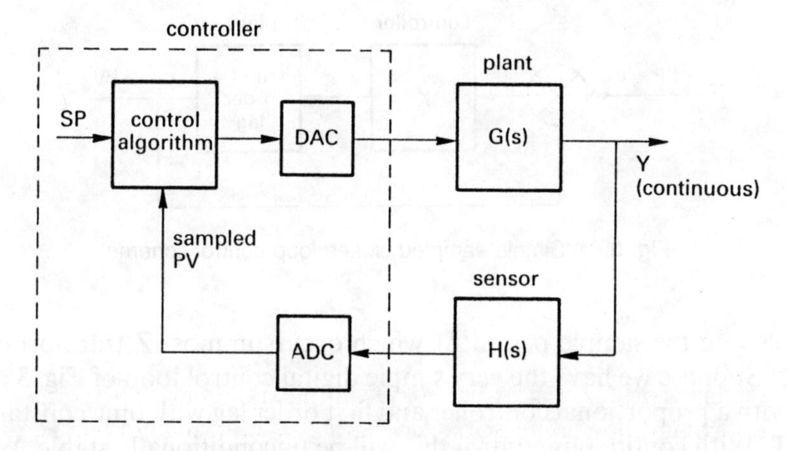

Fig. 3.60 Closed loop digital control.

3.7.8. Closed loop transfer functions and the Z plane

A digital closed loop control system can be represented by Fig. 3.60 where we have a digital controller connected to a continuous plant with forward transfer function G(s) and feedback transfer function H(s). As mentioned earlier, because there is no sampling between G and H we cannot obtain the Z transforms G(Z) and H(Z) and multiply them together to get the combined Z transform. Instead we must obtain the combined Z transform GH(Z).

In a similar way, if the plant is composed of several blocks in series, these must be combined and a common Z transform obtained for the combination.

With this observation, the sampled closed loop transfer function is very similar to the continuous closed loop transfer function, and

$$\frac{P(Z)}{D(Z)} = \frac{KG(Z)}{1 + K.GH(Z)} \tag{3.52}$$

Note that GH(Z) is used in the denominator, rather than G(Z).H(Z).

In the s-plane, poles are formed where the denominator is zero. The values of s can be complex with real and imaginary parts. The behaviour of the plant can be deduced from the pole positions.

Similar ideas apply in the Z plane. Poles occur where the denominator of equation 3.52 is zero. For a quadratic and higher polynomial, these poles can also be complex and drawn on the Z plane with real and imaginary values for each pole. One fundamental difference is the pole positions depend not only on the gain K, but

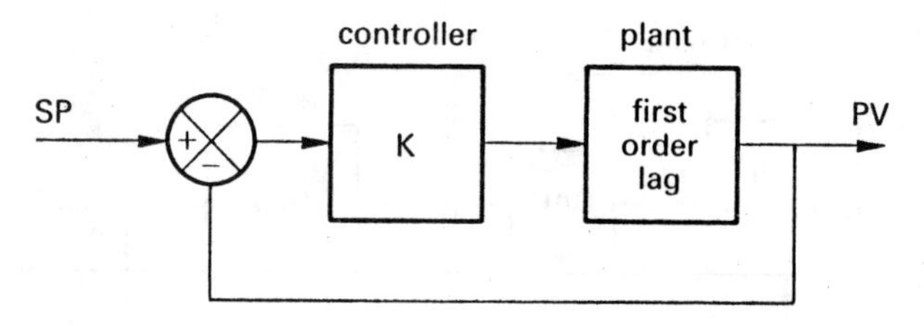

Fig. 3.61 Simple sampled closed loop control scheme.

also on the sample period Δt which occurs in most Z transforms.

Suppose we have the very simple digital control loop of Fig. 3.61 with a proportional controller and first order lag with time constant T. With continuous control this will be unconditionally stable. We will assume a sample period of Δt which gives the open loop transfer function:

$$\frac{Ka(1 - e^{-a\Delta t})}{a(Z - e^{-a\Delta t})}$$

where K is the combined controller/plant gain and a is 1/T. For simplicity we will set T = 1, giving the closed loop transfer function:

$$\frac{K(1 - e^{-\Delta t})}{Z - e^{-\Delta t} + K(1 - e^{-\Delta t})}$$

Note that both of these relationships depend on K and Δt.

Choosing a sample period of 0·1 s simplifies the closed loop transfer function to:

$$\frac{0{\cdot}1K}{Z - 0{\cdot}9 + 0{\cdot}1K}$$

The pole positions for various values of K are plotted on Fig. 3.62a. Suppose we choose a value of K which puts the pole at Z = 0·5. Ignoring the numerator (which has no effect on the transient response in this case), the transfer function has the form 1/(Z − 0·5) which can be written as

$$\frac{Z^{-1}}{1 - 0{\cdot}5Z^{-1}}$$

This is a geometric series which can be expanded using equation 3.24 to give the series

$$1 + 0{\cdot}5Z^{-1} + (0{\cdot}5)^2Z^{-2} + (0{\cdot}5)^3Z^{-3} + \ldots$$

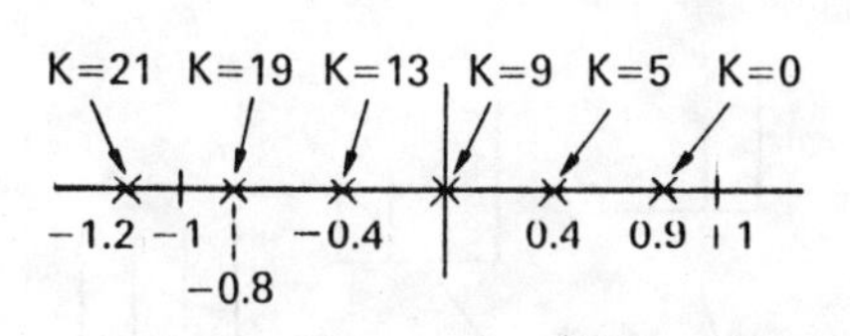

(a) Pole positions for various gains

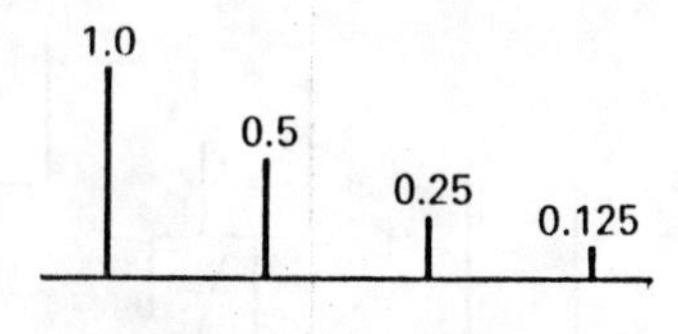

(b) Response to unit pulse with pole at +0.5 corresponding to gain of 6

Fig. 3.62 The effect of gain on pole position.

This produces the sample sequence of Fig. 3.62b and corresponds to the response to a unit pulse.

Similar calculations for different values of pole positions give the responses shown on Fig. 3.63. As can be seen, for stability the pole must lie between $Z = -1$ and $Z = +1$. If the pole is positive, the response follows a simple exponential rise or decay. If the pole is negative the system is oscillatory.

Pole positions depend not only on the gain K, but also on the sample period Δt. The Z transform, and hence the pole positions, depend not only on the gain, but also on the sample time. Increasing the sample time will alter the pole position, and may eventually lead to instability.

So far we have considered poles purely on the real axis. The pole positions are determined by the roots of the closed loop transfer response as K and Δt are varied. As this is a polynomial the roots are likely to occur as complex pairs as Fig. 3.64a. To interpret the significance of the pole positions these poles are best viewed as a polar plot of magnitude c and angle θ.

It can be shown that the Z transform of an exponentially varying sine wave is

$$Z[e^{-\sigma t} \sin \omega t] = \frac{AZ}{(Z - ce^{j\theta})(z - ce^{-j\theta})} \tag{3.53}$$

where $c = e^{-\sigma\Delta t}$

$\theta = \omega\Delta t$

$A = e^{-\sigma\Delta t} \sin \omega t$

It follows that the value of c gives us the change of the sinsusoidal envelope in one sample period. If $c < 1$ the envelope and hence the signal will decay away to zero. If $c > 1$ then the signal will grow exponentially to infinity. With $c = 1$ stable oscillations (or a stable

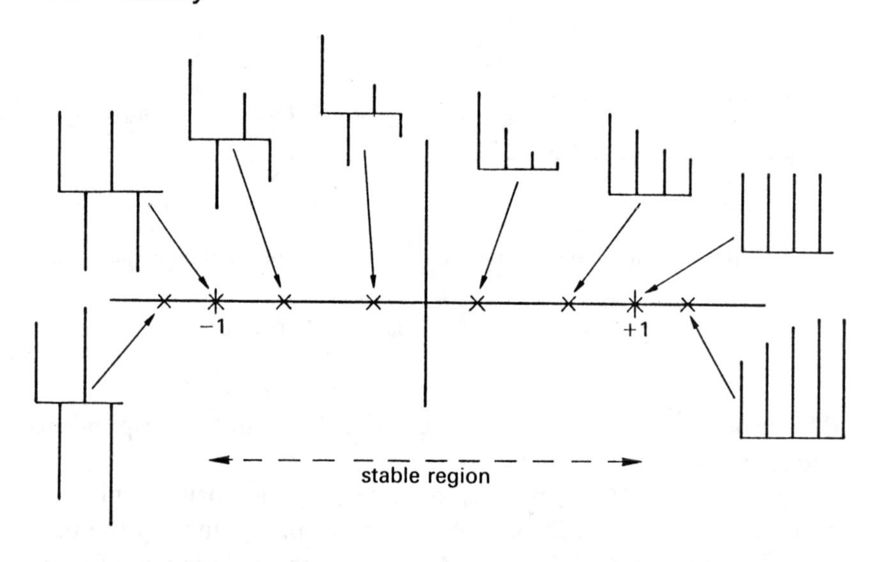

Fig. 3.63 Significance of pole position on real axis.

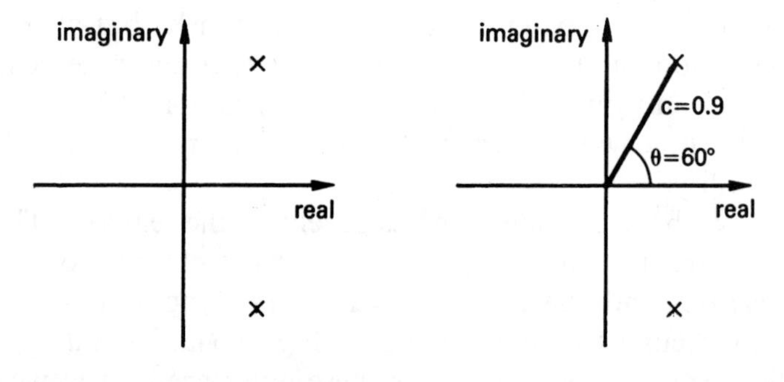

(a) Complex conjugate roots (b) Magnitude and angle, drawn for c=0.9 and θ=60°

Fig. 3.64 Complex roots of polynomials.

value) result. The value of c thus tells us the stability of the system, and for stability all poles must lie within the unity radius circle.

The value of θ depends on the sampling frequency and the frequency of the sinusoidal waveform:

$$\frac{\theta}{2\pi} = \frac{\text{sinusoidal frequency}}{\text{sampling frequency}} \tag{3.54}$$

where the sinusoidal frequency $= \omega/2\pi$ and the sampling frequency is $1/\Delta t$.

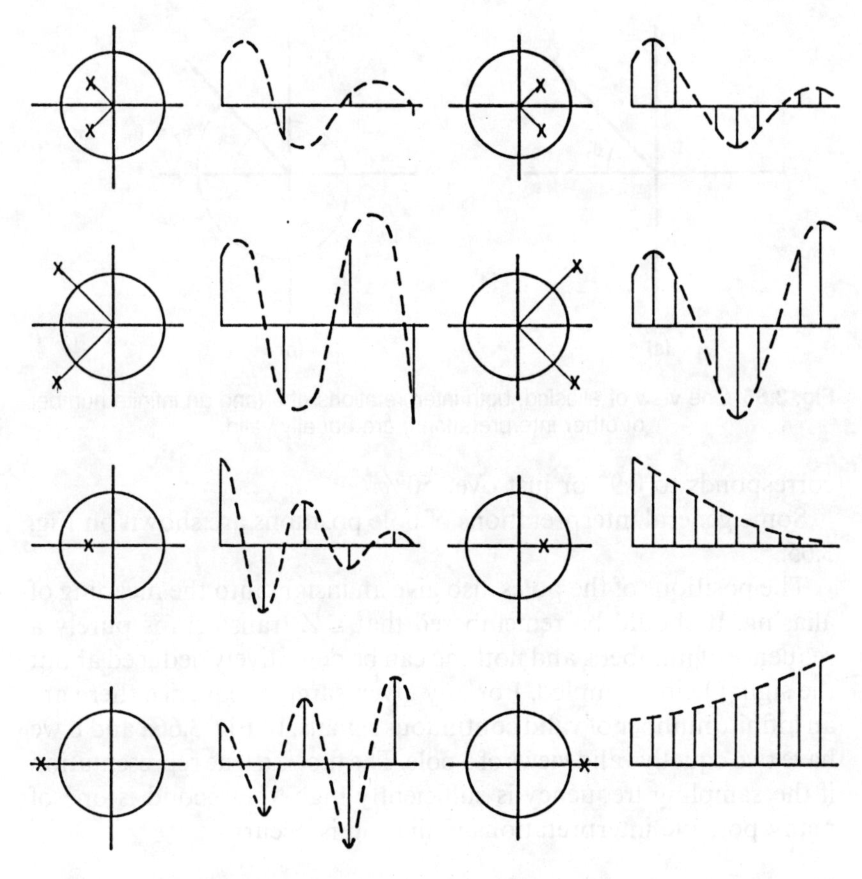

Fig. 3.65 Interpretation of response from pole positions on the Z plane.

For example, if $\theta = 45°$, or $\pi/4$ radians, the sinusoidal frequency is one eighth of the sampling frequency, and there will be eight sample values per cycle.

In Fig. 3.64b, we have two poles arranged such that $c = 0{\cdot}9$ and θ is 60° or 1·05 radians. The sampling frequency is 10 Hz. The sinusoidal frequency will be given by

$$fs = \frac{1{\cdot}05 \times 10}{2\pi} = 1{\cdot}67\,\text{Hz}$$

The amplitude after a given time can be found from

$$c = e^{-\sigma \Delta t}$$

After 0.6 s, for example, there will have been six samples, so the amplitude of the sinusoid will have fallen to c^6 of its initial value. This

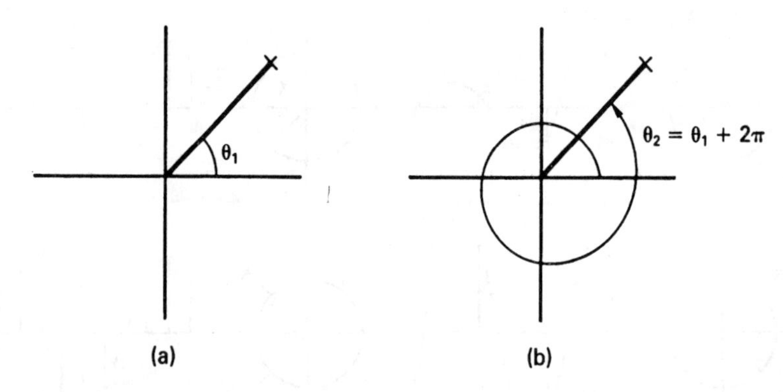

Fig. 3.66 One view of aliasing; both interpretations of θ (and an infinite number of other interpretations) are equally valid.

corresponds to $0{\cdot}9^6$ or just over 50%.

Some general interpretations of pole positions are shown on Fig. 3.65.

The positions of the poles also give an insight into the meaning of aliasing. It should be remembered that a Z transform is purely a sequence of numbers, and nothing can be definitively deduced about the signal being sampled. For any given sample sequence there are an infinite number of valid continuous signals. In Fig. 3.66a and b we have two equally valid views of a pole. The first is a true representation if the sampling frequency is sufficiently high, the second is one of many possible interpretations if aliasing is occurring.

Chapter 4
Controllers

4.1. Controller basics

4.1.1. Introduction

It is, perhaps, surprising that the three basic controller types of Fig. 4.1 are used for the vast majority of control loops despite the availability of more modern, and better, control algorithms. The P + I + D controller has evolved from pneumatics, through discrete component amplifiers and operational amplifier ICs to digital implementations in microprocessor based instruments and computer systems.

There are many reasons for the continuing popularity of three term control and its variants. It can deal adequately with most control problems, and is cheap and easy to implement. Its longevity has also allowed process control engineers and technicians to build up a solid base of knowledge.

This chapter discusses features found on commercial controllers (and control packages found on computers) plus practical aspects of using controllers on real plants.

4.1.2. Definitions of terms

Although the three term control algorithm of Fig. 4.1c is universally accepted, there is a wide variation in the terms used to describe it. This can lead to considerable confusion when reading different manufacturer's data sheets.

The gain term K can be described as a straight gain or as the inverse 100/K%, termed the proportional band. A gain of 2 thus corresponds to a proportional band of 50%.

The integral action is often referred to as 'Reset'; a term arising out

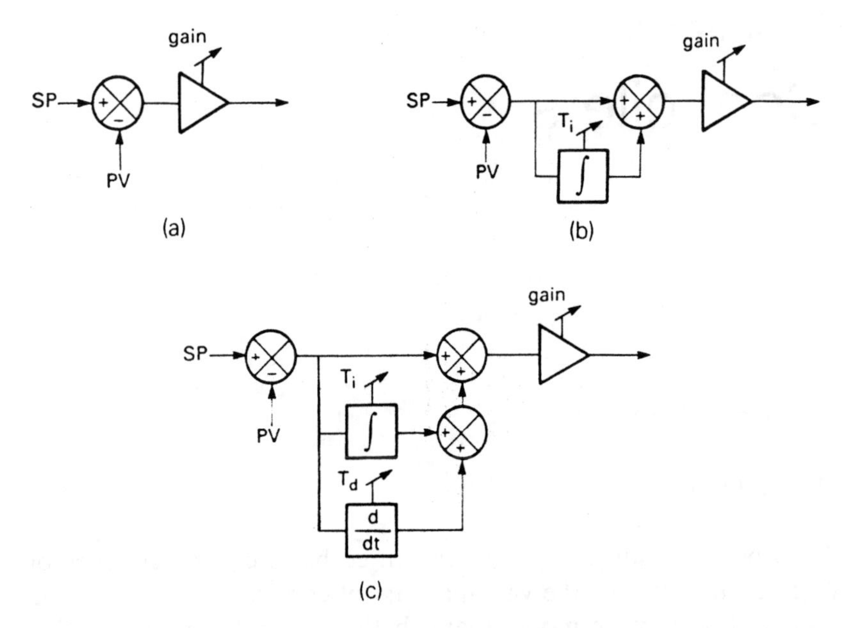

Fig. 4.1 Basic controller types, (a) Proportional controller; (b) Proportional plus integral (P + I) controller; (c) Three term (P + I + D) controller.

of the fact that integral action removes the proportional offset (see Section 1.3.1) and resets the operating point to the set point. Rather confusingly a few manufacturers use the term 'Reset' to describe the feedback signal PV.

The constant T_i, the integral time, determines the contribution of the integral term. The shorter T_i the faster the system will return to the set point after a disturbance or set point change, but the more unstable the system may become.

Integral action on many controllers is adjusted via a potentiometer labelled 'repeats per minute'. This is the inverse of T_i (an integral time of 20 seconds is 3 repeats per minute). Increasing T_i is the same as decreasing repeats per minute.

The effect of derivative action is controlled by T_d, the derivative time. This control is often labelled 'rate', but the units are universally time (seconds or minutes). Increasing T_d increases responsiveness to set point changes or disturbances, and usually (but not always) increases the stability of a system. Increasing T_d, however, gives a system more high frequency gain which makes it noise prone, often resulting in continual actuator 'dither' and premature wear of moving parts.

4.1.3. Frequency response of controllers

Figure 4.2 shows the frequency response of a P + I (Fig. 4.2a) and a P + I + D (Fig. 4.2b) controller plotted on a Bode diagram. The P + I + D controller is shown for a setting of $T_i = 4T_d$, a recommended ratio as this produces a double zero at a predictable position on the s-plane (see Section 3.5.4).

It can be seen that at low frequencies both controllers have an increasingly high amplitude ratio (infinite at DC) and a phase shift of $-90°$. The P + I amplitude ratio is 3 dB at $\omega = 1/T_i$ rads/sec; at this frequency the phase shift is $-45°$. The phase shift has fallen to zero at $\omega = 10/T_i$ rads/sec.

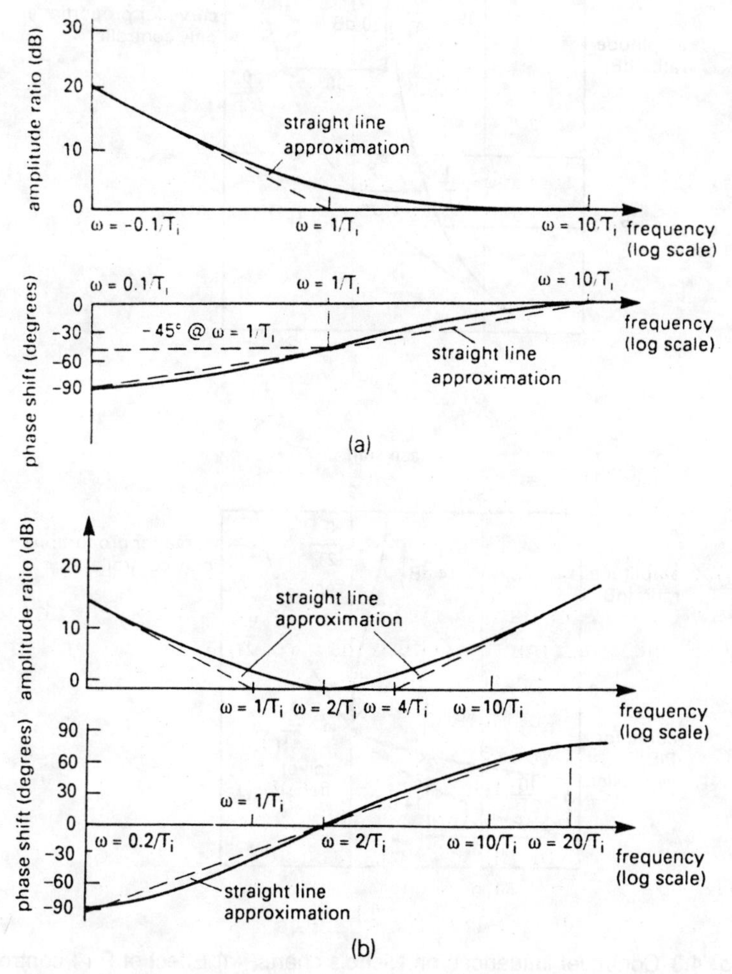

Fig. 4.2 Bode diagrams for (a) P + I and (b) P + I + D controllers.

The P + I + D amplitude ratio falls to 0 dB at $\omega = 2/T_i$ rads/sec then starts to rise again, theoretically to infinity but practically to a limit usually of around +20 – 30 dB. The phase shift tends towards −90° at frequencies below $\omega = 0{\cdot}01/T_i$ rads/sec, and towards +90° at frequencies above $\omega = 10/T_i$ rads/sec. The phase shift is zero at $\omega = 2/T_i$.

The effect of a controller can be seen by considering Nichols charts for systems controlled first by a proportional only controller, and then by a P + I and P + I + D controller.

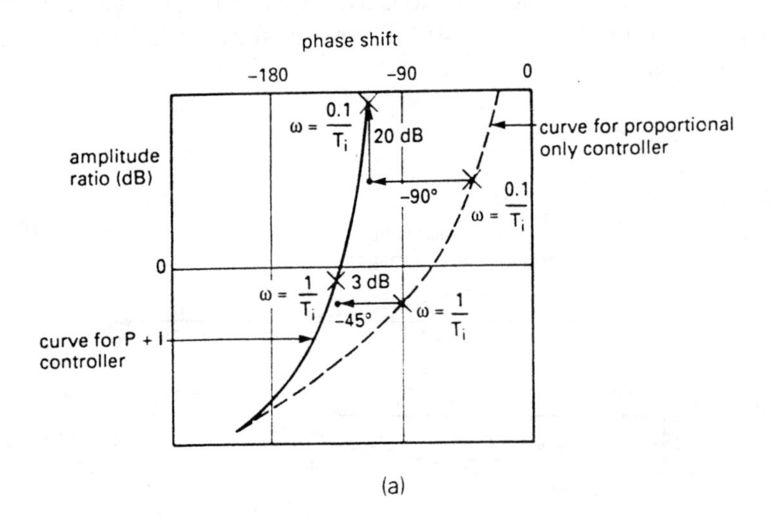

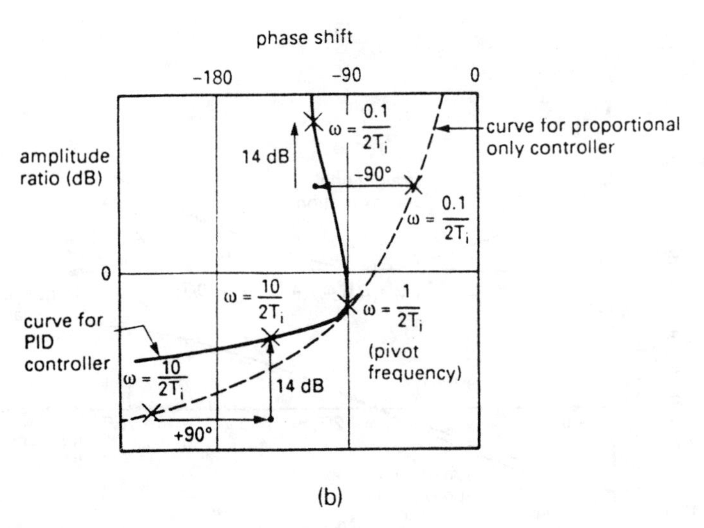

Fig. 4.3 Controller influences on Nichols charts, (a) Effect of P + I controller; (b) Effect of P + I + D controller.

We can consider the frequency at which the controller has least effect as a form of pivot point. For the P + I controller, this occurs at $\omega = 1/T_i$ rads/sec where the amplitude ratio is 3 dB and the phase shift −45°. At this point, the Nichols chart is shifted up by 3 dB and to the left by 45° as shown in Fig. 4.3a. Below the pivot frequency, the gain and the phase shift increase, both of which stretch and bend the trace to the left as shown. Above the pivot frequency the trace approaches that for the proportional only control. The general effect is a reduction of the gain and phase margin. The removal of offset has been obtained via a reduction in stability.

A P + I + D controller's pivot point is $\omega = 1/2T_i$ rads/sec. At this frequency the amplitude ratio is 0 dB and the phase shift 0°, and the P + I + D and proportional only curves cross or touch. At low frequencies the curve is shifted to the left by 90° and stretched in a similar manner to the P + I controller. At high frequencies positive phase shift occurs along with increasing gain. The effect of this will vary according to the fall off of gain and increase in phase shift of the system. Figure 4.3b shows a typical result.

4.1.4. Effect of controllers on root locus

A P + I controller introduces a pole at the origin plus a zero at $\sigma = -1/T_i$ as shown in Fig. 4.4a. Because one pole and one zero is introduced into the system the number of asymptotes are unchanged, although the asymptote's origin may be moved.

A P + I + D controller introduces a pole at the origin plus two zeros whose position can be controlled by T_i and T_d (albeit with interaction). If, as is conventional, T_i is set to $4T_d$ a double zero is produced at $\sigma = -1/2T_d$. With two zeros and one pole being added the number of asymptotes is reduced by one (see Section 3.5.4).

The effect of the controller on various systems is shown on Fig. 4.5

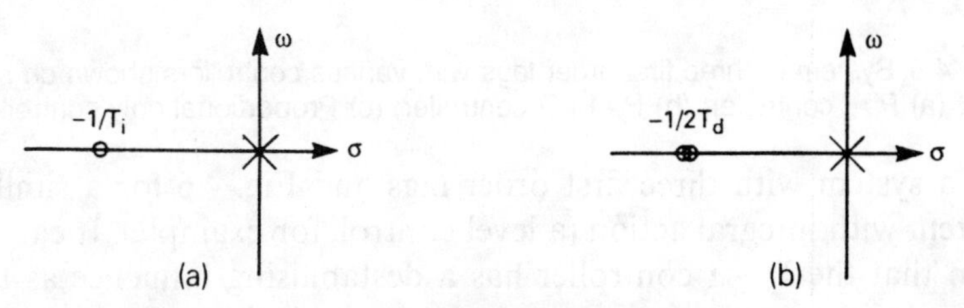

Fig. 4.4 Open loop pole and zero position on s-plane for controllers, (a) P + I controller; (b) P + I + D controller.

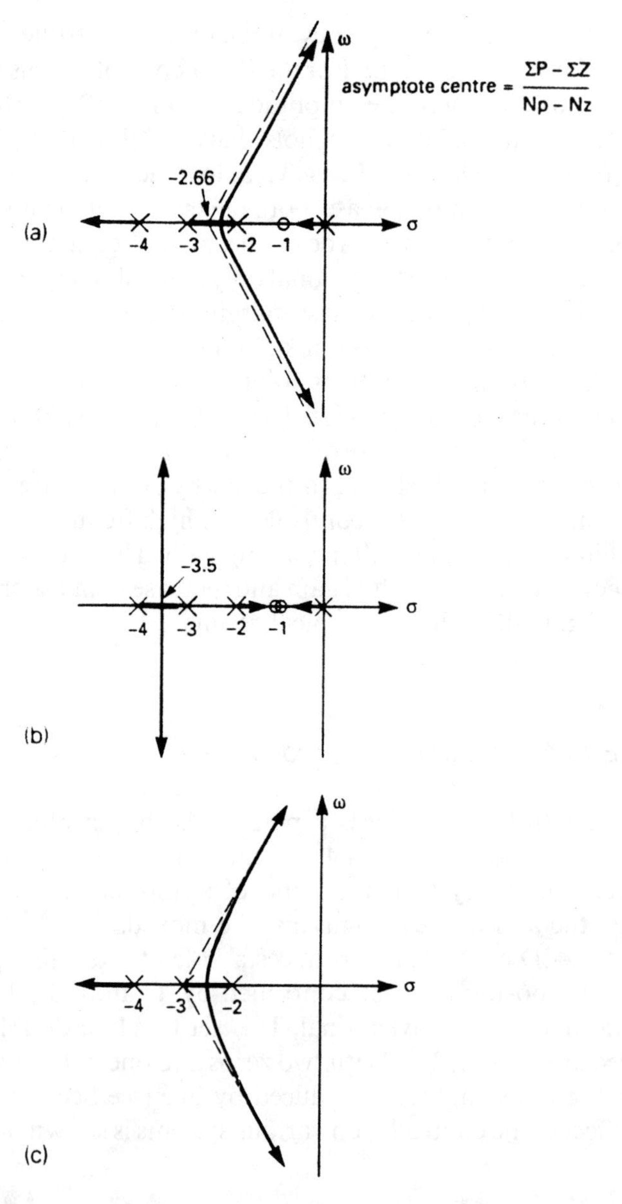

Fig. 4.5 System of three first order lags with various controllers shown on root loci, (a) P + I controller; (b) P + I + D controller; (c) Proportional only controller.

for a system with three first order lags and Fig. 4.6 for a similar system with integral action (a level control, for example). It can be seen that the P + I controller has a destabilising influence as the origin of the asymptotes is pushed progressively to the right for decreasing T_i.

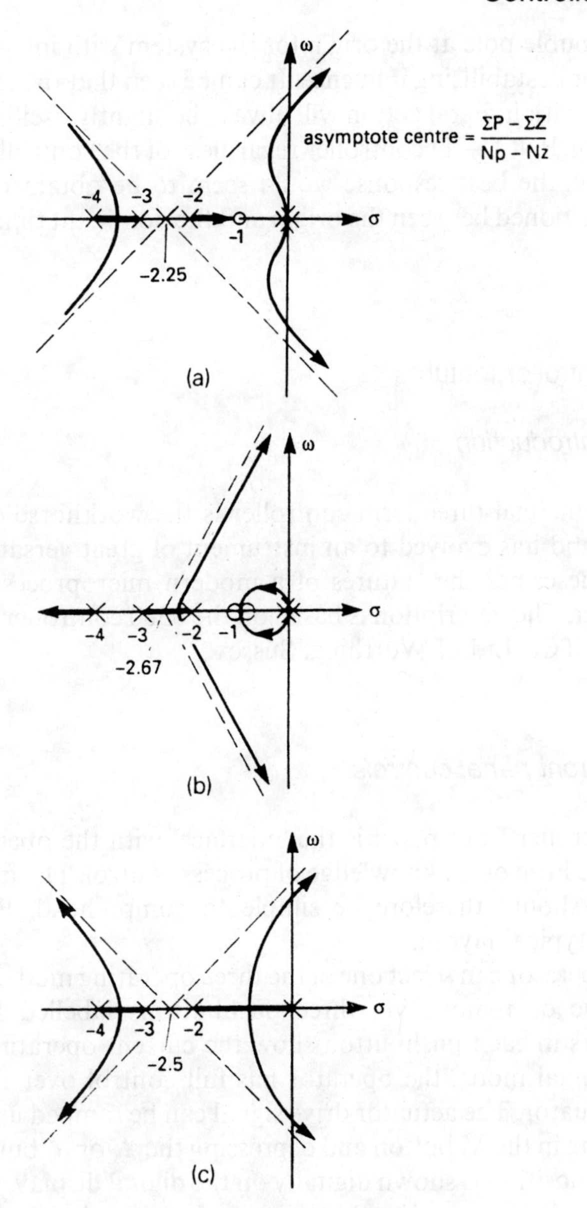

Fig. 4.6 System of integral action and three first order lags with various controllers shown on root loci, (a) P + I controller; (b) P + I + D controller; (c) Proportional only controller.

The P + I + D controller does not in general, destabilise the system provided the double zero is to the right of the dominant time constant. For the simple three lag system the loop is stable for all gains unlike the proportional only controller.

The double pole at the origin for the system with integral action is a major destabilising influence. It can be seen that the response of a system with integral action will always be slightly oscillatory with a P + I or P + I + D controller regardless of the controller setting. As before, the best response would seem to be obtained with the zeros positioned between the origin and the dominant time constant pole.

4.2. Controller features

4.2.1. Introduction

The commercial three term controller is the workhorse of process control and has evolved to an instrument of great versatility. This section describes the features of a modern microprocessor based controller. The description is based on the 6360 controller manufactured by TCS Ltd of Worthing, Sussex.

4.2.2. Front panel controls

The controller front panel is the 'interface' with the operator who may have little or no knowledge of process control. The front panel controls should therefore be simple to comprehend. Figure 4.7 shows a typical layout.

The operator can select one of the three operating modes; manual, automatic or remote, via three pushbuttons labelled M, A, R. Indicators in each pushbutton show the current operating mode.

In manual mode, the operator has full control over the driven plant actuator. The actuator drive signal can be ramped up or down by holding in the M button and depressing the ▲ or ▼ buttons. The actuator position is shown digitally on the digital display, whilst the M button is depressed and continuously in analog form on the horizontal bargraph.

In automatic mode the unit behaves as a three term controller with a set point loaded by the operator. The unit is scaled into engineering units (i.e. real units such as °C, psi, litres/min) as part of the set up procedure so the operator is dealing in real plant variables. The digital display shows the set point value when the SP button is depressed and the value changed with the ▲ and ▼ buttons. The set

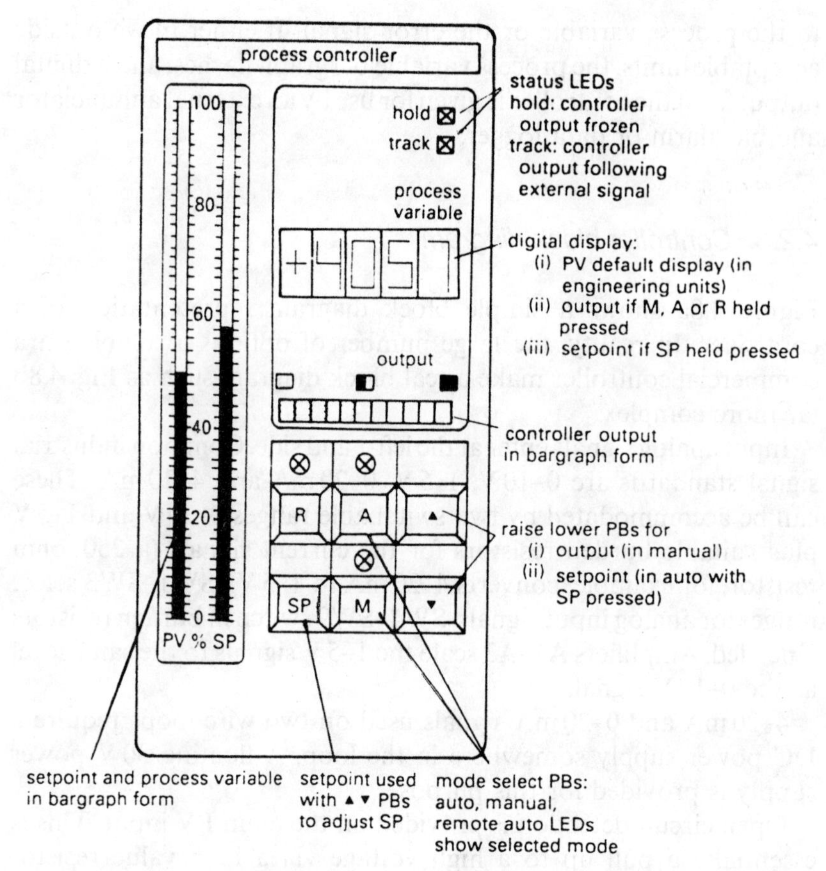

Fig. 4.7 Controller front panel controls (based on TCS 6360 Controller).

point is also displayed in bargraph form on the right-hand side of the dual vertical bargraph.

Remote mode is similar to automatic mode except the set point is derived from an external signal. This mode is used for ratio or cascade loops and batch systems where the set point has to follow a predetermined pattern (annealing furnaces are a common application). As before the set point is displayed in bargraph form and the operator can view, but not change, the digital value by depressing the SP button.

The process variable itself is displayed digitally when no push button is depressed, and continually on the left-hand bargraph. In automatic or remote modes the height of the two left-hand bargraphs should be equal; a very useful quick visual check that all is under control.

Alarm limits (defined during the controller set up) can be applied to the process variable or the error signal. If either move outside acceptable limits, the process variable bargraph flashes, and a digital output from the controller is given for use by an external annunciator audible alarm or data logger.

4.2.3. Controller block diagram

Figure 4.8a shows a simple block diagram representation of a controller. In reality, the large number of options available on a commercial controller make a real block diagram such as Fig. 4.8b far more complex.

Input analog signals enter at the left-hand side. Common industrial signal standards are 0–10 V, 1–5 V, 0–20 mA and 4–20 mA. These can be accommodated by two switchable ranges 0–10 V and 1–5 V plus suitable burden resistors for the current signals (a 250 ohm resistor, for example, converts 4–20 mA to 1–5 V). SW1–SW3 select ranges for analog input signals. SW4–SW9 switch in burden resistors if needed. Amplifiers A1–A3 scale the 1–5 V signals to the same level as the 0–10 V signal.

4–20 mA and 0–20 mA signals used on two wire loops require a DC power supply somewhere in the loop. A floating 30 V power supply is provided for this purpose.

Open circuit detection is provided on the main PV input. This is essentially a pull up to a high voltage via a high value resistor (typically several hundred K ohms). A comparator signals an open circuit input when the voltage rises. Short circuit detection can also be applied on the 1–5 V input (the input voltage falling below 1 V). Open circuit or short circuit PV is usually required to bring up an alarm and trip the controller to manual, with the output signal driven high, held at last value, or driven low according to the nature of the plant being controlled. The open circuit trip mode is determined by switches as part of the set up procedure.

The PV and remote SP inputs are scaled to engineering units and linearised. Common linearisation routines are thermocouples, platinum resistance thermometers and square root (for flow transducers). A simple adjustable first order filter can also be applied to remove process or signal noise. The set point for the P + I + D algorithm is selected from the internal set point or the remote set point by the front panel auto and remote push button contents A, R.

The purpose of the track analog input is described below in Section 4.2.4.

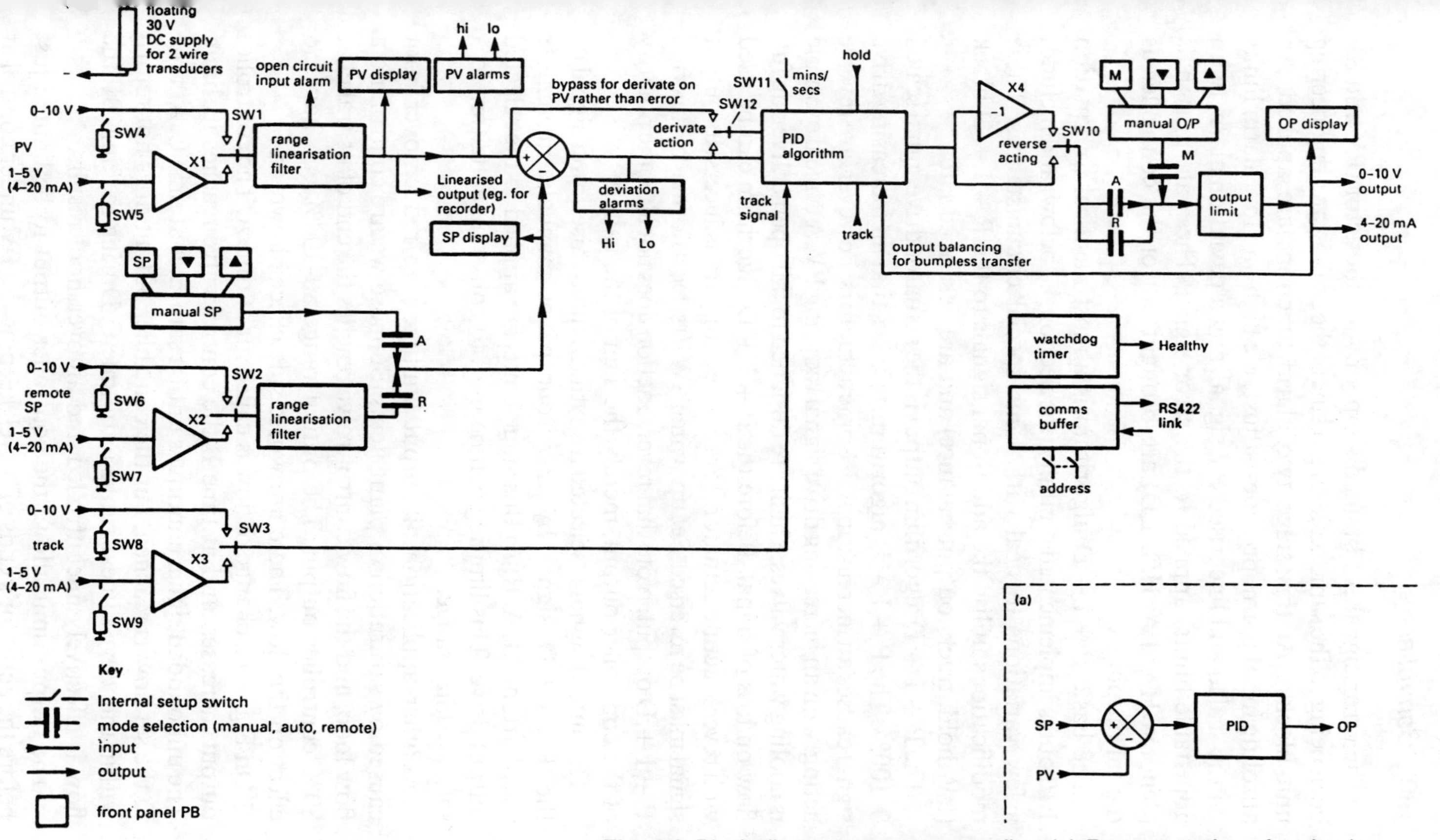

Fig. 4.8 Block diagram process controller, (a) Representation of a simple controller; (b)Representation of commercial controller.

The error signal is obtained simply by a subtractor (PV and SP both being to the same scale as a result of the scaling and engineering unit blocks). At this stage two alarm functions are applied. An absolute input alarm provides adjustable high and low alarm limits on the scaled and linearised PV signal, and a deviation alarm (with adjustable limits) applied to the error signal. These alarm signals (denoted HA, LA, HD, LD) are brought out of the controller as digital outputs.

The basic P + I + D algorithm is almost identical to equation 1.10 but is implemented digitally (see Section 4.4 below) and includes a few variations to deal with some special circumstances. These modifications utilise the additional signals to the P + I + D block (PV, hold, track, output balance) and are described later.

The P + I + D algorithm output is the actuator drive signal scaled 0–100%. The P + I + D algorithm assumes that an increasing drive signal causes an increase in PV. Some actuators, however, are reverse acting, with an increasing drive signal reducing PV. A typical example is cooling water valves which are designed to fail open delivering full flow on loss of signal. Before the P + I + D algorithm can be used with reverse acting actuators (or reverse acting transducers) its output signal must be reversed. Set up switch SW10 selects normal or inverted P + I + D output. Note that reverse action does not alter the polarity of the controller output, merely the sign of the gain.

The output signal is selected from the manual raise/lower signal or the P + I + D signal by the front panel manual/auto/remote pushbuttons M, A, R. At this stage limits are applied to the selected output drive. This limiting can be used to constrain actuators to a safe working range.

Another application of output limiting is to reduce overshoot caused by saturation of plant items. Suppose we are controlling a flow loop, and the flow control valve reaches the end of its travel at 80% controller output. The output range 80–100% can have no effect on the flow. There are worse side effects, however.

On Fig. 4.9 a change of flow is called for at time A. The controller output increases, and at time B the control valve reaches its limit. The unbounded signal continues until it saturates at time C. At time D the set flow is reached, but the controller output does not return within the valve's linear range until time E. During the time BE the flow is effectively uncontrolled and an overshoot results.

The output limit allows the controller output to be limited just before the actuator's ends of travel, keeping the PV under control at all times.

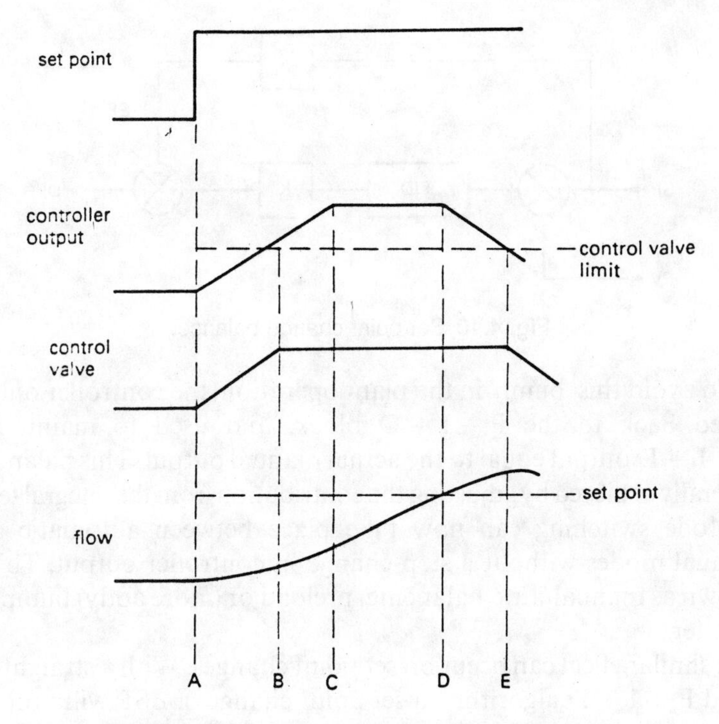

Fig. 4.9 The need for controller output limiting.

Two controller outputs are provided, 0–10 V and 4–20 mA for use with voltage and current driven actuators. The linearised PV signal is also retransmitted as a 0–10 V signal for use with separate external indicators and recorders.

4.2.4. Bumpless transfer and track mode

The output from the P + I + D algorithm is a function of time and the values of the set point and the process variable. When the controller is operating in manual mode it is highly unlikely that the output of the P + I + D block will naturally be the same as the demanded manual output. In particular the integral term will probably cause the output from the P + I + D block to eventually saturate at 0% or 100% output.

If no precautions are taken, therefore, switching from auto to manual, then back to auto again some time later will result in a large step change in controller output at the transition from manual to automatic operation.

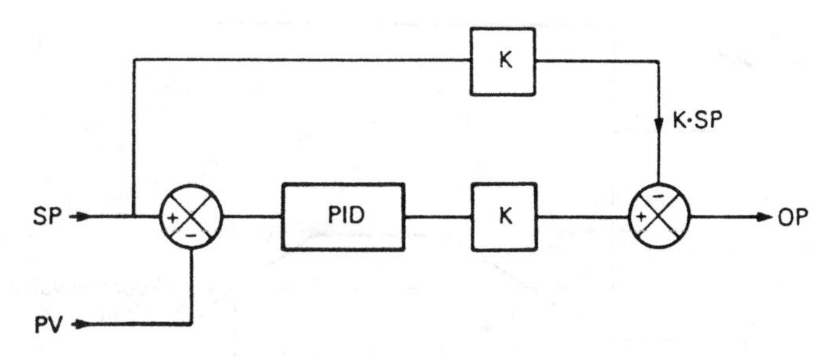

Fig. 4.10 Set point change balance.

To avoid this 'bump' in the plant operation, the controller output is fed back to the P + I + D block, and used to maintain a P + I + D output equal to the actual manual output. This balance is generally achieved by adjusting the contribution from the integral term.

Mode switching can now take place between automatic and manual modes without a step change in controller output. This is known as manual/auto balancing, preload or (more aptly) bumpless transfer.

A similar effect can occur on set point changes. With a straightforward P + I + D algorithm, a set point change of ΔSP will produce an immediate change in controller output of K.ΔSP where K is the controller gain. In some applications this is unacceptable. In Fig. 4.10 a term K.SP is subtracted from the P + I + D block output. The controller now responds to errors caused by changes in PV in the normal way, but only reacts to changes in SP via the integral and derivative terms. Changes in SP thus result in a slow change in controller output. This is known as set point change balance, and is a switch selectable set up option.

This balance signal fed back from the output to the P + I + D block is also used when the controller output is forced to follow an external signal. This is called track mode.

A typical example is shown in Fig. 4.11 which represents a fume extraction system. The control loops are required to maintain a steady duct negative pressure by adjusting the position of the fan dampers. Dependent on the fume flow required, one or both fans may be run. If the two pressure loops are left independent when both fans are running, unpredictable and possibly unstable results may occur as the loops interact. Independent pressure control loops are required, however, when only one fan is run.

In Fig. 4.11, fan 1 is declared to be the master fan, and its controller output is fed as a track signal to the fan 2 controller. A digital signal informs fan 2 controller when both fans are running, forcing fan 2

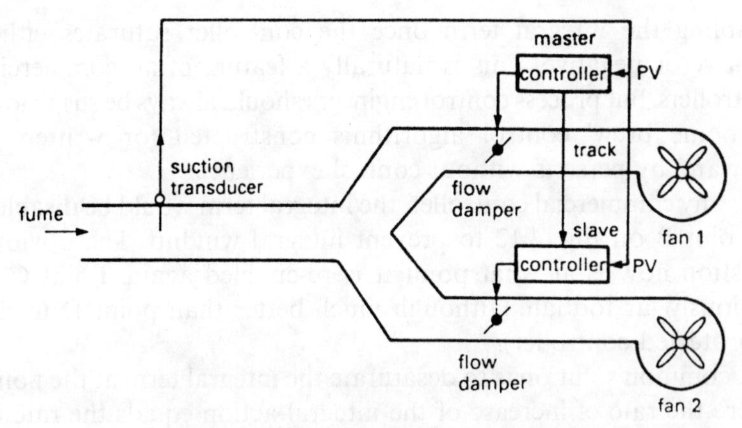

Fig. 4.11 Typical application of track mode.

vane to follow fan 1 vane. When fan 1 or fan 2 are run on their own the controllers act independently as normal.

As before, the P + I + D algorithm needs to be balanced to avoid a bump when transferring between track mode and automatic mode. The feedback output signal achieves this balance as described previously.

4.2.5. Integral windup and desaturation

Large changes in SP or large disturbances to PV can lead to saturation of the controller output or a plant actuator (in Section 4.2.3 it was shown to be desirable to have the controller saturate first). Under these conditions the integral term in the P + I + D algorithm can cause problems.

Figure 4.12 shows the probable response of a system with unrestricted integral action. At time A a step change in set point occurs. We assume the scheme in Fig. 4.10 is not in use, so the OP rises first in a step (K × set point change) then rises at a rate determined by the integral time. At time B the controller saturates at 100% output, but the integral term keeps on rising. With fixed controller output PV rises in an approximately linear manner.

At time C PV reaches, and passes, the required value, and as the error changes sign the integral term starts to decrease, but it takes until time D before the controller desaturates. Between times B and D the plant is uncontrolled, leading to an unnecessary overshoot and possibly even instability.

This effect is called 'integral windup' and is easily avoided by disabling the integral term once the controller saturates either positive or negative. This is naturally a feature of all commercial controllers, but process control engineers should always be suspicious of 'home brew' control algorithms constructed (or written in software) by persons without control experience.

In any commercial controller, the integral term would be disabled at point B on Fig. 4.12 to prevent integral windup. The obvious question now is at what point it is re-enabled again. Point C is obviously far too late (although much better than point D in the unprotected controller).

A common solution is to desaturate the integral term at the point where the rate of increase of the integral action equals the rate of decrease of the proportional and derivative terms. This occurs when the slope of the P + I + D output is zero, i.e. when

$$e = -T_i\left(\frac{de}{dt} + T_d\frac{d^2e}{dt^2}\right) \tag{4.1}$$

with e being the error and T_i, and T_d the controller constants.

Equation 4.1 brings the controller out of saturation at the earliest possible moment, but this can, in some cases, be too soon leading to an unnecessarily damped response. Some controllers allow adjustment of the desaturation point by adding an error limit circuit as Fig. 4.13. The error limit delays the balance point to equation 4.1, forcing the controller to remain in saturation for a longer time. The speed of desaturation and the degree of overshoot can thus be adjusted by the commissioning engineer.

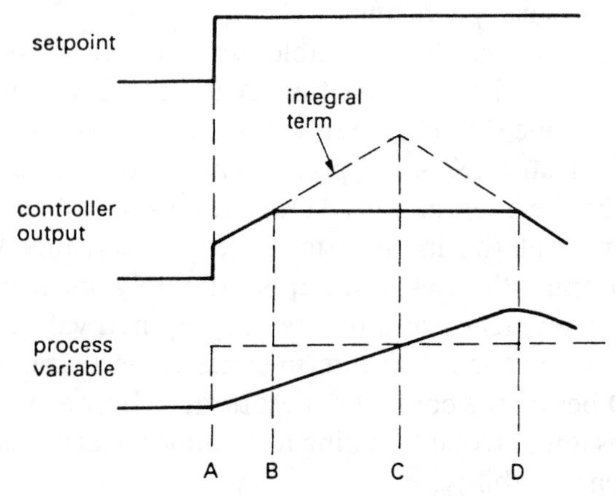

Fig. 4.12 The effect of integral windup.

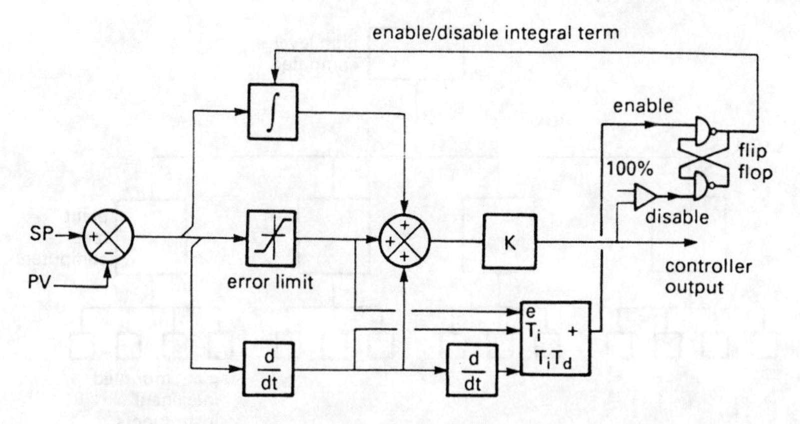

Fig. 4.13 Common method of avoiding integral windup.

4.2.6. Selectable derivative action

The term $T_d$de/dt in the three term controller algorithm can be rearranged as

$$T_d\left(\frac{d(SP)}{dt} - \frac{d(PV)}{dt}\right)$$

where SP is the set point and PV the process variable. The derivative term thus responds to changes in both the set point and the plant feedback signal.

This is not always desirable; in particular a step change in set point leads to an infinite spike controller output and a vicious 'kick' to the actuator. Commercial controllers therefore include a selectable option for the derivative term to be based on true error (SP–PV) or purely on the value of PV alone.

There is generally no noticeable difference in plant performance between these options; stability or the ability to deal with disturbances or load changes are unaffected, and derivative on PV is normally the preferred choice. The only occasion when true derivative on error is advantageous is where the PV is required to track a continually changing SP (military gunnery control is one obvious example).

4.2.7. Miscellaneous features

A multi-level control system is increasingly common, with plant control being undertaken by intelligent controllers or programmable controllers, with one or more levels of computer supervision above

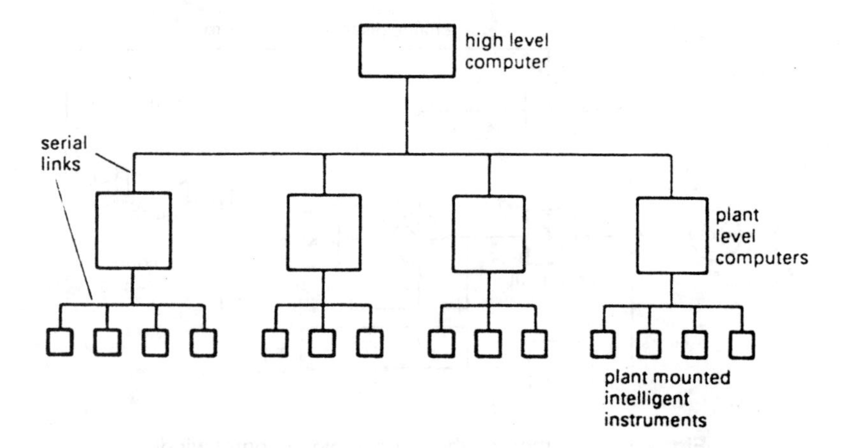

Fig. 4.14 Intelligent instruments as part of a distributed computer system.

them as Fig. 4.14. This arrangement gives protection against total plant failure whilst retaining the advantages of computer control.

Modern controllers are invariably provided with a mechanism for providing serial communications to a higher level supervisory system. This link, usually arranged as a drop line to several controllers, can be used by the supervisory host computer to download set points and read the values of process variable and instrument status (e.g. operating modes and alarm flags).

Typically an instrument is allocated an address by which it is uniquely identified, allowing the host computer to say 'Set point for instrument 127 is 1500 litres/min'.

The failure of a controller can be potentially dangerous, so commercial controllers are fitted with extensive self-checking features. Typically, in digital instruments, a watchdog circuit will be fitted to initiate external safety shutdown circuits in the event of a controller failure. The controller will, at the same time, shut its own outputs down. Usually the controller is arranged to trip to manual mode, and the user can select what value the output takes on the transfer to manual. The usual options are hold last valid output, drive high (e.g. cooling control) or drive low (e.g. feedstock control).

4.2.8. Controller data sheet

A data sheet for a typical modern controller (the TCS 6360 instrument) is given in Fig. 4.15.

Facing page and following two pages

Fig. 4.15 Extract from the data sheet for the TCS 6360 Controller. (Courtesy of TCS Limited, Worthing, Sussex.)

3-term and ratio controller: Features

- No options.
- All controllers are identical and interchangeable.
- Single loop integrity.
- Built-in diagnostic routines.
- PID, Ratio or ON/OFF control.
- Microprocessor technology and solid state displays.
- Remote monitoring and supervision via a simple serial link.
- Field proven unit with a two-year warranty.
- Fully compatible with the TCS range of advanced instrumentation.

Description

The 6360 single loop process controller combines the flexibility of modern microprocessor technology with the integrity associated with conventional analogue instruments.

A microprocessor is incorporated in every 6360 enabling a user to characterise each device for any 3-term, ratio or ON/OFF control loop function using a simple plug-in hand-held terminal. As the loop characteristics are defined by easily changed parameters all 6360 controllers are identical and interchangeable regardless of application. One 6360 is a spare for all the others on a plant as its function is defined and changed by the technician via the hand-held terminal. Use of the terminal ensures security of the settings which are retained indefinitely when the device is powered and for at least five years if unpowered.

Functionally the controller operates as a conventional analogue unit providing facilities to raise/lower the setpoint, ratio or output, via front panel push-buttons and to change control mode to Manual, Automatic, Remote (for cascade connection) or Ratio. The 6360 will interface to 4-20mA signals from plant mounted equipment or 0-10V signals from the System 6000 range of signal conditioners and output drivers.

Each controller has a suite of input linearisation routines and TCS will provide custom linearisation at an additional cost. As well as providing a control output the 6360 generates 0-10V signals representing the linearised process variable and the setpoint or error deviation.

Supervision and monitoring of the 6360 is made particularly simple by the provision of a communications interface. This allows an intelligent device to monitor or update any of the control parameters of a network of 6360 controllers via an RS422 serial bus using a standard ANSI protocol. The use of a TCS 8245 Communications Buffer Unit enables RS232, TTL and fibre optic interfaces to be implemented.

The solid state technology offers high levels of reliability while the diagnostic procedures built into each instrument provide further output integrity.

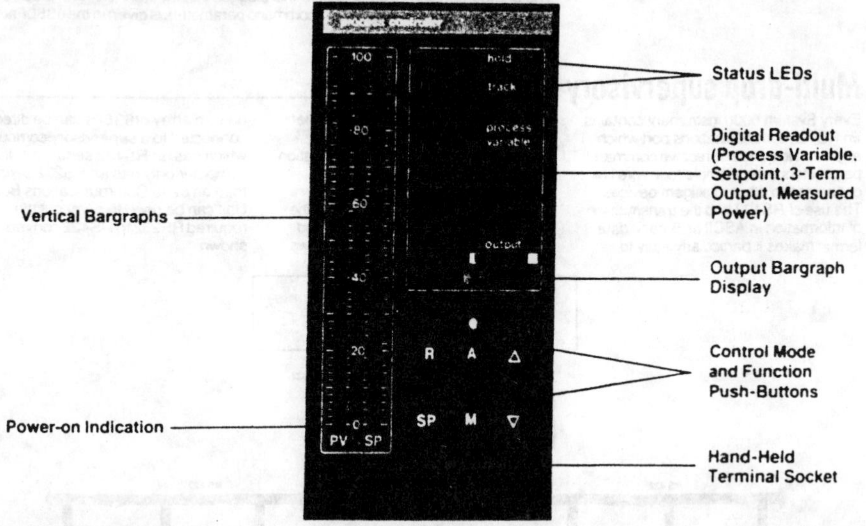

Operator displays and controls

Operator displays

Vertical Bargraph
Two 101 segment bargraphs displaying process variable and set point.

Digital Readout
(for process variable, setpoint, 3-term output and measured power indication) 4-digit, orange LED display with sign and decimal point.

Alarm Indication
High or low alarms indicated by flashing the process variable bargraphs.

Power-on Indication
Lowest segment of both bargraphs illuminated.

3-Term Output or Measured Power Display
Horizontal yellow LED bargraph with 10 segments to indicate 0-100% output.

Status Indicators
2 yellow rectangular LEDs to indicate TRACK and HOLD status.

Operator controls

Control Mode Selection
3 illuminated push-buttons:
Manual (M) with integral yellow LED.
Local Auto (A) with integral green LED.
Remote Auto or Ratio (R) with integral green LED.

Function Selection
2 non-illuminated push-buttons.
Raise (▲) increments the output when Manual (M) is pressed, or increments the setpoint when (SP) is pressed.
Lower (▼) decrements the output when Manual (M) is pressed, or decrements the setpoint when (SP) is pressed.

Display Selection
1 non-illuminated push-button (SP) causes the digital readout to display the current setpoint while pressed.

NOTE: Pressing the Manual, Auto or Remote buttons causes the digital readout to display the current output level or the measured power. Pressing the Raise or Lower buttons alone causes the alarm levels to be displayed on the bargraphs.

Communications

Every System 6000 microprocessor based instrument is fitted with an RS232 port and an RS422 port for serial data communications. The RS232 port is available via a front-panel socket and is used for the 8260 Hand-held programming terminal. The RS422 port is available on the module rear connector pins and is bussed onto the supervisory data link common to all modules. All parameters that can be monitored via the 8260 terminal can also be accessed and updated via the supervisory data link.

Hand-held terminal link

Each System 6000 instrument can be set up using a plug-in 8260 Hand-held terminal. Every parameter is accessed by means of a simple 2 character command mnemonic and all data is entered directly in engineering units. This technique ensures the accuracy and security of parameter settings.

Specification

Transmission Standard
2-wire RS232/V24 (± 12V).

Data Rate
300 baud.

Character Length
10 bits made up of:
1 start + 7 data + 1 parity (even) + 1 stop.

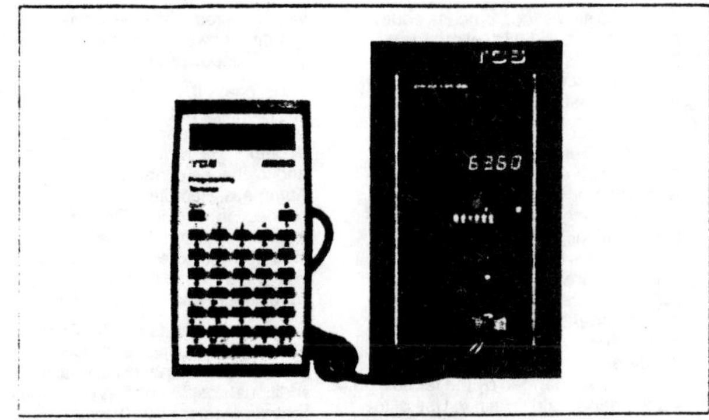

The photograph shows an 8260 terminal plugged into the front-panel of a 6360 controller. A full list of the available command parameters is given in the 6360 Facts Card.

Multi-drop supervisory link

Every System 6000 instrument contains an RS422 communications port which enables it to send and receive command parameters over a simple four-wire link connected to other intelligent devices. The use of RS422 and the transmission of information in ASCII or Binary data format makes it particularly easy to communicate with the 6360 controller. To hook the 6360 into a distributed control system requires no modification to the instrument and no further expenditure on options. The four-wire link is simply connected up so that the 6360 becomes part of the distributed control system. The illustration shows how an array of 6360s can be directly connected to a supervisory computer which has an RS422 serial port. If the computer only has an RS232 serial port then an 8245 Communications Buffer Unit can be used to carry out the required RS232 to RS422 conversion as shown.

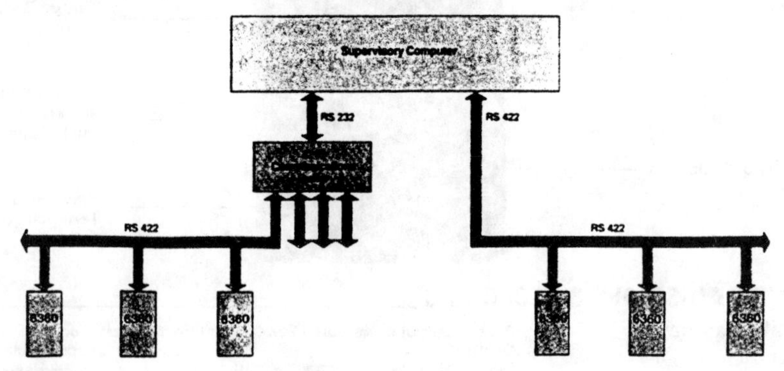

Specification

Transmission Standard
4-wire RS422 (0-5V).

Line Impedance
120-240Ω twisted pair.

Line Length
4000 ft max. (at 9600 baud).

Number of Units/Line
16.

Data Rate
Selectable from 110, 300, 600, 1200, 2400, 3600, 4800, or 9600 baud.

Character Length (ASCII/Binary)
10/11 bits – 300 to 9600 baud.
11/12 bits – 110 baud (2 stop).

Protocol

All microprocessor based instruments in the System 6000 range employ a standard ANSI protocol known as BI-SYNCH. The exact form of BI-SYNCH implemented within System 6000 corresponds with the American National Standard specification:

ANSI – X3.28 – 2.5 – A4 Revision 1976

TCS have implemented both an ASCII and Binary version of this protocol within each instrument.

The ASCII mode is simplest to use as all data is transmitted in ASCII characters. The Binary mode offers a 4 to 1 increase in transmission speed by compressing the data into a binary format, and also supports additional features like Multi-Parameter and Enquiry Polling.

Input/Output signals

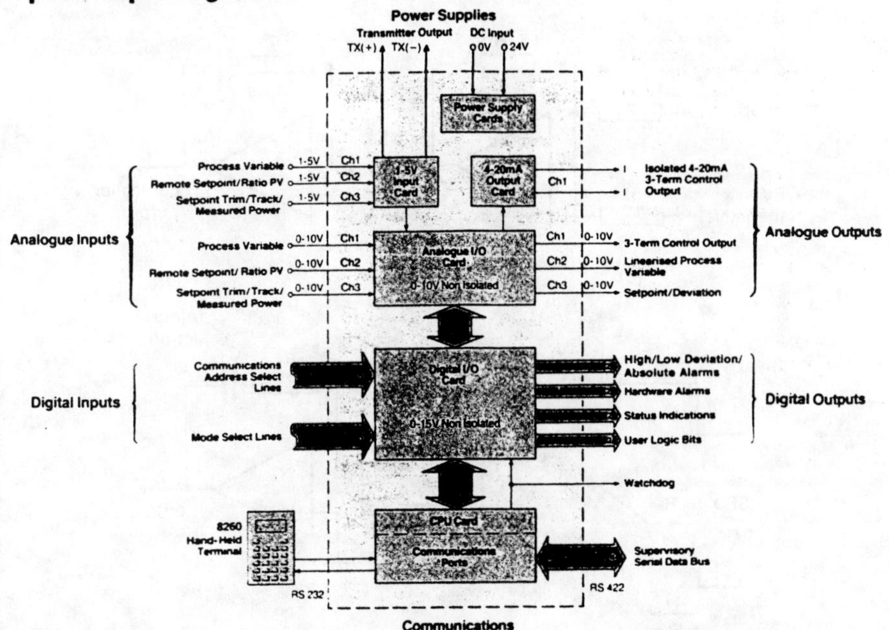

Analogue inputs

Number of Channels
3 direct non-isolated inputs or 3 conditioned non-isolated inputs.

Channel Functions
Channel 1 = Process Variable input.
Channel 2 = Remote Setpoint/Ratio Process Variable input.
Channel 3 = Setpoint or Ratio Trim/Track/Measured Power input.

Input Signal Levels
Direct inputs are 0-10V range. conditioned inputs are 1-5V or 4-20mA range with external sense resistors.

Resolution
12 bit binary ADC (.025%) hardware applied to inputs.
15 bit binary representation obtained after digital input filtering and signal averaging giving resolution of 1 digit in ± 9999.

Accuracy
± 1 LSB max. over 0 to 50°C range for hardware
± 1 digit of reading for 0-4000 range.
± 2 digits of reading for 0-8000 range.
± 3 digits of reading for 0-9999 range. after input filtering.

Sampling Rate
ADC samples 1 channel every 12ms, i.e. any one channel is sampled once every 36ms.

Input Impedance
1MΩ pull-down to −5V on channel 1.
1MΩ pull-down to 0V on channels 2 and 3.

Input Signal Processing
Linear (normal or inverse).
Normalised square root.
Type J, K, T, S, R, E, B thermocouples.
Platinum resistance thermometers.
User specified linearisation functions.

Analogue outputs

Number of Channels
3 direct non-isolated outputs plus 1 isolated output.

Channel Functions
Channel 1 = 3-Term control output.
Channel 2 = Process Variable output.
Channel 3 = Setpoint output or amplified deviation (error).

Output Signal Levels
Direct outputs are 0-10V range.
Isolated output is 4-20mA (channel 1 only).

Output Circuit Type
Medium-term analogue sample-and-hold circuits preceded by DAC.

Output Resolution
12 bit binary (.025%) giving minimum analogue voltage steps of 2.5mV

0-10V Output Accuracy
± 1 LSB max. over 0 to 50°C range.

Isolated Output Accuracy
± 0.5% of full scale.

Sample and Hold
DAC updates 1 channel every 12ms, i.e. any one channel is refreshed once every 36ms.

Output Drift Rate Under Watchdog Failure Conditions
½mV/sec maximum (equivalent to 1% of full scale in 3 minutes).

Output Drive Capability
± 5mA for direct voltage outputs.

Isolation Voltage
± 50V minimum with respect to system ground.

Digital inputs

Number of Inputs
8 external non-isolated inputs.

Input Functions
4 communications unit address select lines.
4 mode select lines

Input Voltage Levels
15V = logic one.
0V = logic zero.

Input Impedance
100kΩ pull-down to 0V (gives 150μA logic one current).

Digital outputs

Number of Outputs
8 external non-isolated outputs plus Watchdog.

Output Functions
2 deviation or absolute alarms.
2 hardware alarms.
2 status indications.
2 user logic bits.

Output Voltage Levels
15V = logic one.
0V = logic zero.

Output Drive Capability
2k2 open-collector pull-up to +15V supply, maximum logic zero sink current = 16mA.

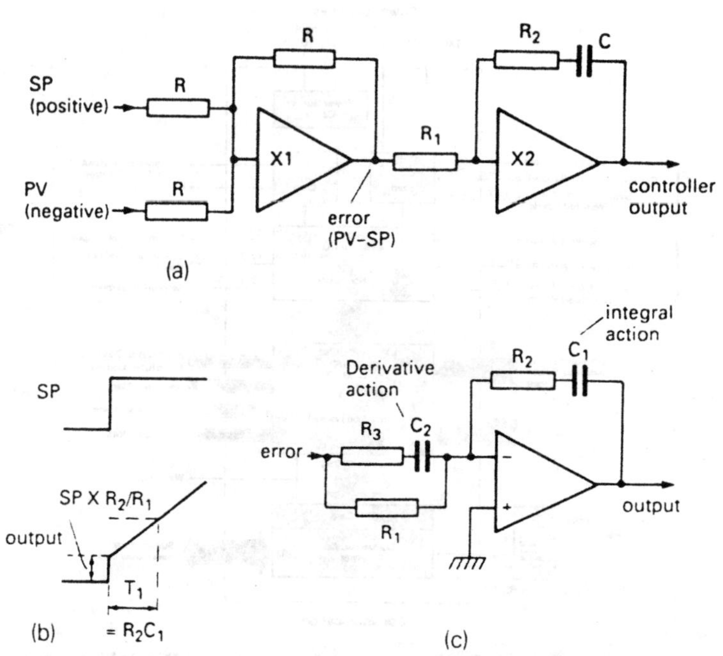

Fig. 4.16 A simple analogue P + I and P + I + D controller, (a) Circuit diagram; (b) Setting of gain and T_i; (c) P + I + D controller.

4.3. Analog controller and compensator circuits

4.3.1. Three term controllers

Although most commercial controllers are now digital/microprocessor based, analog controllers are still to be found at the cheaper end of the market and as integral parts of other equipment (e.g. the speed amplifier/controller in a DC drive). Analog P + I and P + I + D controllers are constructed from DC amplifiers,

Figure 4.16 shows various implementations of P + I and P + I + D controllers using DC amplifiers. Figure 4.16a, for example, is a simple P + I controller. Amplifier X1 performs the error subtraction (SP–PV) with SP positive and PV negative. The output of X1 is the error signal.

Consider a 1 V step change in SP with the controller previously balanced. The output of the amplifier X2 will experience a step change of R_2/R_1 V, followed by a ramp of slope $1/R_1.C$ V/sec as shown in Fig. 4.16c. This is the response of a P + I controller.

The gain is set by the ratio R_2/R_1. The integral time is defined as the time taken to repeat the initial step. This time is $(R_2/R_1) \times (R_1C_1) = R_2C_1$. Resistor R_1 thus sets the gain (increasing resistance

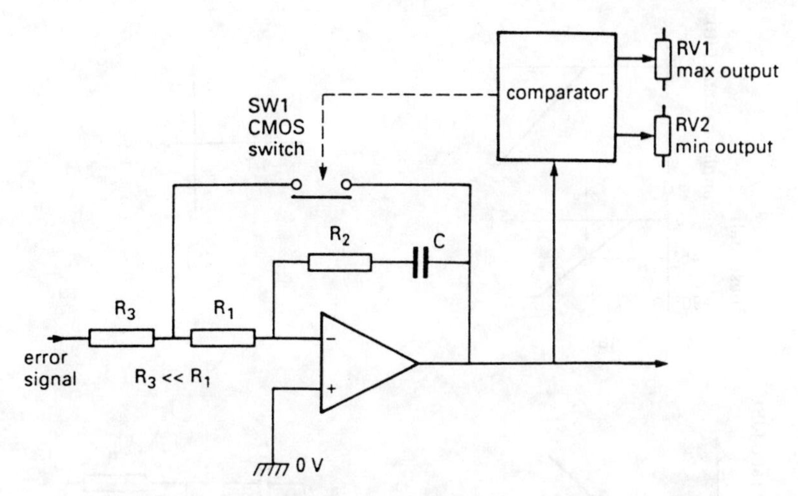

Fig. 4.17 Anti-integral windup and integral balance circuit.

increases the proportional band, decreasing the gain) and R_2 the integral time.

Integral windup protection and integral balance is normally achieved by switching circuits similar to the anti-windup circuit of Fig. 4.17. The CMOS switch SW1 closes if the controller output tries to move outside the values set on the presets RV1, RV2. SW1 applies a signal opposing the input of sign such that the output will move away from the limit. Once the output has just cleared the limit, SW1 opens and normal action is resumed. In an extended period of saturation the controller output will just remain at the limit value. Note that the integral term contribution from the capacitor is held at the value corresponding to the saturation limit.

4.3.2. Compensator circuits

A first order lag has the Bode diagram of Fig. 4.18a and the transfer function $1/(1 + sT)$ where T is the time constant. This provides a pole on the s-plane at $\sigma = -1/T$ as Fig. 4.18b. Related to the first order lag is the first order lead, which has the transfer function $(1 + sT)$ and the Bode diagram of Fig. 4.18c. The lead circuit contributed a zero on the s-plane at $\sigma = -1/T$ as Fig. 4.18d. The lead circuit can be approximated by Fig. 4.18e, but limitations of the amplifier prevent the gain increasing indefinitely as implied by Fig. 4.18c.

The lead and the lag circuits allow the designer to adjust the frequency response of a system to give the required behaviour. An

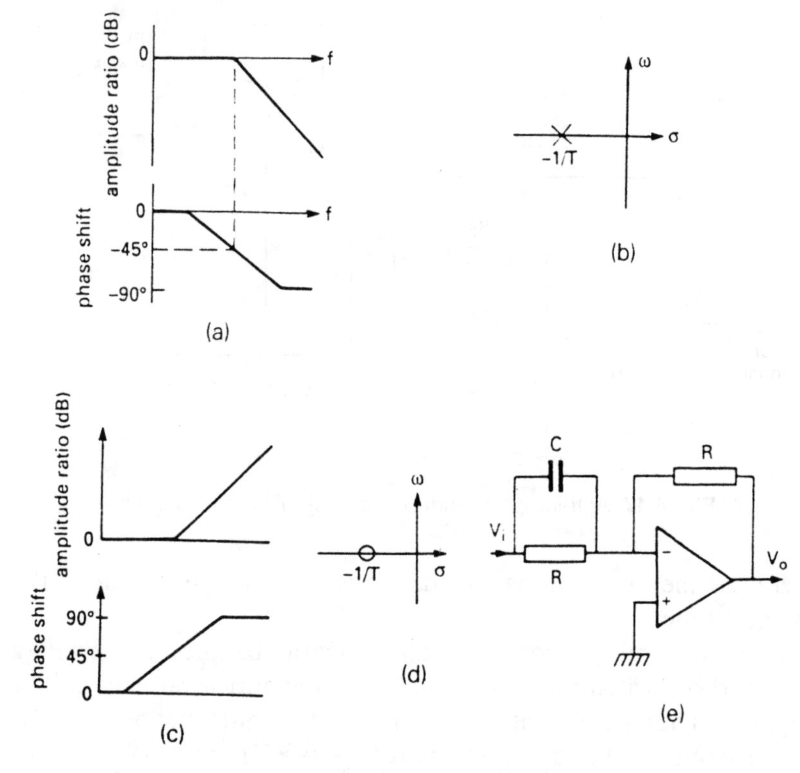

Fig. 4.18 First order lead compensator, (a) Bode diagram of first order lag; (b) First order lag on s-plane; (c) Bode diagram of first order lead; (d) First order lead on s-plane; (e) First order lead compensator.

alternative, but equally valid, view is they allow the designer to add poles and zeros to the root locus. Circuits of this type are termed 'compensators'.

Practical compensators have transfer functions of the form

$$\frac{(1 + sT_1)}{(1 + sT_2)} \tag{4.2}$$

and contribute a pole at $-1/T_2$ and a zero at $-1/T_1$. Combining Fig. 4.18a and 4.18c it is possible to build up the Bode diagram for the above function. There are two possible combinations, $T_1 > T_2$ and $T_1 < T_2$.

Figure 4.19 shows the case where $T_1 > T_2$. This is called a lead compensator. The Bode diagram is shown on Fig. 4.19a, with amplitude ratio of 0 dB at low frequencies, rising at 6 dB/octave between $1/T_1$ and $1/T_2$ rads/sec. The gain is constant above $1/T_2$

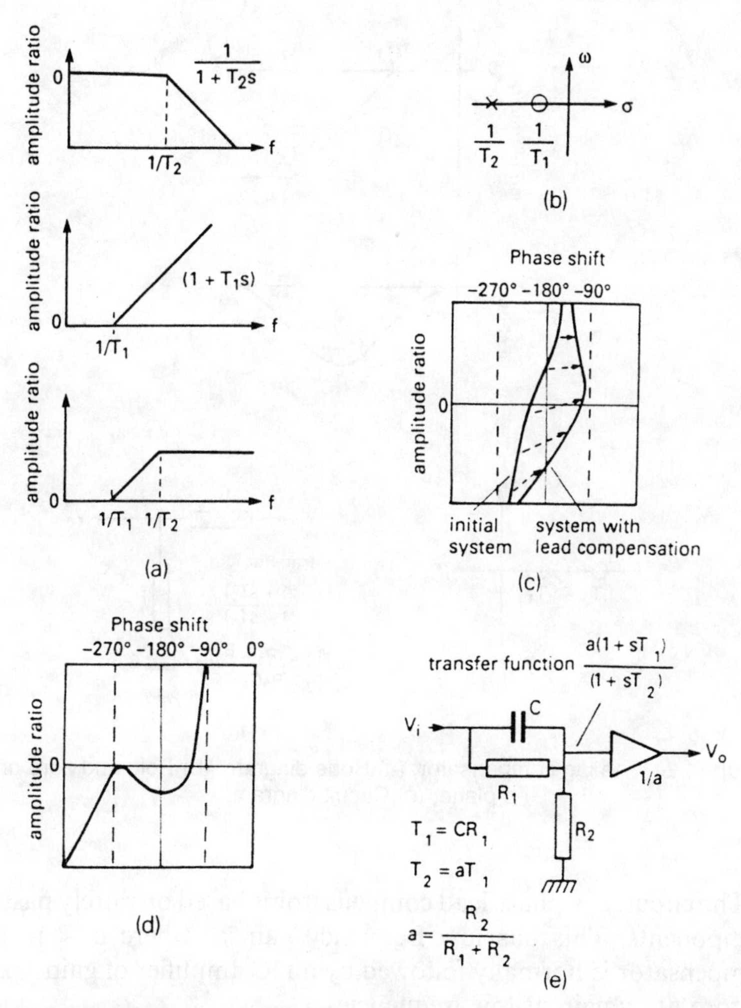

Fig. 4.19 The lead compensator, (a) Formation of Bode diagram; (b) Pole and zero on s-plane; (c) Effect of lead compensation on Nichols chart; (d) System which could become unstable with lead compensation; (e) Circuit diagram.

rads/sec. 90° of phase lead is introduced between frequencies $1/T_1$ and $1/T_2$.

On the s-plane the lead compensator adds a zero and a pole with the zero nearest the origin as Fig. 4.19b. In the most general of terms the zero increases the system stability.

A lead compensator increases the phase margin of a system and hence the stability as shown on the Nichols chart of Fig. 4.19c. The gain margin may be reduced if the open loop gain rises again after the 180° phase shift frequency. Figure 4.19d would probably become unstable with the addition of a phase lead circuit.

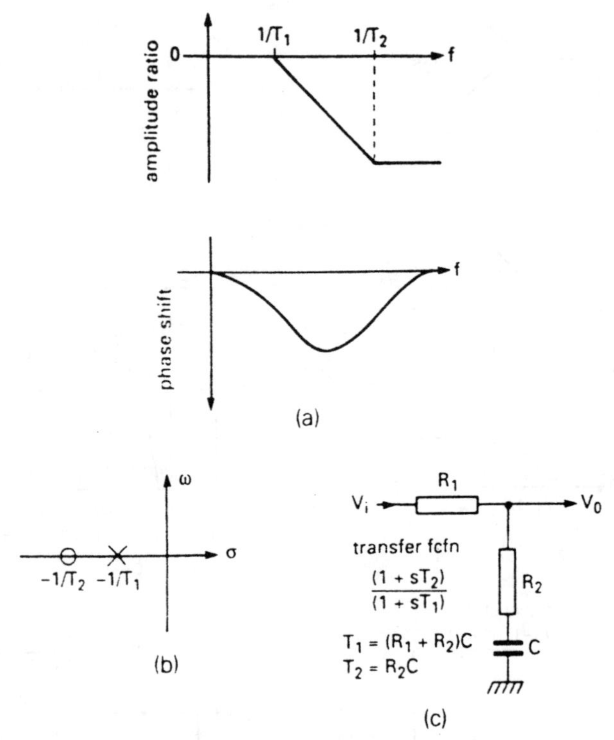

Fig. 4.20 The lag compensator, (a) Bode diagram; (b) Pole and zero on s-plane; (c) Circuit diagram.

The circuit of a phase lead compensator is based on purely passive components. This has low frequency gain a, where $a < 1$. The compensator is normally followed by a DC amplifier of gain 1/a to restore unity gain at low frequencies.

The lag compensator (where $T_1 < T_2$) has the Bode diagram of Fig. 4.20a and introduces a pole and zero as Fig. 4.20b. Lag compensation reduces the gain (and hence the gain margin) at high frequencies allowing a high controller gain to be used. Lag compensation is less common than lead compensation. A typical passive circuit is shown in Fig. 4.20c. This has unity gain at low frequencies, and does not need to be followed by an amplifier.

The circuit of Fig. 4.21a is called a lead–lag circuit and has the approximate transfer function

$$\frac{(1 + sT_1)(1 + sT_2)}{(1 + s\alpha T_1)\left(1 + s\dfrac{T_2}{\alpha}\right)} \tag{4.3}$$

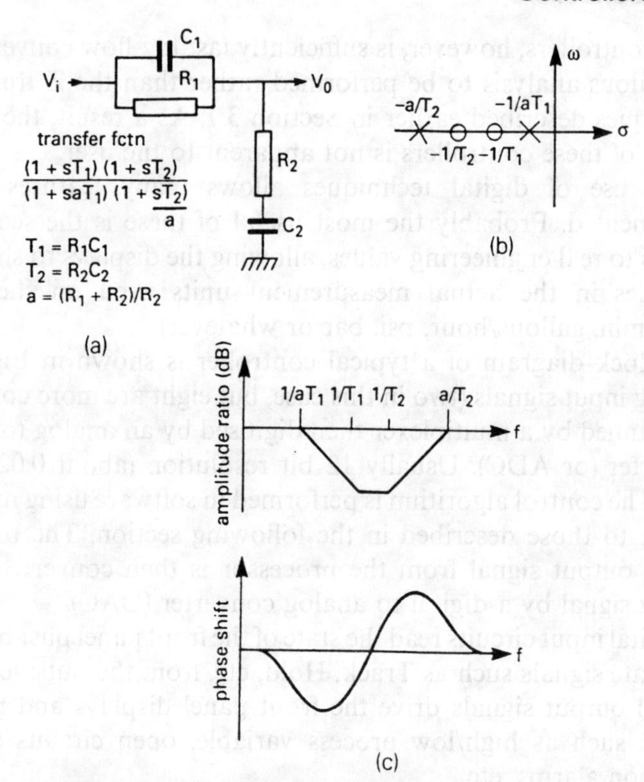

Fig. 4.21 The lead/lag compensator, (a) Circuit diagram; (b) Poles and zeros on s-plane; (c) Bode diagram.

if $T_1 > T_2$ and $\alpha > 1$. This introduces two pole/zero pairs onto the s-plane as Fig. 4.21b, and produces the Bode diagram of Fig. 4.21c.

A general observation is necessary on compensators. By their nature, compensators raise the gain quite dramatically over a range of frequencies. Normal DC amplifier circuits operate over a range of ± 10 V. Care should be taken to ensure that the output of a compensator circuit does not drive subsequent stages into saturation, thereby making the circuit non-linear.

4.4 Digital systems

4.4.1. Digital controllers

The cheapness and versatility of microprocessors has led to modern controllers being based on digital techniques rather than the traditional analog circuits of previous sections. The sample time of these controllers, however, is sufficiently fast to allow conventional continuous analysis to be performed rather than the Z transform techniques described earlier in Section 3.7. As a result, the digital nature of these controllers is not apparent to the user.

The use of digital techniques allows many features to be implemented. Probably the most useful of these is the scaling of signals to real engineering values, allowing the displays to show the variables in the actual measurement units used on the plant (litres/min, gallons/hour, psi, bar or whatever).

A block diagram of a typical controller is shown in Fig. 4.22. Analog input signals (two in this case, but eight are more common) are scanned by a multiplexer then digitised by an analog to digital converter (or ADC). Usually 12 bit resolution (about 0·025%) is used. The control algorithm is performed in software using methods similar to those described in the following section. The resulting digital output signal from the processor is then converted to an analog signal by a digital to analog converter (DAC).

Digital input circuits read the state of the front panel pushbuttons, and state signals such as Track, Hold, etc. from the outside world. Digital output signals drive the front panel displays and provide alarms such as high/low process variable, open circuits signals, deviation alarms, etc.

A typical modern controller will have eight analog inputs, four analog outputs, eight digital inputs and eight digital outputs. Increasingly digital controllers are being made user-configurable. The supplier provides the controller containing predefined, but unconnected, software blocks. Typical blocks are the analog/digital inputs and outputs, three term controllers, maths blocks, filter blocks, alarm blocks, ratio blocks, totalisation blocks manual stations, display blocks and so on.

The designer can then link these blocks to build any required control strategy. An analog input signal from an orifice plate, say,

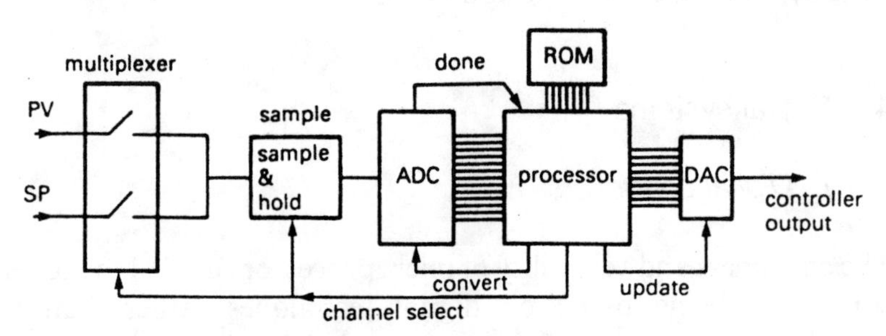

Fig. 4.22 Block diagram of digital controller.

can be connected to a square root block, scaled to real flow units then connected to the PV input of a three term control PID block. Usually this building is done with a very simple computer based pick/place/link drawing package. Figure 4.23 shows a slightly simplified gas follow air burner control scheme designed in this manner.

4.4.2. Digital algorithms

The algorithms used in digital systems are generally based on difference equations rather than differential equations, with the sample time Δt being substituted for dt. In the continuous graph of Fig. 4.24a, for example, the slope is dy/dt.

The digital representation is sampled (a topic discussed below) and the equivalent waveform is shown on Fig. 4.24b. If y(t) is the value of y at time t, and $y(t - \Delta t)$ the value of y at the previous sample Δt seconds earlier, the slope at time t approximates to

$$\frac{y(t) - y(t - \Delta t)}{\Delta t}$$

An alternative representation is to say y_n is the 'nth sample', then the slope becomes

$$\frac{y_n - y_{n-1}}{\Delta t}$$

Figure 4.25a shows the integration of a continuous waveform as being equivalent to the area under the curve. The equivalent digitised representation is shown in Fig. 4.25b. The area under the curve from sample Y_0 at $t = 0$ to sample Y_n at $t = n\Delta t$ is the sum of the individual rectangular areas, i.e.

$$Y_1\Delta t + Y_2\Delta t + \ldots + Y_n\Delta t$$

or more concisely

$$\sum_{1}^{n} Y_n\Delta t$$

A more accurate integration algorithm is the trapezoid representation of Fig. 4.25c, where each rectangle has area

$$\Delta t(Y_n + Y_{n-1})/2$$

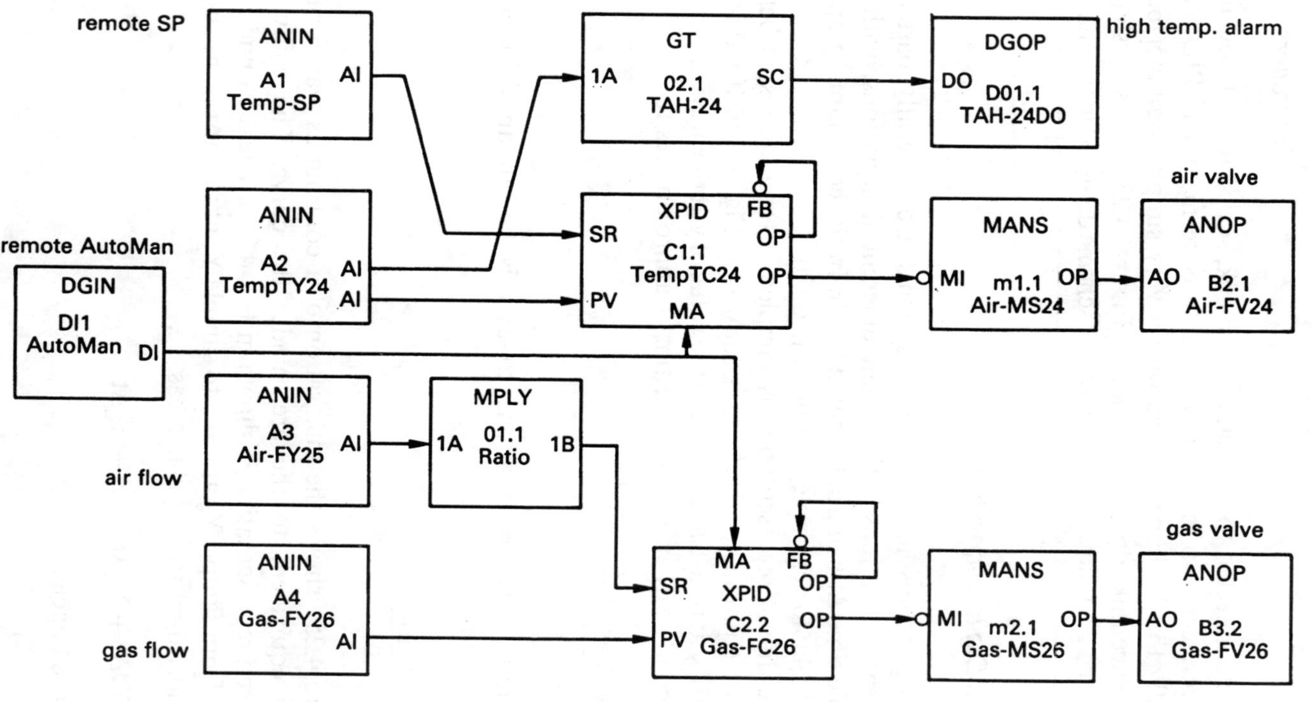

Fig. 4.23 Gas follow air temperature controller.

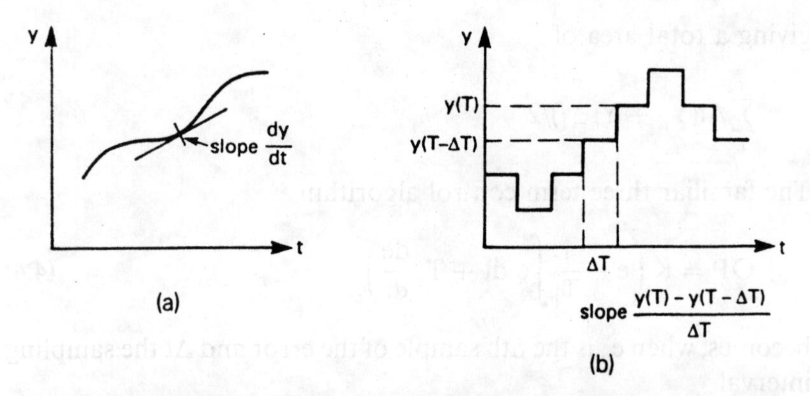

Fig. 4.24 Difference equations and digital systems, (a) Analog system; (b) Digital system.

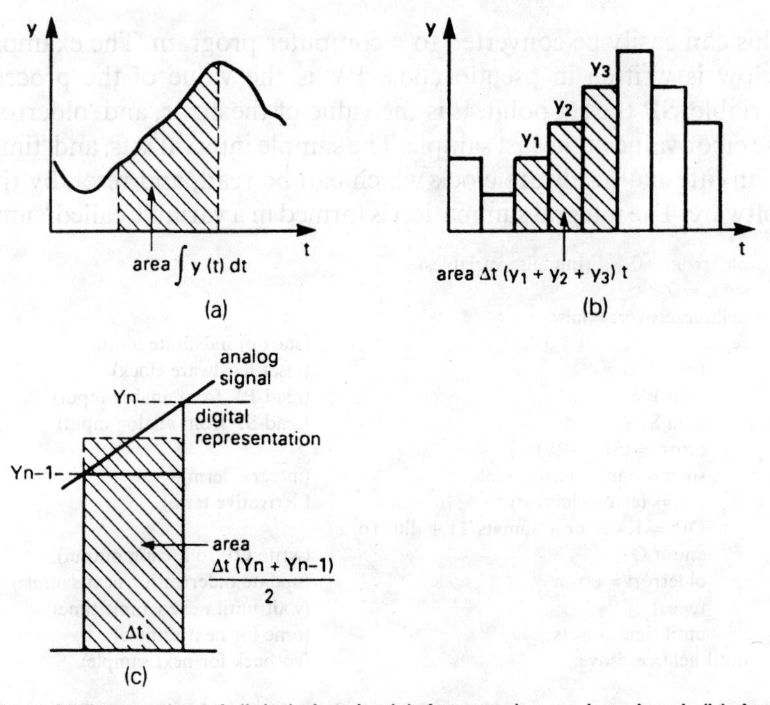

Fig. 4.25 Integration of digital signals, (a) Area under analog signal; (b) Area under a digital signal; (c) More accurate integration of digital signal.

giving a total area of

$$\sum_{1}^{n} \Delta t(Y_n + Y_{n-1})/2$$

The familiar three term control algorithm

$$OP = K\left(e + \frac{1}{T_i}\int e\,dt + T_d\frac{de}{dt}\right) \tag{4.4}$$

becomes, when e_n is the nth sample of the error and Δt the sampling interval

$$OP = K\left(e_n + \frac{1}{T_i}\Sigma\Delta t\frac{(e_n + e_{n-1})}{2} + \frac{T_d}{\Delta t}(e_n - e_{n-1})\right) \tag{4.5}$$

This can easily be converted to a computer program. The example below is written in pseudo code. PV is the value of the process variable, SP the set point. E is the value of the error, and 'olderror' the error value at the last sample. The sample interval is ts, and 'time' is an internal hardware clock which can be read, and reset, by the software. The integral summation is formed in a variable called 'sum'.

```
olderror: = 0      (Initialise variables)
sum: = 0
hellfreezesover: = false
repeat                                          (start of indefinite loop)
    time: = 0                                   (reset hardware clock)
    anin PV                                     (read PV from analog input)
    anin SP                                     (read SP from analog input)
    error: = (SP − PV)
    sum: = sum + (error + olderror)/2           (integral term)
    diff: = (error-olderror)                    (derivative term)
    OP: = K*(error + sum*ts/TI + diff*Td/ts)
    anout OP                                    (write OP to analog output)
    olderror: = error                           (update olderror for next sample)
    repeat                                      (wait until next sample time)
    until time > = ts                           (time for next sample)
until hellfreezesover                           (go back for next sample)
```

The above program, which loops indefinitely, needs some refinement. It works in 'percent of range' rather than engineering units, and has no protection against integral windup, for example. Additional routines for driving displays and auto/manual changeover would also be needed.

Similar digital techniques can also be applied to produce first order lags, lead–lag compensators, etc. The difference equation for a first order lag with time constant T seconds becomes

$$V_f = \text{oldvf} + \frac{\Delta t}{T}(V_n - \text{oldvf}) \quad (4.6)$$

where V_f is the current filtered value based on the raw unfiltered value V_n, and oldvf is the last filtered value. As before Δt is the sample interval. The difference equation assumes a rate of change of $(V_n - \text{oldvf})$ for the next sample interval.

A first order filter can thus be simulated by the listing below. T is the time constant and ts the sample time.

```
oldvfilter: = 0
repeat
    time: = 0
    anin V
    vfilter: = oldvfilter + (V – oldvfilter)*ts/T
    oldvfilter: = vfilter
    repeat: until time > = ts
until hellfreezesover
```

This is represented by the signal diagram of Fig. 4.26a. By similar reasoning, a second order system can be represented by Fig. 4.26b.

A straight time delay can be represented by an analog bucket chain moving analog values along every sample as Fig. 4.26c. With a sample time of Δt seconds, a pure delay of T seconds is equivalent to taking the value from the $\text{INT}(T/\Delta t + 0{\cdot}5)$ position.

The accuracy of a digital controller obviously depends on the resolution and accuracy of the ADC/DAC circuits. Less obviously, the sample time Δt appears in all the difference equations, so the consistency and accuracy of the sample clock is of equal importance to the analog circuitry. Digital controllers generally employ a hardware crystal controlled clock which uses interrupts to initiate controller routines at regular intervals. A typical controller will perform a complete input/compute/output cycle every 50 mS (i.e. 20 samples per second).

4.5. Pneumatic controllers

In the age of high technology electronics and microcomputers it is perhaps surprising that in many industries the traditional process controller still uses pneumatic signal standards and techniques. Despite many obituaries, pneumatics is still alive and well and living in industry with a complete range of pneumatic control devices (P + I + D controllers, lead–lag compensators, root extractors for flow measurements, ratio stations, etc.) readily available.

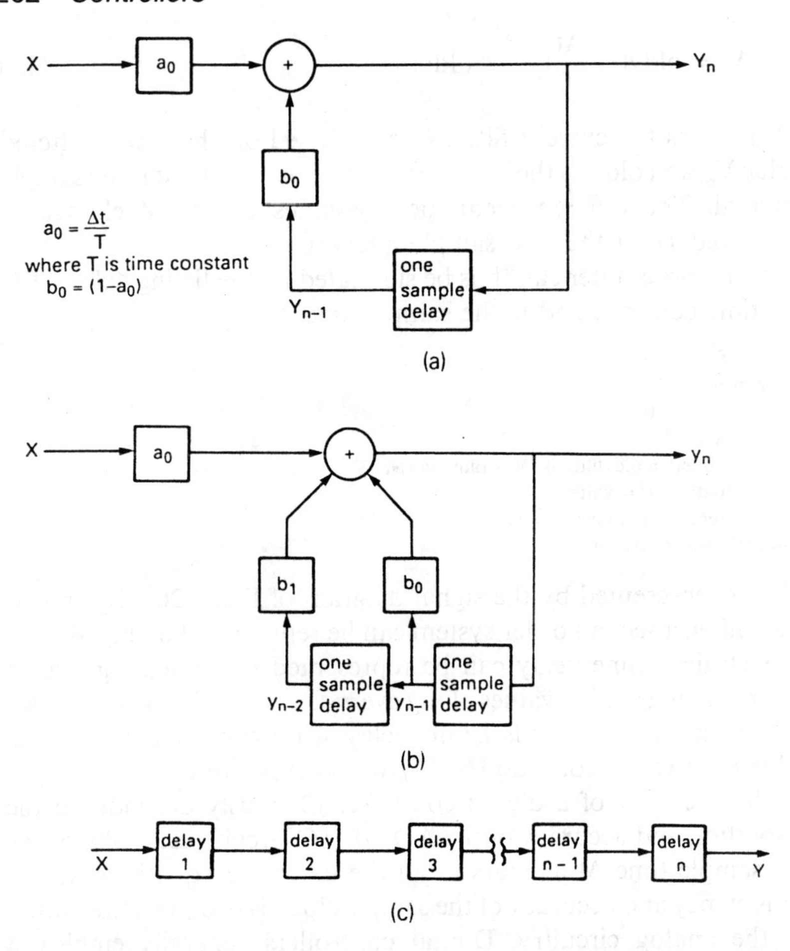

Fig. 4.26 Digital representation of common analog blocks, (a) First order lag; (b) Second order system; (c) A transit delay.

In plants with explosive atmospheres pneumatics are often the only solution. Many industries have large zone 1 areas (obvious examples are petrochemical plants and, less obviously, flour mills and whisky distilleries) where it would be prohibitively expensive to provide safe control areas local to the plant. There is also a large base of knowledge and experience built up around pneumatics that ensures its continuance.

Electronic systems generally represent signals as a 4–20 mA current or a 1–5 V or 0–10 V voltage. The equivalent pneumatic signal is a pressure with range 3–15 psig (or the equivalent metric signal 0·2–1 bar). The offset zero gives protection against signal loss (0 psi representing a negative signal) and increases response speed on

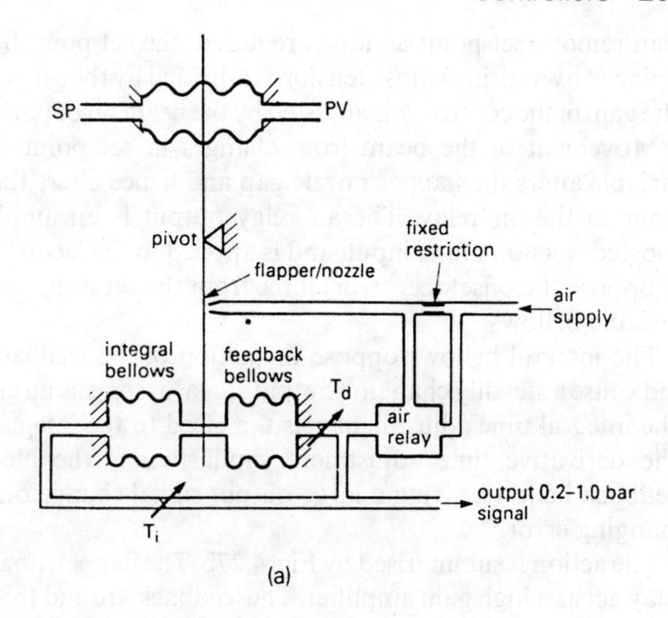

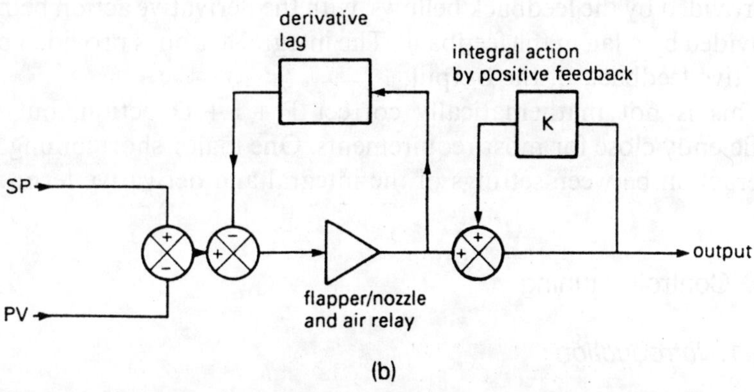

Fig. 4.27 Pneumatic controllers, (a) Block diagram of pneumatic P + I + D controller; (b) Representation of controller action.

a falling signal (which follows an exponential decay towards 0 psi as the signal line vents to atmosphere).

Speed of response is the major limitation of pneumatic systems. At best, transmission is limited to the speed of sound. Practically attainable speeds are far lower, typically 1 sec/100 m. Large volume actuators also act as a first order lag of several seconds time constant as the enclosed air is pressurised or depressurised.

Figure 4.27a shows the construction of a typical three term pneumatic controller. The set point and process variable signals apply opposing forces to one end of the force beam. (If local, rather than remote, set point action is required, the set point bellows are replaced by a spring whose tension is adjusted by the set point knob). The gain of the controller is adjusted by the beam pivot point position.

Movement of the beam from changes in set point or process variable alters the flapper nozzle gap and hence alters the pressure input to the air relay. The air relay output is an amplified, and boosted, version of the input, and is applied to the feedback bellows to oppose the offset (i.e. error) force from the set point and process variable bellows.

The integral bellows oppose the action of the feedback bellows and cause a steadily changing output signal as long as an error exists. The integral time adjustment sets the bleed to the integral bellows. The derivative time adjustment similarly sets the bleed to the feedback bellows, giving a large output signal change on a rapidly changing error.

The action is summarised by Fig. 4.27b. The flapper/nozzle and air relay act as a high gain amplifier. The feedback around this amplifier is provided by the feedback bellows, with the derivative action being provided by a lag in the feedback. The integral action is provided by positive feedback of the output.

This is not mathematically correct $P + I + D$ action, but is sufficiently close for most requirements. One major shortcoming is interaction between settings of the integral and derivative terms.

4.6. Controller tuning

4.6.1. Introduction

Before a three term controller can be used, values must be set for the gain (or proportional band), integral time and derivative time. In theory, if a model of the controlled plant is available, suitable values can be found in advance from Nichols charts or Nyquist diagrams using empirical methods based on hill climbing techniques.

In the usual situation, however, the plant model is not known (except in the most general terms) and the controller has to be tuned based on measurements which can be made on the plant with simple equipment.

This section describes practical methods of tuning a control loop. It should be noted that all involve disturbing the plant and some involve driving the plant into instability. The safety considerations of the tests should be clearly understood by all concerned.

Most of the tests aim to give a quarter cycle delay and assume the plant consists of a transit delay in series with a second order block (or two first order lags) plus possible integral action. A useful first step (if only to gain confidence) is to attempt to control the plant manually. This should identify the broad outline of the plant's characteristics (e.g. the rough value of the dominant time constant, if the plant exhibits a transit delay and any obvious non-linearities such as hysteresis).

In conducting the tests, it is useful to have a two pen recorder connected as shown in Fig. 4.28. The ranges of both pens should be the same.

In the tests below, controller gains are referred to in proportional band (PB = 100%/gain), with time used for integral and derivative action. Conversion to gain or repeats per minute is obviously straightforward.

4.6.2. Ultimate cycle methods

The aim of these methods is to determine the controller gain which will just sustain continuous oscillation; i.e. the point A and the gain K on the Nichols chart and Nyquist diagrams of Fig. 4.29a, b. The technique is based on a paper 'Optimum Settings for Automatic Controllers' published by J.G. Ziegler and N.B. Nichols in 1942.

The integral and derivative terms are disabled (to give a P only controller) and the gain slowly increased. Step disturbances are introduced and the system response observed. (One easy way to do this is by switching the controller to manual, changing the output slightly, say by 5–10%, and switching back to auto). As the gain is

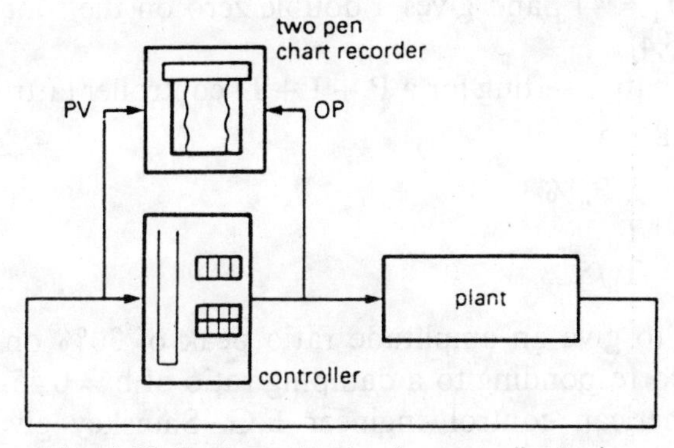

Fig. 4.28 Equipment setup for controller tuning.

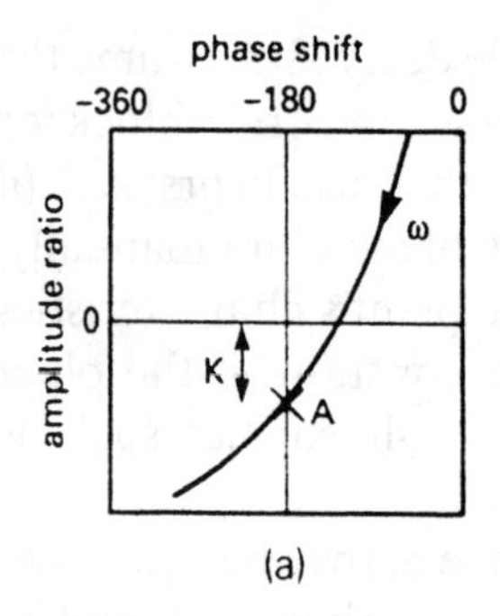

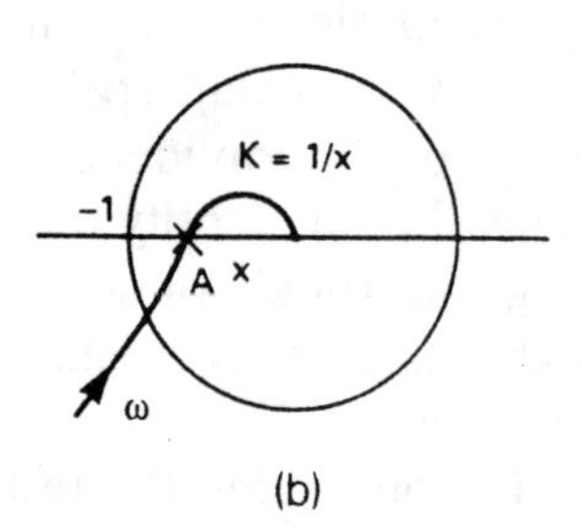

Fig. 4.29 Basis of ultimate cycle tests. Point A denotes the frequency at which continuous oscillations will occur, and K is the required gain, (a) Nichols Chart; (b) Nyquist diagram.

increased, the response will become progressively underdamped, and finally continuous oscillation will result. Care must be taken (and patience exercised) to allow transients to die away before each new value of gain is tried. The ultimate gain for stability is denoted by P_u. The period of oscillation T_u should now be measured from the chart recorder.

The required controller settings are now

P only

PB	$2\ P_u\%$

P + I

PB	$2{\cdot}2\ P_u\%$
T_i	$0{\cdot}8\ T_u$

P + I + D

PB	$1{\cdot}67\ P_u\%$
T_i	$T_u/2$
T_d	$T_u/8$

This sets $T_i = 4T_d$ and gives a double zero on the root locus (see Section 3.5.4).

An alternative setting for a P + I + D controller (attributable to Atkinson) gives

PB	$2\ P_u\%$
T_i	T_u
T_d	$T_u/5$

This aims to give an amplitude ratio peak of 30% on the Bode diagram, corresponding to a damping ratio of $b = 0{\cdot}45$.

The American control engineer F.G. Shinskey gives slightly different values

P + I

PB $2\ P_u\%$

T_i $0{\cdot}43T_u$

P + I + D

PB $2P_u\%$

T_i $0{\cdot}34T_u$

T_d $0{\cdot}08T_u$

The slight differences between all these methods are not really significant; all only aim to set the controller in the right area with final adjustment being made by trial and error. In general, it is advisable to leave T_i and T_d alone, and adjust the gain to give the required damping.

Shinskey gives a useful rule of thumb for identifying the plant characteristics. If the manual test (suggested earlier) shows a transit delay T_t to be present, then if

$T_u/T_t = 2$	The plant is pure transit delay
$2 < T_u/T_t < 4$	The plant is dominated by transit delay
$T_u/T_t = 4$	There is a single dominant first order lag
$T_u/T_t > 4$	There are several first order delays of similar magnitude

4.6.3. Decay method

This is a variation on the ultimate cycle method. With the controller in P only mode, trial and error methods similar to those above are used to give a decay ratio of approximately 4 : 1 (see Fig. 4.30). If the PB which achieves this is denoted P_q and the period of the decaying oscillations is T_q the required controller settings are

P only

PB $P_q\%$ (obviously)

P + I

PB $1{\cdot}5P_q\%$

Ti T_q

P + I + D

PB $P_q\%$

Ti $0{\cdot}67T_q$

Td $0{\cdot}17T_q$

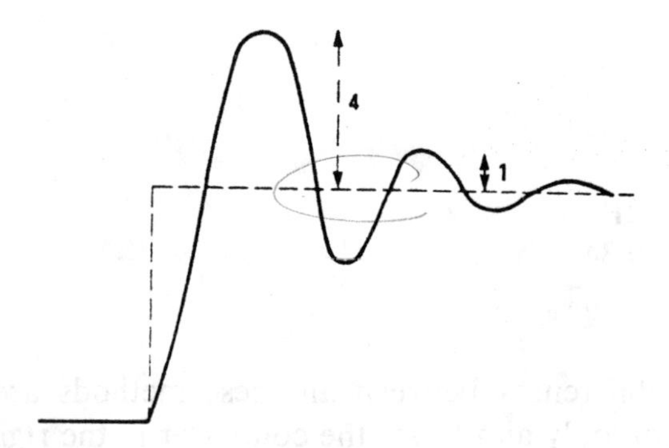

Fig. 4.30 Signal waveform used in decay method.

4.6.4. Bang/bang oscillation test

This is the fastest tuning method as it requires only a single test. It is, however, the most vicious of the methods, and can be misleading if the plant is non-linear. The controller is set for P only mode, with gain as high as the controller will allow (ideally infinite gain). This turns the controller into a bang/bang servo.

When the controller is switched to auto, oscillation will result as Fig. 4.31. The period of the oscillations T_o and the peak to peak height of the oscillations noted H_o are shown as a percentage of the controller input range.

The required controller settings are

P only

$PB = 2H_o\%$

P + I

$PB = 3H_o\%$

$T_i = 2T_o$

P + I + D

$PB = 2H_o\%$

$T_i = T_o$

$T_d = T_o/4$

If the PV oscillations themselves saturate a PB > 100% is required (i.e. gain less than unity). The range of the controller output swing should be progressively limited until the PV signal is within the input range. The controller output swing OP and the PV swing PV can then be read from the chart recorder (both as a percentage of full scale). H_o then is simply $100 \times PV/OP\%$ allowing the above relationships to be used.

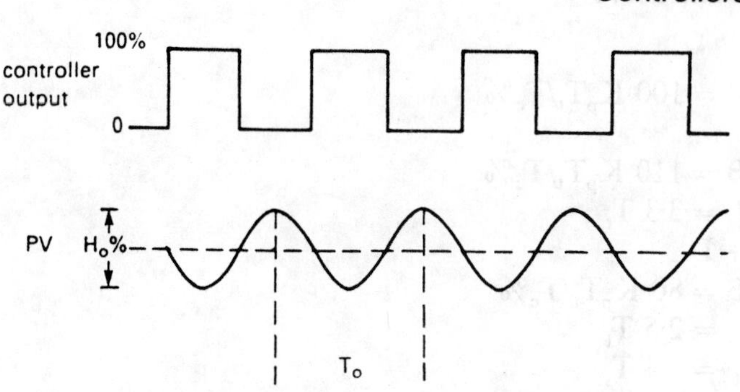

Fig. 4.31 The ultimate cycle method.

4.6.5. Reaction curve test

This open loop test was proposed by the American control engineers G.H. Cohen and G.A. Coon in 1953 (based on earlier work by Ziegler and Nichols). It assumes a measurable transit delay, and cannot be applied to plants with integral action (e.g. level controls).

The controller output is adjusted manually to bring the plant to near the desired operating point. After transients have died away, a small manual output step ΔOP is applied, which results in a change ΔPV in the process variable as Fig. 4.32.

The process gain K_p is then simply ΔPV/ΔOP.

A tangent is drawn to the process variable curve at the steepest point from which an apparent transit delay T_t and apparent time constant T_c can be inferred. Controller settings are then

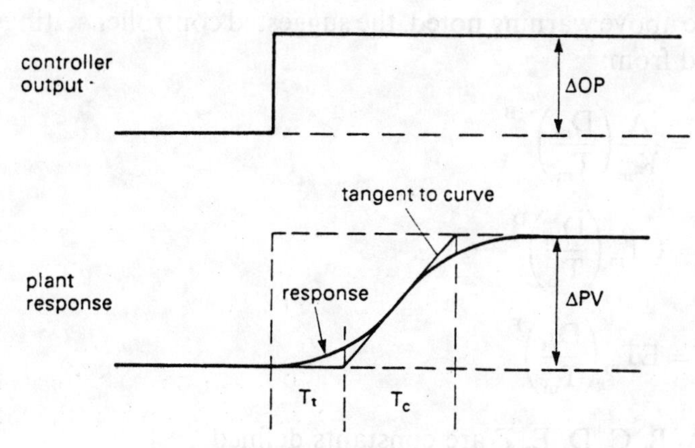

Fig. 4.32 The reaction curve test.

P only

$PB = 100\ K_p T_t / T_c \%$

P + I

$PB = 110\ K_p T_t / T_c \%$

$T_i = 3{\cdot}3\ T_t$

P + I + D

$PB = 80\ K_p T_t / T_c \%$

$T_i = 2{\cdot}5\ T_t$

$T_d = 0{\cdot}4\ T_t$

The reaction curve test is the gentlest of the tuning methods.

4.6.6. A model building tuning method

The closed loop tuning methods described so far require the plant to be pushed to (and possibly beyond) the edge of instability in order to set the controller. An interesting gentle tuning method was described by Yuwana and Seborg in the journal AIChE (American Institute of Chemical Engineers), Vol. 28, no. 3 in 1982.

The method assumes the plant behaves as a dominant first order lag in series with a transit delay, i.e.

$$\frac{K_m e^{-D_m s}}{1 + T_m s}$$

This assumption can give gross anomalies with plants with integral action such as level or position controls, but is commonly used for many manual and automatic/adaptive controller tuning methods. With the above warning noted, the suggested controller settings can be found from:

$$K_c = \frac{A}{K_m}\left(\frac{D_m}{T_m}\right)^{-B}$$

$$T_i = CT_m\left(\frac{D_m}{T_m}\right)^{D}$$

$$T_d = ET_m\left(\frac{D_m}{T_m}\right)^{F}$$

where A, B, C, D, E, F are constants defined:

Mode	*A*	*B*	*C*	*D*	*E*	*F*
P	0·490	1·084				
PI	0·859	0·997	1·484	0·680		
PID	1·357	0·947	1·176	0·738	0·381	0·99

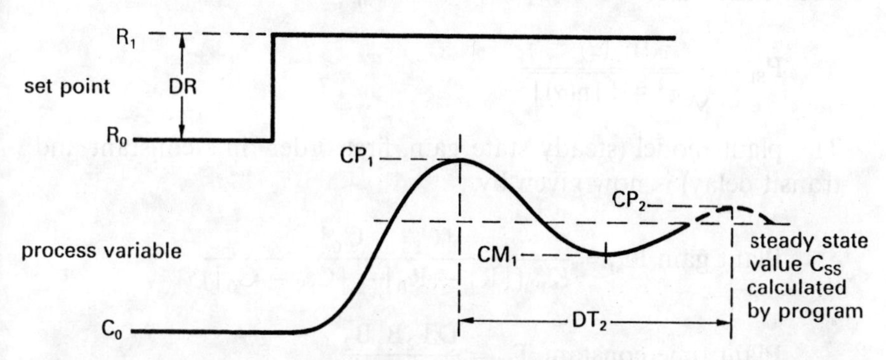

Fig. 4.33 Data used by tuning program. Note that R_0 and C_0 need not be the same, and DR can be positive (as shown) or negative.

These apparently random equations and constants come from experimental work described by Miller *et al.* in *Control Engineering*, Vol. 14, no. 12.

The method of finding the plant gain, time constant and transit delay is based on a single quick test with the plant operating under closed loop control. The test is performed on the plant operating under proportional only control, with a gain sufficient to produce a damped oscillation as Fig. 4.33 when a step change in set point from R_0 to R_1 is applied. The subsequent process maximum CP_1, minimum CM_1 and next maximum CP_2 are noted along with the time DT_2 between CP_1 and CP_2. The controller proportional band used for the test, PB, is also recorded, from which the controller gain $K_{PB} = 100/PB$ is found.

Given the values from the test the method estimates the value of the plant steady state gain K_M, lag time constant T_M and the transit delay time D_M. The background mathematics is given at length in the original paper, but the steps are as follows.

The final steady state value, denoted C_{SS}, is first predicted by:

$$C_{SS} = \frac{CP_1.CP_2 - CM_1^2}{CP_1 + CP_2 - 2CM_1}$$

and an intermediate damping factor α is given by

$$\alpha = \frac{CP_1 - C_{SS}}{C_{SS} - CM_1}$$

from which the true damping factor P_{SI} is obtained

$$P_{SI} = \frac{1n\ (\alpha)}{\sqrt{\pi^2 + [1n(\alpha)]^2}}$$

The plant model (steady state gain, first order time constant and transit delay) is now given by

$$\text{Plant gain } K_M = \frac{|C_{SS} - C_0|}{K_{PB}(|R_1 - R_0| - |C_{SS} - C_0|)}$$

$$\text{Plant time constant } T_M = \frac{DT_2B_1B_2}{2\pi}$$

$$\text{Plant transit delay } D_M = \frac{DT_2B_2}{\pi B_1}$$

where B_1 and B_2 are two intermediate values given by

$$B_1 = P_{SI}\sqrt{K+1} + \sqrt{P_{SI}^2[K+1] - 1 + K}$$

$$B_2 = \sqrt{(1 - P_{SI}^2)(K+1)}$$

with $K = K_{PB}.K_M$, the apparent loop gain for the test.

Note that if the real plant contains integral action (e.g. level or position control) the method will fall down when calculating the steady state gain giving (correctly) a gain of infinity and a divide by zero error.

The values for K_M, T_M and D_M can now be plugged back into the earlier equations to give the required controller settings for K_C, T_I and T_D.

The equations above are not very practical for use on site, so the original paper was developed into a program for the Hewlett Packard HP-67 calculator by Jutan and Rodriguez and published in the magazine *Chemical Engineering*, September 1984. The nomenclature used in the above equations is based on this article.

A Pascal version of this method suitable for use on a notebook computer is given later in Chapter 7, and an autotuner based on the technique is also included with the SIMULATE program.

4.6.7. General comments

The results of the initial tuning exercises above should not be viewed as rules of law written in stone. At best they put the engineer in the right area, at worst they can actually mislead. They should be viewed as a starting point for further manual adjustment of tuning constants.

It is always useful to have the proportional gain as high as possible to give large initial control action to changes and disturbances (unless the measured variable signal is noisy) so the gain should be adjusted first with integral and derivative setting left at their original settings. The gain should be adjusted to give the desired overshoot and damping.

Integral action is best adjusted next to give best removal of offset error. It may be necessary to decrease the gain again as the integral time is decreased. It is useful to keep a fixed ratio between T_i and T_d ($T_i = 4T_d$ is useful as this introduces a double pole into the root locus) whilst adjusting T_i. A useful rule of thumb is that T_i gain is an 'index' of stability for a given system, i.e. a T_i of 12 sec and a gain of 2 will give similar performance to a T_i of 24 sec and a gain of 4.

The derivative action should be adjusted last as a final stability/overshoot adjustment. Too much derivative action, however, can make the system noise prone and lead to 'twitchy' actuators. Care should be taken with derivative action in systems with significant transit delays. Here, derivative action can be a destabilising influence.

It should be remembered that the loop conditions will change with time and temperature and as items in the loop fail and get replaced with spares. Loop tuning should always be rather conservative, aiming for acceptable rather than perfect control. A finely tuned loop may become unstable under very slight changes in operating conditions.

Cascade control, described later, has an inner loop controlling an inner variable within an outer loop as Fig. 4.34 where an air flow loop is contained within a temperature control loop. In such cases,

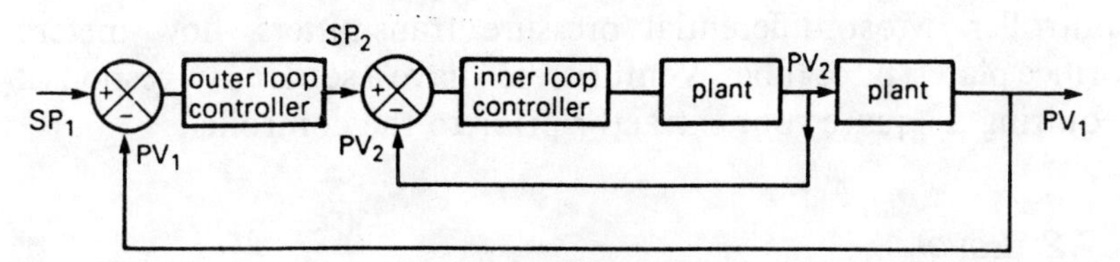

Fig. 4.34 Cascade control. The inner loop must be faster than the outer loop and must be tuned first.

the inner loop must be faster than the outer loop, and it is essential to tune the inner loop first and the outer loop last.

4.7. Characteristics of real loops

4.7.1. Flow

Flow is the property of a moving fluid, and the characteristics of flow loops are determined by mass and kinetic energy. The flow of fluid in a pipe is determined when pressure drop along the pipe balances the resistance to flow. If a flow control valve is opened, this changes the pressure drop, causing an imbalance between pressure and resistance force. The mass of fluid now accelerates until a new balance is achieved.

The change in pressure and the accelerating force propagates through the fluid at the speed of sound, so transit delays are negligible unless the flow control valve and flow transducer are widely separated. Transit delays can, however, occur in pneumatic signal pipes. The change in flow takes place in an exponential manner, with time constant

$$T = \frac{K.L.A}{F} \text{sec}$$

where L is the length of the pipe containing the fluid, A the cross sectional area and F the flow rate. K is a constant related to the fluid type, density and pressure. The time constant is therefore inversely related to the flow. A typical range of values is 0·5–4 seconds.

There is usually a non-linear relationship between valve stem position and flow, even when the valve is operated with a constant pressure differential (which is how valve characteristics are specified).

The net effect of both of these phenomena is to make flow loops distinctly non-linear, making it desirable to tune the loop at the expected flow rate, with checks for stability at high and low flows.

Flow turbulence causes signal noise, which is exaggerated by the use of derivative action. Flow is usually controlled by a P + I controller. Most differential pressure transmitters, flow meters (orifice plates, pitot tubes, Venturis, etc.) have a square law response, requiring a square root extractor prior to the controller.

4.7.2. Level

It is important to establish the control requirements for a level control system at an early stage. Figure 4.35 shows a typical level

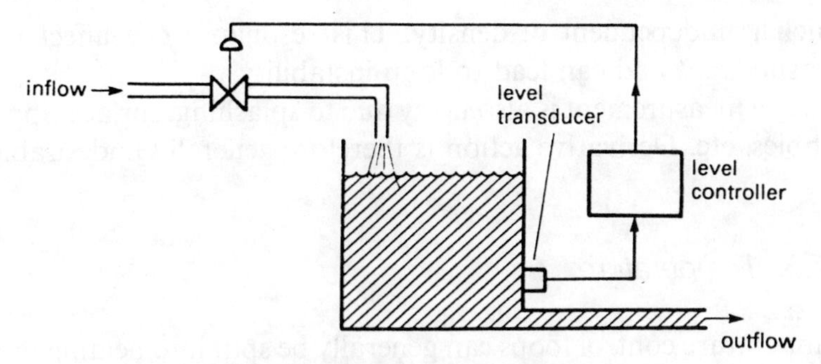

Fig. 4.35 A level control system.

control system with a metered input flow. A surge tank, by its nature, serves to decouple input and output flow rates and its purpose would be defeated by tight level control.

A tank integrates differences between inflow and outflow to give level. A simple P only control will give adequate control for a surge tank. The tank will slowly change until outflow and inflow match, tank level being inversely proportional to outflow (tank level falls for increasing outflow).

A P + I controller will give true level control but the combination of the tank integral action and the P + I controller leads to a double pole at the origin of the root locus as Fig. 4.36 and a response which will always be somewhat oscillatory.

Enclosed liquids can exhibit resonance (the bathtub effect). The resonant period of a tank of diameter d is

$$T = 2\pi \sqrt{\frac{d}{2g}}$$

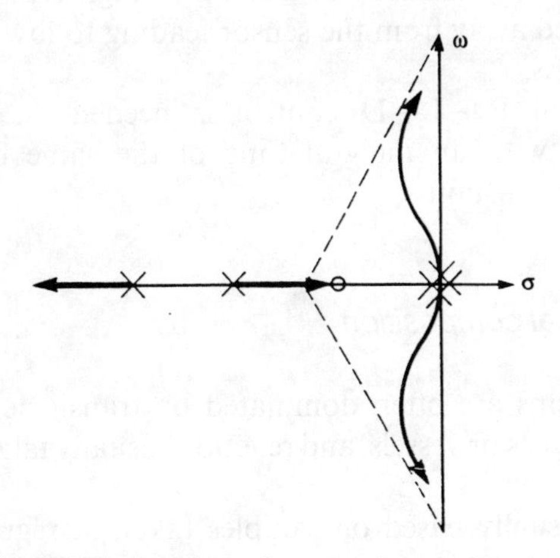

Fig. 4.36 Root locus for a level control system.

which is independent of density. This resonance can affect level measurement and can lead to loop instability.

Level measurement is also noisy due to splashing, surface ripples, bubbles, etc. Derivative action is therefore generally undesirable.

4.7.3. Temperature

Temperature control loops can generally be split into heating loops where the temperature of a material is to be maintained by the addition of energy, and cooling loops where the temperature of an exothermic reaction is to be controlled by a cooling jacket.

Temperature control loops are dominated by heat capacity and heat transfer. When heat is added, or removed from mass we have

$$\text{rate of temperature rise} = \frac{\text{heat in} - \text{heat out}}{\text{heat capacity}}$$

where heat capacity is a function of mass and specific heat.

Heat transfer through a mass is a function of temperature difference across opposing surfaces and the thermal conductivity.

Temperature control systems are therefore modelled as a series of interacting first order lags. Heat transfer through thick layers can have the properties of a transit delay, as can liquid heating systems where mixing is by natural convection. Transit delays can also appear with closed loop cooling systems where the temperature of the secondary fluid is used for control.

It can be difficult to make good temperature measurements; flame impingement on thermocouples leading to high readings, or heat being conducted away from the sensor leading to low readings are typical problems.

Generally full P + I + D control is needed on temperature control loops, with an integral time of the same order as the dominant time constant.

4.7.4. Chemical composition

Chemical systems are often dominated by transit delays. Mixing takes place in pipes or vessels, and reactions usually take a fixed time to complete.

Control is usually based on samples taken at regular intervals

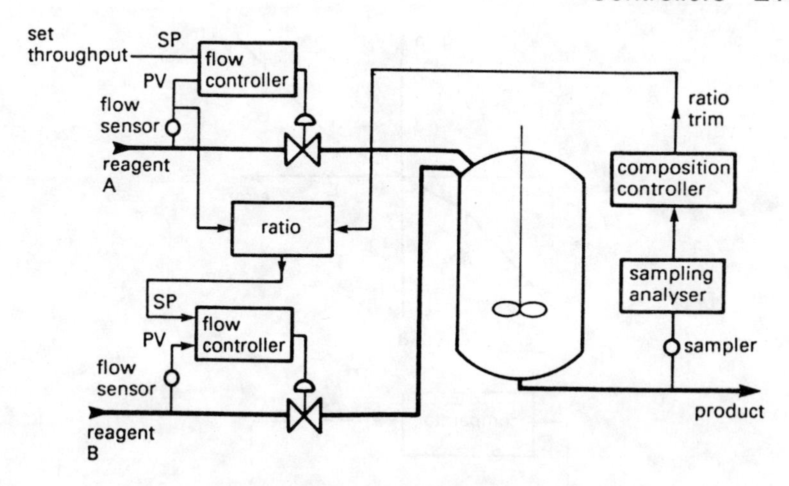

Fig. 4.37 A typical chemical composition control system.

which introduce the characteristics of a digital sampled system (see Section 3.7). Figure 4.37 shows a typical continuous process where the input of reagents are flow controlled, the second being ratioed to the first. A regular sampler/analyser measures the output product composition and corrects the ratio controller at regular intervals. Often, in such systems, a knowledge of the throughput rate, pipe size and vessel dimensions can give an initial guesstimate of transit delays existing in the system.

4.8. Other control algorithms

4.8.1. Variable gain controllers

Process variable noise occurs in many loops; level and flow being possibly the worst offenders. This noise causes unnecessary actuator movement, leading to premature wear and inducing real changes in the plant state. Noise can, of course, be removed by first or second order filters, but these reduce the speed of the loop and the additional phase shift from the filters can often act to destabilise a loop.

A controller with gain K will pass a noise signal K.n(t) to the actuator where n(t) is the noise signal. One obvious way to reduce the effect of the noise is to reduce the controller gain, but this degrades the loop performance.

Usually the noise signal has a small amplitude compared with the signal range (if it has not the process will be practically uncontrollable).

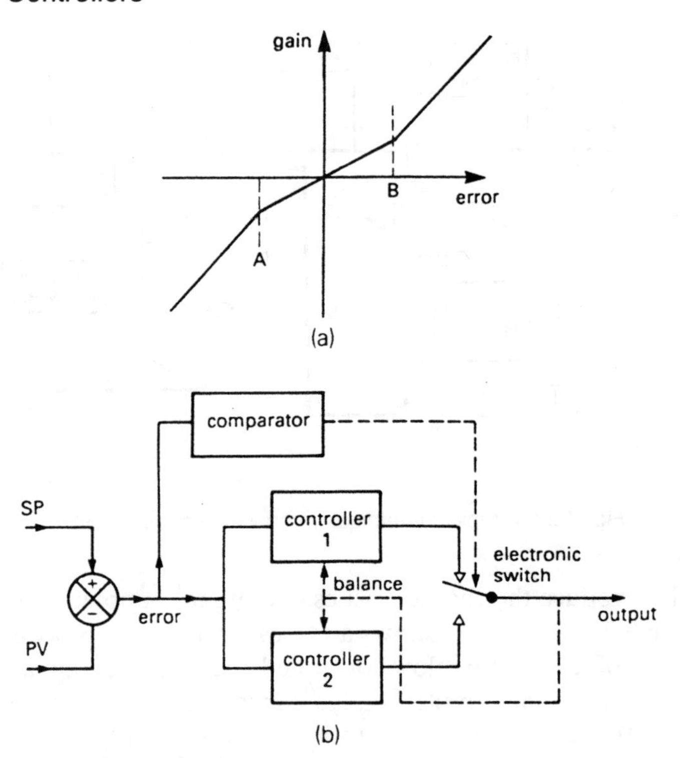

Fig. 4.38 Variable gain controller, (a) System response; (b) Block diagram.

What is intuitively required is a low gain when the error is low, but a high gain when the error is high.

Figure 4.38a shows how such a scheme operates. The noise amplitude lies in the range AB, so this is made a low gain region. Outside this band the gain is much higher. The gain in the region AB should be low, but not zero, to keep the process variable at the set point. With a pure deadband (i.e. zero gain in AB) the process variable would cycle between one side of the centre band and the other.

Figure 4.38b shows a possible implementation. A comparator switches between a low gain and high gain controller according to the magnitude of the error. Note that integral balancing is required between the two controllers to stop integral windup in the unselected controller.

Figure 4.38 has two gain regions. It is possible to construct a controller whose gain varies continuously with error. Such a controller has a response

$$\mathrm{OP} = \mathrm{K}f(\mathrm{e})\left(\mathrm{e} + \frac{1}{\mathrm{T_i}}\int \mathrm{e}\,\mathrm{dt} + \mathrm{T_d}\frac{\mathrm{de}}{\mathrm{dt}}\right) \tag{4.7}$$

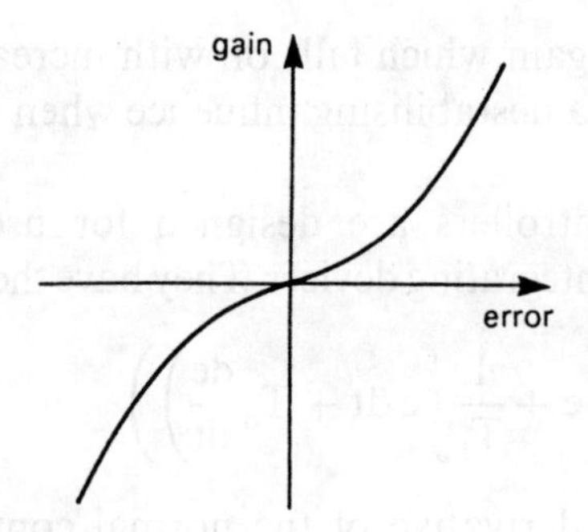

Fig. 4.39 Variable gain controller with parabolic response.

where f(e) is a function of error.

A common function is

$$f(e) = \text{ABS}\left(m + (1 - m)\frac{e}{100}\right) \tag{4.8}$$

where e is expressed as a percentage (0–100%) and m is a user set linearity adjustment ($0 < m < 1$). The ABS operation (which always returns a positive sign) is necessary to prevent the controller action changing sign on negative error.

With $m = 1$, $f(e) = 1$ and equation 4.7 behaves as a normal three term controller. With $m = 0$, the proportional part of equation 4.7 follows a square law with error, giving a response similar to Fig. 4.39. Like Fig. 4.38a, this has low gain at small error (zero gain at zero error) but progressively increasing gain as the error increases.

Position control systems often need a fast response but cannot tolerate an overshoot. These often use equation 4.8 with m at a low value approximating to the quadratic curve. This gives a high take off speed, but a low speed of approach.

4.8.2. Incremental controllers

Diaphragm operated actuators can be arranged to fail open or shut by reversing the relative positions of the drive pressure and return spring. In some applications a valve will be required to hold its last position in the event of failure. One way to achieve this is with a motorised actuator, where a motor drives the valve via a screw thread.

Such an actuator inherently holds its last position but the position is now the integral of the controller output. An integrator introduces

90° phase lag and gain which falls off with increasing frequency. A motorised valve is a destabilising influence when used with conventional controllers.

Incremental controllers are designed for use with motorised valves and similar integrating devices. They have the control algorithm

$$OP = \frac{d}{dt}\left(K\left(e + \frac{1}{T_i}\int e\,dt + T_d\frac{de}{dt}\right)\right) \tag{4.9}$$

which is the time derivative of the normal control algorithm. In difference equation terms, equation 4.9 becomes

$$OP_n = K\left((e_n - e_{n-1}) + \frac{1}{T_i}e_n + T_d(e_n - 2e_{n-1} + e_{n-2})\right)$$

where OP_n and e_n are the current output and error sample, e_{n-1} the previous error sample, and e_{n-2} the error two samples ago.

Incremental controllers are sometimes called boundless controllers or velocity controllers because the controller output specifies the actuator rate of change (i.e. velocity) rather than actual position.

Incremental controllers cannot suffer from integral windup per se, but it is often undesirable to keep driving a motorised valve once the end of travel is reached. End of travel limits are often incorporated in motorised valves to prevent jamming. The controller also has no real 'idea' of the valve true position, and hence cannot give valve position indication. If end of travel signals are available, a valve model can be incorporated into the controller to integrate the controller output to give a notional valve position. This model would be corrected whenever an end of travel limit is reached. Alternatively a position measuring device can be fitted to the valve for remote indication.

Pulse width modulated controllers are a variation on the incremental theme. Split phase motor drive valves require logic raise/lower signals, and by using time proportional raise/lower outputs normal proportional control can be simulated.

4.8.3. Inverse plant model

The ideal control strategy, in theory, is one which mimics the plant behaviour. Given a totally accurate model of the plant, it should be possible to calculate what controller output is required to follow set point change, or compensate for a disturbance. The problem here is, of course, having an accurate plant model, but even a rough

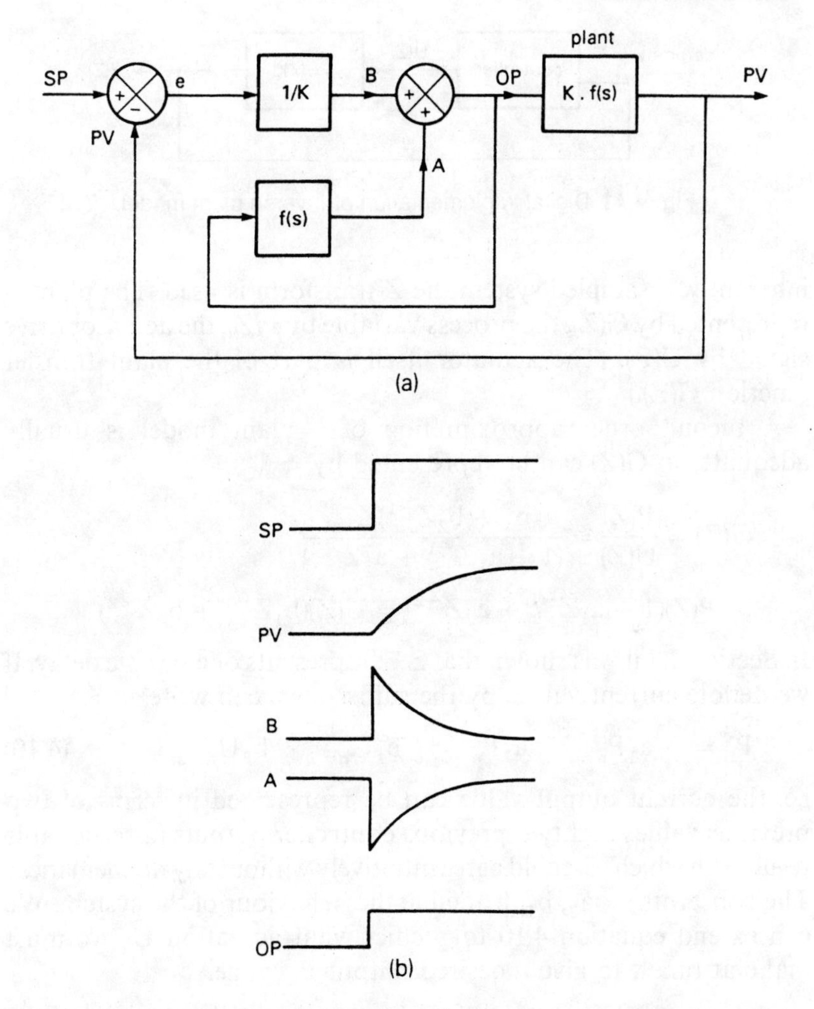

Fig. 4.40 The inverse plant model, (a) Block diagram; (b) System operation.

approximation should suffice as the controller output will converge to the correct value eventually.

One possible solution is shown in Fig. 4.40. The process is represented by a block with transfer function K.f(s) where K is the DC (low frequency) gain. Following a change in set point, the signal A should mimic exactly the inverse changes in the variable B, leading to a constant output from the controller which is exactly correct to bring the plant to the set point without overshoot. With a perfect model, the change at A should match the inverse change at B as the set point is approached.

A digital implementation is shown in Fig. 4.41. Since this is,

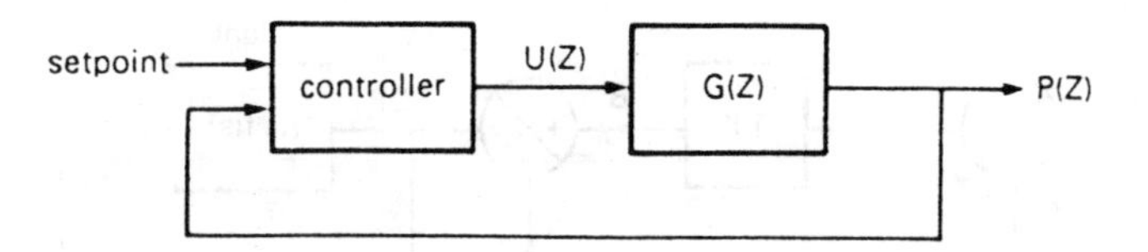

Fig. 4.41 Digital implementation of inverse plant model.

inherently, a sampled system the Z transform is used. The plant is represented by G(Z), the process variable by P(Z), the actuator drive signal by U(Z). (The actuator itself is part of the plant transfer function G(Z).)

A second order approximation of a plant model is usually adequate, so G(Z) can be represented by

$$G(Z) = \frac{P(Z)}{U(Z)} = \frac{(b_1 + b_2Z^{-1})Z^{-1}}{(1 + a_1Z^{-1} + a_2Z^{-2})}$$

$$\text{or } P(Z)(1 + a_1Z^{-1} + a_2Z^{-2}) = U(Z)(b_1Z^{-1} + b_2Z^{-2})$$

In Section 3.7 it was shown that Z^{-1} represents one sample delay. If we denote current values by the suffix n, we can write

$$P_n = -a_1P_{n-1} - a_2P_{n-2} + b_1U_{n-1} + b_2U_{n-2} \tag{4.10}$$

i.e. the current output value can be represented in terms of two previous values and two previous controller outputs (a reasonable result with which we could agree intuitively without any mathematics). The constants a_1, a_2, b_1, b_2 define the behaviour of the system. We can extend equation 4.10 to predict what actuation U_n we must make at time t to give a desired output P_{n+1}, i.e.

$$P_{n+1} = -a_1P_n - a_2P_{n-2} + b_1U_n + b_2U_{n-1} \tag{4.11}$$

or the desired controller output to give set point $SP = P_{n+1}$, i.e.

$$U_n = \frac{1}{b_1}(SP + a_1P_n + a_2P_{n-2} - b_2U_{n-1})$$

The controller thus looks at the current process variable, the required process variable, the last sampled value of process variable and actuation and computes directly the next actuation to bring the error to zero. Figure 4.42 shows a computational block diagram for this approach.

The problem with this simple, and apparently ideal, controller is that it will probably demand actuation signals which will drive the controller output, the actuator or parts of the plant into saturation.

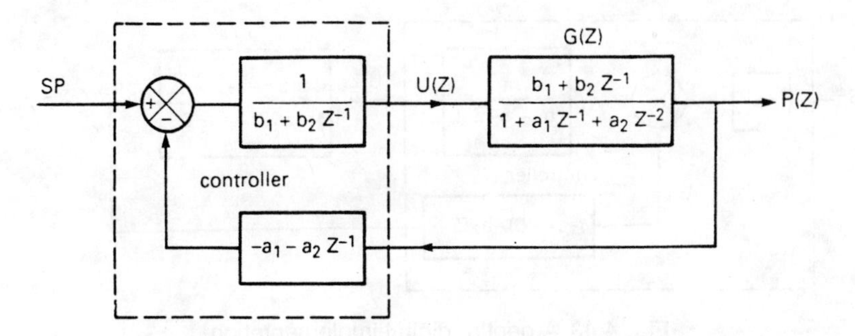

Fig. 4.42 Digital implementation of inverse plant model.

A more gentle version of this technique aims to get a fraction α of the way from the current value to the desired value of the process variables at each step. This approximates to an exponential response. We now have

$$\underset{\text{prediction}}{P_{n+1}} = \underset{\text{current}}{P_n} + \alpha(\underset{\text{set point}}{SP_n} - \underset{\text{current}}{P_n}) \tag{4.12}$$

Applying the Z transform gives

$$P(Z) = \frac{SP(Z)(1-\beta)Z^{-1}}{(1-\beta Z^{-1})}$$

where $\beta = 1 - \alpha$.

This is the Z transform of a first order lag, as we intuitively deduced earlier.

Substituting equation 4.12 for P_{n+1} into the earlier equation 4.11 and rearranging to give the required actuation U_n gives

$$U_n = \frac{1}{b_1}(P_n(1 + a_1 - \alpha) + a_2 P_{n-1} + \alpha SP_n - b_2 U_{n-1}) \tag{4.13}$$

which can be represented by the computational block diagram of Fig. 4.43.

Tuning of a controller based on Fig. 4.43 required the mathematical building of a plant model (i.e. determining values for a_1, a_2, b_1, b_2) and choosing a value for α that does not lead to undue saturation of any component in the loop.

A variation on the technique (called the Dahlin design) uses a controller with a compensator whose transfer function C(Z) is designed to combine with the plant transfer function G(Z) to give a first order closed loop response. If K(Z) denotes the required closed loop response

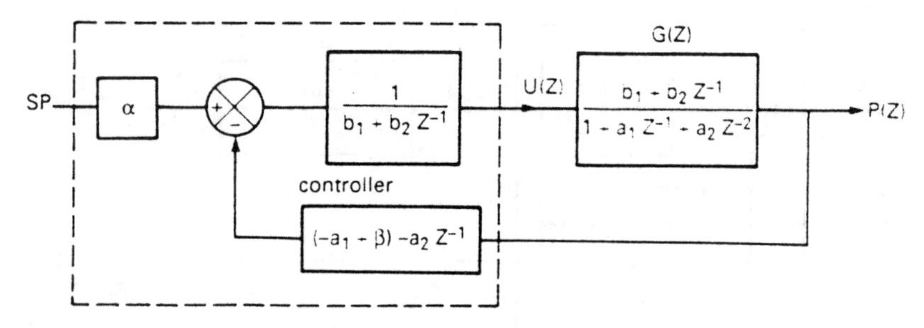

Fig. 4.43 A gentler digital implementation.

$$K(Z) = \frac{C(Z).G(Z)}{1 + C(Z)G(Z)}$$

Since G(Z) is the known plant model, and K(Z) is the required response, the required compensator function is

$$C(Z) = \frac{1}{G(Z)} \cdot \frac{K(Z)}{(1 - K(Z))}$$

Tuning of the controller of Fig. 4.44 again requires knowledge of a plant model, and choice of a required response which does not drive plant components into saturation.

The compensator of Fig. 4.44 attempts to cancel the plant transfer function. The technique can also be applied to analog systems using the s-transform. Knowledge of a plant model must again be available, and care taken to ensure impossibly large, or fast, actuations are not demanded.

4.9. Autotuning controllers

4.9.1. Introduction

Tuning a controller can be more of an art than an exact science, and can be unbelievably time consuming. Time constants of tens of minutes are common in temperature control, and lags of hours can

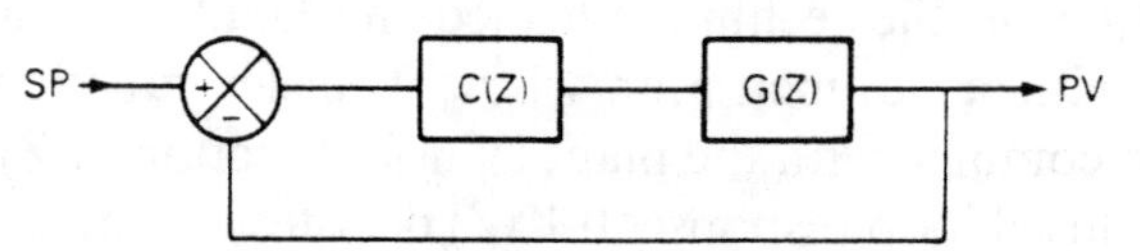

Fig. 4.44 The Dahlin design of controller.

occur in chemical processes. In a blast furnace, for example, there can be a lag of over twenty-four hours before a change in the feed ore affects the output iron. As a result a great deal of patience and time is needed to perform the manual tests described in the previous sections. The introduction of digital controllers based on microprocessors has allowed tuning algorithms to be built into the controllers. Such devices are known as self-tuning, autotuning or adaptive controllers.

Process control engineers tend to have two distinct views of self-tuning controllers. Many see them solely as a commissioning aid which can take much of the tedium out of setting up control loops. They allow the commissioning engineer to tune many loops at the same time, reducing the work and significantly shortening commissioning time. Once the loops are tuned the autotune feature is disabled, and the controllers left to run with a fixed setup or replaced by a conventional controller. The term 'autotuner' is sometimes used to describe this approach. In many plants, however, the open loop characteristics are not fixed, but change and drift with time and outside influences. Under these circumstances a self-tuning controller can be allowed to adjust itself continually to compensate for the changes in the plant. The term 'adaptive controller' is commonly applied. The terms 'autotuner' and 'adaptive' are really describing the use rather than the controllers themselves. This section will use the term self-tuning controllers to describe both applications.

There are, though, two distinct types of self-tuning controllers. Modelling, or explicit, self-tuners try to build a model of the plant then determine controller parameters to suit this model. Implicit self-tuners use tests similar to those in Section 4.6.

4.9.2. Model identification (explicit self-tuners)

Model identification is based on the principle of Fig. 4.45. A control action OP is applied to the plant and the model. The plant returns a real change of PV and the model a predicted change. These changes are compared and the resulting error used to update the model. The controller parameters are updated based on the model. The steps of nudge the plant, see the change, update the model, set controller parameters are followed continuously.

The manual tuning method described in Section 4.6.6 is a simple example of an explicit tuning method. The technique applies a small step change of set point to the controller, and observes the response

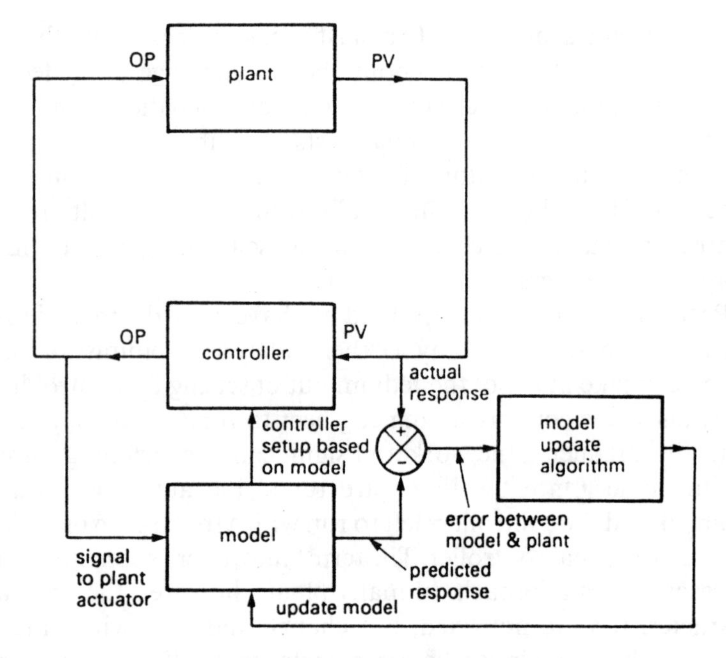

Fig. 4.45 A modelling self tuning controller.

of the plant. A first order plus transit delay model is deduced, from which the controller settings are calculated.

A more detailed method of setting the controller parameters from the model is based on the s-plane used to draw root loci. As explained earlier in Section 3.5 the closed loop response is determined by the position of the poles. These must be in the left-hand part of the s-plane for a stable system, and the rate of decay of the transient response is determined by their position along the σ axis.

A technique called 'pole placement' chooses controller parameters to position the closed loop poles to give the required response. Suppose our model building suggests the plant behaves as a single first order lag. This will have the transfer function A/(1 + sT) where T is the time constant and A the steady state gain. The transfer function has a single open loop pole at

$$s = -\frac{1}{T}$$

Let us assume our lag has a time constant of 4 seconds, and a steady state gain of 0·5. Its transfer function will be 0·5/(1 + 4 s). We will first try to make it behave as first order lag of time constant 0·8 s by using

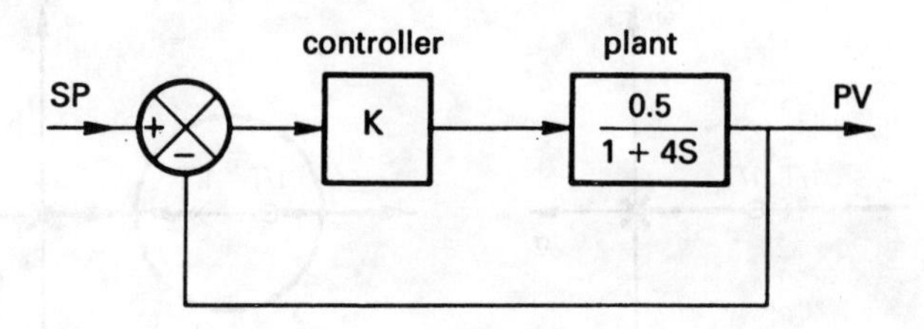

Fig. 4.46 Pole placement with first order plant and proportional only controller.

a proportional only controller of gain K as shown on Fig. 4.46. Our self-tuner at this stage can only adjust K.

The closed loop transfer function is CG/(1 + CG) where C and G are the transfer functions of the controller and plant. Poles occur where 1 + CG = 0, which for our system is at

$$1 + \frac{0{\cdot}5\mathrm{K}}{1 + 4\,\mathrm{s}} = 0$$

from which

$$\mathrm{s} = -\frac{(1 + 0{\cdot}5\mathrm{K})}{4}$$

If the closed loop response is to be a first order lag of time constant 0.8 s, we want a pole at

$$\mathrm{s} = -\frac{1}{0{\cdot}8} = -1{\cdot}25$$

hence

$$-\frac{1 + 0{\cdot}5\mathrm{K}}{4} = -1.25$$

or

$$\mathrm{K} = 8$$

The required controller gain is therefore 8 (a PB of 12·5%).

A proportional only controller will not reach the set point or compensate exactly for disturbances. We can improve the performance by using a P + I controller. A P + I controller introduces a pole at the origin of the s-plane ($\sigma = 0$, $\omega = 0$) and a zero at $-1/T_i$. The transfer function of the controller is $K(1 + 1/sT_i)$ or more conveniently $K(1 + sT_i)/sT_i$.

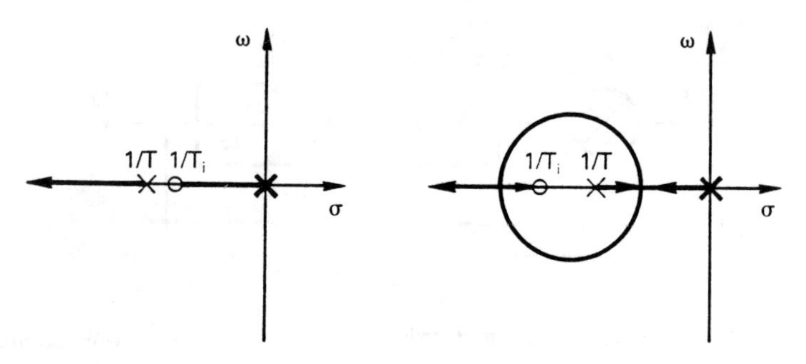

Fig. 4.47 The two different root loci for a first order lag plus a PI controller.

If our P + I controller is used with our first order lag, two shapes of root loci are possible dependent on where the controller zero is placed. These two loci are shown on Fig. 4.47. The first behaves as two first order lags in series. The dominant pole is nearest the origin, so the apparent response can never be better than a time constant of T_i.

The second behaves as a dominant second order response. With this second locus we can place our poles at any required position by varying K and T_{ni} as shown on Fig. 4.48. Let us assume we want to place the dominant poles at $s = -1 \pm j$, a slightly underdamped response.

As before, the closed loop transfer function is CG/(1 + CG), which is

$$\frac{\dfrac{K(1+sT_i)}{sT_i} \cdot \dfrac{0{\cdot}5}{(1+4s)}}{1+\dfrac{K(1+sT_i)}{sT_i} \cdot \dfrac{0{\cdot}5}{(1+4s)}}$$

This simplifies to

$$\frac{0{\cdot}5K(1+sT_i)}{4T_is^2+(1+0{\cdot}5K)T_is+0{\cdot}5K} \tag{4.14}$$

The poles occur where the denominator is zero, i.e. where

$$4s^2+s(1+0{\cdot}5K)+0{\cdot}5K/T_i=0$$

from which

$$s^2+s(1+0{\cdot}5K)/4+0{\cdot}125K/T_i=0 \tag{4.15}$$

We require the poles to be at $s = -1 \pm j$, which is the solution of the quadratic equation

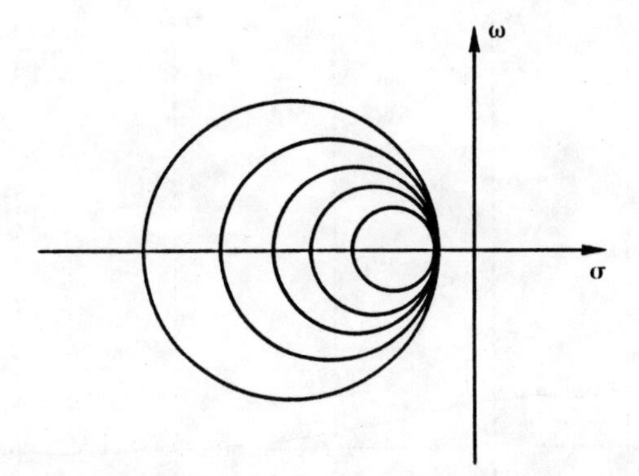

Fig. 4.48 The effect of varying T_i and K. The value of T_i determines which curve is used, and the value of K the position of the pole pairs allowing the pole position to be placed as required.

$$(s + 1 + j)(s + 1 - j) = 0$$

or

$$s^2 + 2s + 2 = 0 \tag{4.16}$$

The left-hand side of equations 4.15 and 4.16 must be identical, allowing us to equate the coefficients for s,

$$\frac{(1 + 0{\cdot}5K)}{4} = 2$$

giving K = 14 and

$$\frac{0{\cdot}125K}{T_i} = 2$$

giving $T_i = 0.875$ seconds

Substituting K and T_i back into equation 4.14 produces

$$\text{closed loop transfer function} = \frac{7(1 + 0{\cdot}875s)}{3{\cdot}5s^2 + 7s + 7}$$

Factorising of the quadratic denominator shows the poles to be at the required positions. The predicted step response, drawn with the program DIGSIM from Chapter 7, is shown on Fig. 4.49.

Higher order systems can be dealt with in a similar way. Essentially pole placement entails solving simultaneous equations. In the simple example above we had two unknowns, K and T_i, and two equations equating the coefficient of s and the constant term.

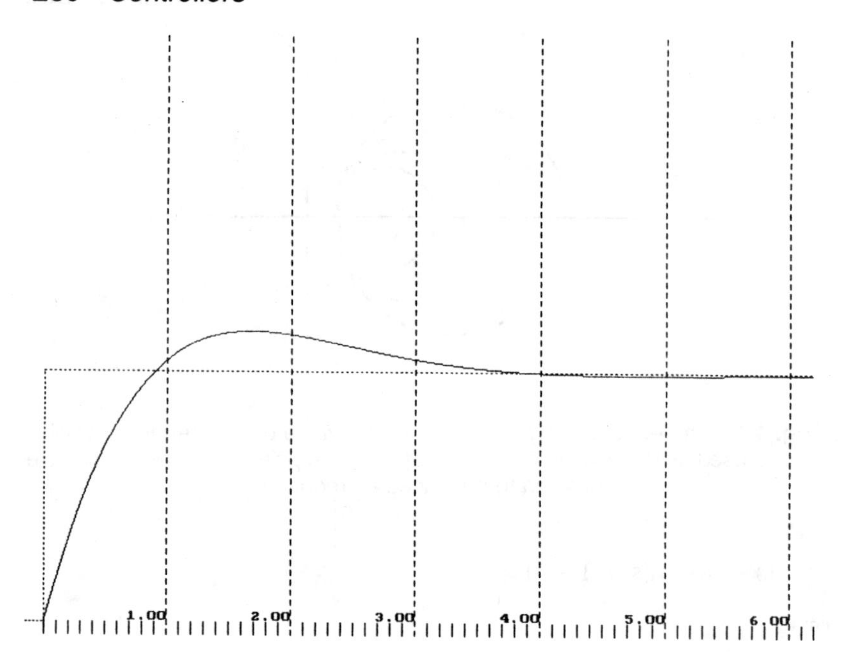

Fig. 4.49 Predicted step response drawn with computer program DIGSIM described in Chapter 7.

With three unknowns, K, T_i and T_d say, we would have three equations allowing the controller parameters to be found.

Pole placement implies that any required response can be obtained from a plant. A blast furnace with a 24 hour first order dominant response can, according to pole placement, be made to respond in less than a second. This is possible in theory, but in practice demands ridiculous control actions. It may be possible to bring a 100 kg vat of liquid from 10 to 100°C in 0·1 sec with the application of, say, 100 MW of heating power, but such a power will be beyond the ability of the heaters, or will result in instant failure from thermal shock. The results of a pole placement must therefore always be checked to see if the settings are within the capabilities of the plant.

The two methods above have used a continuous control algorithm. Self-tuning methods can be developed for discrete sampled data systems by considering pole positions in the Z plane rather than the s-plane. As shown earlier in Section 3.7.8 the poles on the Z plane must lie within the unit circle for stability, with the pole positions showing the response as Fig. 4.50.

A generalised block diagram for a sampled system is shown on Fig. 4.51. This will have the closed loop transfer function

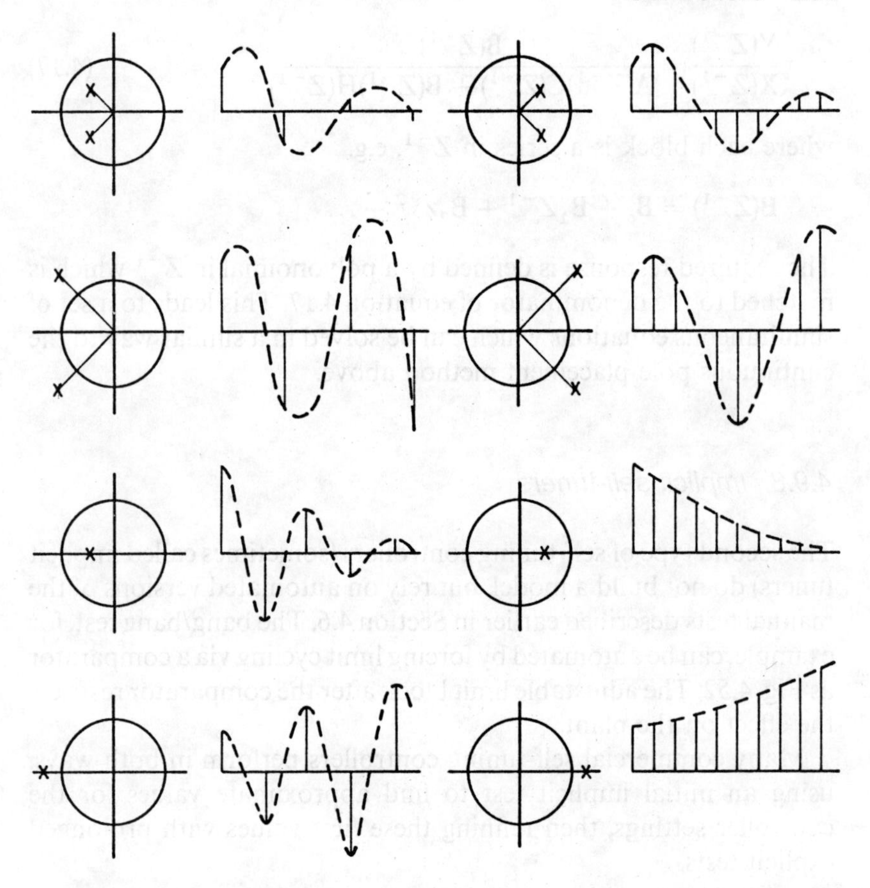

Fig. 4.50 Interpretation of response from pole positions on the Z plane.

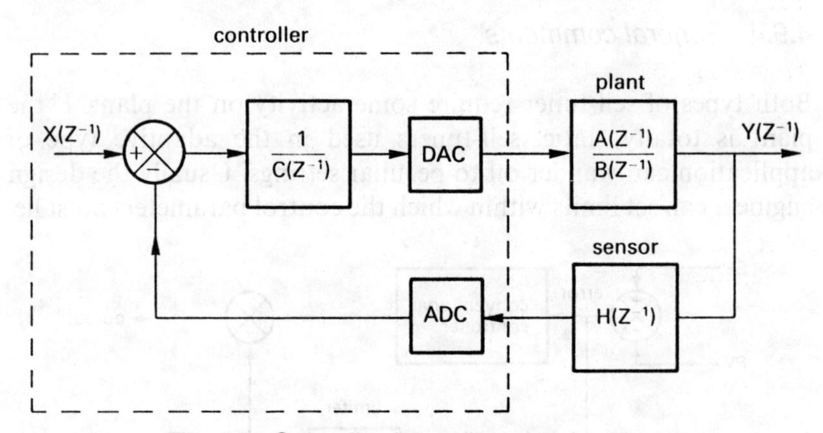

Fig. 4.51 Generalised sampling control system.

$$\frac{Y(Z^{-1})}{X(Z^{-1})} = \frac{B(Z^{-1})}{A(Z^{-1})C(Z^{-1}) + B(Z^{-1})H(Z^{-1})} \tag{4.17}$$

where each block is a series in Z^{-1}, e.g.

$$B(Z^{-1}) = B_o + B_1Z^{-1} + B_2z^{-2} + \ldots$$

The required response is defined by a polyonomial in Z^{-1} which is matched to the denominator of equation 4.17. This leads to a set of simultaneous equations which can be solved in a similar way to the continuous pole placement method above.

4.9.3. Implicit self-tuners

The second type of self-tuning controllers (sometimes called implicit tuners) do not build a model, but rely on automated versions of the manual tests described earlier in Section 4.6. The bang/bang test, for example, can be automated by forcing limit cycling via a comparator as Fig. 4.52. The adjustable limit block after the comparator restricts the effect on the plant.

Many commercial self-tuning controllers perform in both ways, using an initial implicit test to find approximate values for the controller settings, then refining these first values with prolonged explicit tests.

4.9.4. General comments

Both types of self-tuner require some activity on the plant. If the plant is totally static, self-tuners used in the adaptive type of application can wander off to peculiar settings. Usually the design engineer can set limits within which the control parameters must lie.

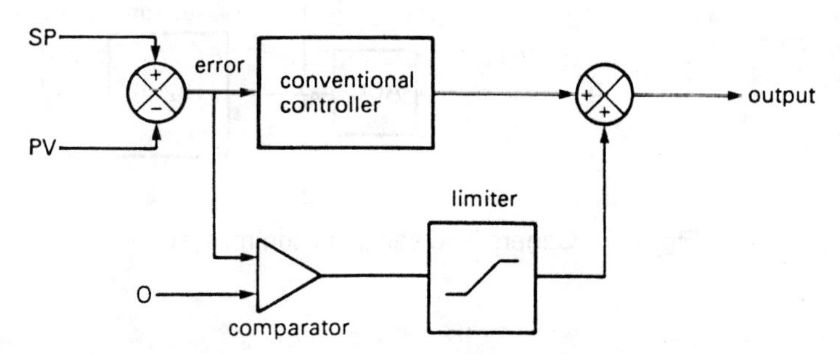

Fig. 4.52 Implicit self tuner based on the bang/bang test.

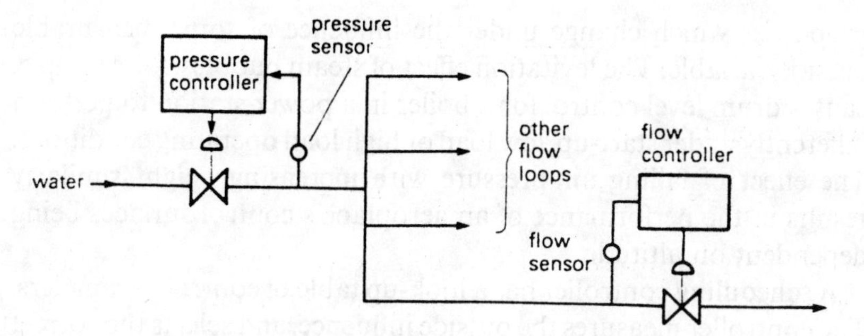

Fig. 4.53 A system which can cause problems for an autotuner. The autotuner on the flow loop attempts to tune the pressure loop.

Many self-tuners in adaptive mode are also designed to deliberately kick the plant at regular intervals and check the response.

Self-tuners can also become confused by changes in the process variable which are caused by outside influences and not the controller itself. Figure 4.53 shows a problem observed by the author. A high pressure water distribution was held at a nominal pressure by a pressure regulating valve. The regulated pressure header was then used for several flow control loops. Changes in flow caused slightly oscillatory pressure changes from the rather crude and not very well set up pressure regulator. The flow controllers in each leg were therefore constantly changing their outputs to maintain constant flow despite the upstream pressure changes. Self-tuners in the flow control loops confused cause and effect and ended up degaining the flow controllers to remove the observed cycling output. Process noise (common in level control systems) can similarly confuse many self-tuners. Users of self-tuners should always be aware of the controller's limited view of the world (obtained from OP, PV and nothing else) and treat the controller's recommended setting with an initial healthy scepticism.

Because all self-tuning controllers are microprocessor based, they are not limited to using the three term control algorithm (setting K, T_i and T_d), although most do. One notable exception is the ABB (previously ASEA) Novatune which uses a variation on Fig. 4.44 and equation 4.13. Plant tuning then becomes a matter of identifying constants a_1, a_2, b_1, b_2, etc.

4.9.5. Scheduling controllers

The scheduling controller is a close relative of self-tuning controllers being used in adaptive mode. There are many loops which have properties which change under the influence of some measurable outside variable. The levitation effect of steam bubbles, for example, causes drum level control for a boiler in a power station to perform differently under start-up, low load or high load operating conditions. The effect of falling air pressure with increasing height similarly results in the performance of an aeroplane's control surfaces being dependent on altitude.

A scheduling controller has a look-up table of control parameters. The controller measures the outside influence, and selects the correct controller parameters for the current operating conditions as shown on Fig. 4.54.

4.10. Closed loop control with PLCs

Programmable controllers, or PLCs, are becoming increasingly powerful, and most medium range PLCs now provide a three term PID function in their instruction set. Figure 4.55 shows a ratio temperature control program written for an Allen Bradley PLC5 processor. In the PLC5, 'N' variables are 16 bit integers, and 'F' variables are floating point numbers.

These three rungs are controlling the temperature in a furnace,

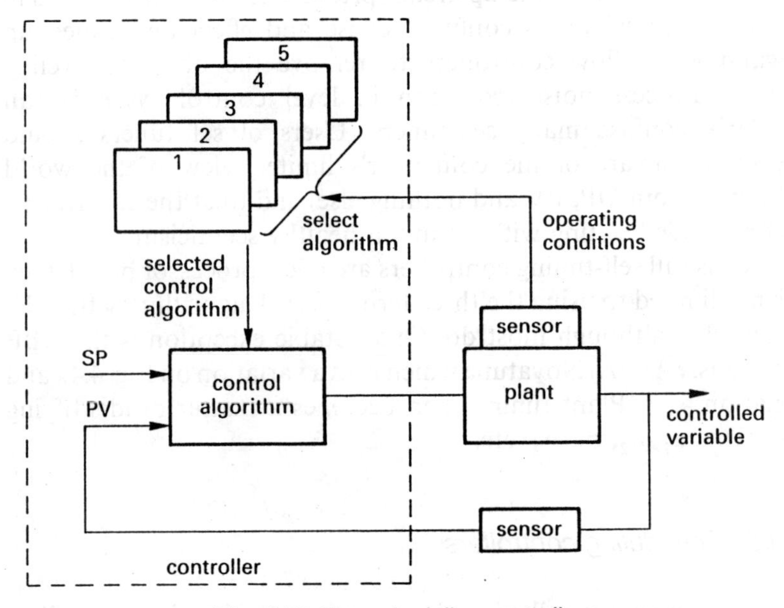

Fig. 4.54 The scheduling controller.

```
        PLC-5 LADDER LOGISTICS Report header (c) ICOM Inc. 1987-1993
                            PLC-5 Ladder Listing
File £2  Proj:PID2             Page:001                         10:07 05/12/95
-------------------------------------------------------------------------------
     |Zone Temperature PID instruction.                                       |
     |Adjusts Air Control Valve                                               |
     |                                                                        |
     |                                                                        |
     |New_AnIn                                  Temprature                    |
     |Data_So                                   PID_Contrl                    |
     |Fire_PID                                  (Air_Flow)                    |
     |    B3                                    +---PID-------------------+   |
   0+---] [-------------------------------------+PID                      +---+
     |    0                                     |Control:          N7:20|   |
     |                                          |Process Variable: N7:100|  |
     |                                          |Tieback:          N7:106|  |
     |                                          |Control Variable: N7:120|  |
     |                                          +-------------------------+   |
     |Multiply Air Flow in N7:105 by F8:6 to get gas setpoint.                |
     |Note that the ratio in F8:6 changes according to                       |
     |post recuperator air temperature and N7:52 is scaled                    |
     |by ten to give reasonable range for PID instruction.                    |
     |                                                                        |
     |                                                                        |
     |New_AnIn                                                                |
     |Data_So                                             Gas_Flow            |
     |Fire_PID                                            Setpoint            |
     |    B3                                              +---MUL-----------+ |
   1+---] [-----------------------------------------------+Mul              +-+
     |    0                                               |A:       N7:105| |
     |                                                    |           1432| |
     |                                                    |B:         F8:6| |
     |                                                    |          1.226| |
     |                                                    |Dest:     N7:52| |
     |                                                    |           1755| |
     |                                                    +-----------------+ |
     |Gas Flow controlled to follow air flow                                  |
     |                                                                        |
     |New_AnIn                                                                |
     |Data_So                                   Gas_Flow                      |
     |Fire_PID                                  PID_Cntrol                    |
     |    B3                                    +---PID-------------------+   |
   2+---] [-------------------------------------+PID                      +---+
     |    0                                     |Control:          N7:50|   |
     |                                          |Process Variable: N7:107|  |
     |                                          |Tieback:          N7:108|  |
     |                                          |Control Variable: N7:121|  |
     |                                          +-------------------------+   |
     |                                                                        |
     |                                                                        |
   3+------------------------------------------------------------------[END]--+

-------------------------------------------------------------------------------
        PLC-5 LADDER LOGISTICS Report header (c) ICOM Inc. 1987-1993
                            PLC-5 Ladder Listing
File £2  Proj:PID2             Page:001                         10:07 05/12/95
```

Fig. 4.55 PID control on an Allen Bradley PLC5.

with the temperature PID block controlling the air valve. The air flow is measured, multiplied by the required ratio and used as the set point for the gas PID block. Ratio control, and this particular form known as lead–lag control, is discussed further in section 5.5.

The control blocks in each PID instruction hold the data and working areas for the PID function; things like auto/man status and the sum for the integral action. The set point is written directly into the third word of the control block. The process variable is obviously the signal being controlled.

```
OFFLINE   Prj:PID2        Mode:Prog              Frc:No     RUNG 2:0/1
                        PID Configuration Data - Addr: N10:0
                 Equation: 0 (0:AB/1:ISA)                 Feed Forward: 0
                     Mode: 0 (0:Auto/1:Manual)       Max Scaled Input: 0
                    Error: 0 (0:SP-PV/1:PV-SP)       Min Scaled Input: 1500
         Output Limiting: 0 (0:NO/1:YES)                     Deadband: 5
         Set Output Mode: 0 (0:NO/1:YES)           Set Output Value %: 0
        Setpoint Scaling: 0 (0:YES/1:NO)             Upper CV Limit %: 99
        Derivative Input: 0 (0:PV/1:Error)           Lower CV Limit %: 0
         Deadband Status: 0                           Scaled PV Value: 0
    Upper CV Limit Alarm: 0                              Scaled Error: 0
    Lower CV Limit Alarm: 0                              Current CV %: 22
   Setpoint out of Range: 0
                PID Done: 0
             PID Enabled: 0                        (Scaled) Setpoint: 0
                                    Proportional Gain (Kp)     [.01]: 100
Last state: 0 (0:zero,1:hold)       Integral Gain (Ki)  [.001/secs]: 15
(only valid on 5/12:A/C,            Derivative Gain (Kd) [.01 secs]: 400
5/15:B/H, 5/25:A/DI)                Loop Update Time     [.01 secs]: 10

Sym:                  Des:
N10:5 =
DT File Name:              Des:
  F1       F2       F3       F4       F5         F6       F7       F8       F9       F10
                                      neWaddr    Des                                 Help
```

Fig. 4.56 Data in PID block.

A three term control algorithm can suffer from integral windup in saturation or manual operation (see Sections 4.4.4 and 4.2.5). The tieback variable is used to give the current value corresponding to the driven actuator output (possibly after auto/manual changeover) and is used to prevent windup and to give bumpless transfer. The controlled output is the signal from the PID algorithm, usually sent to an analog output card via auto/manual changeover logic.

Figure 4.56 shows the typical contents of a PID control block. Each function in this block (e.g. auto/manual mode) is accessible from the program allowing the set point to be varied for tracking applications, the mode changed and the controller parameters (gain, T_i and T_d) to be altered, perhaps to match altering plant conditions allowing the controller to work as a scheduling controller as described in Section 4.9.5.

The PID instruction does not operate continuously, but is fired at regular time intervals; note the values entered for T_i and T_d in Fig. 4.56 are related to the sample interval. In Fig. 4.55 the PID instructions are fired by new data coming from an analog input card.

The three rungs of Fig. 4.56 mask, to some extent, the work that must be done elsewhere in the program. Data from the outside world must be obtained with analog input cards, and the controller output(s) must be written to the actuators with analog output cards. The timing of these reads and writes must be regular and linked to the PID instructions.

Auto/manual changeover logic will also be required, linked into the PID instructions with the tieback variable and the auto/man status flag (which makes the integral term track the tieback in manual).

The operator will also require a link to the control, so pushbuttons, displays and alarms must be provided. Increasingly the operator link will be provided by a computer screen. The Visual Basic programs in Chapter 7 and Fig. 7.20 show one possible approach.

Chapter 5
Complex systems

5.1. Introduction

Most control loops described so far have been relatively simple single input, single output systems, i.e. a single set point and a single controlled variable. It has also been assumed that the systems are linear, i.e. if action A produces result X and action B produces result Y a combined action (A + B) produces result (X + Y). A characteristic of linear systems is that a sine wave input signal produces a sine wave output signal, albeit of possibly different amplitude and shifted in phase.

Systems described so far have behaved in what could be called a sensible manner; a change in input signal always produces a change of output signal in a related sense (turning up the heat always causes the temperature to rise, for example). This is a rather loose description of a mathematical representation called a minimum phase system. Commonly such a system has a transfer function with the highest power in the numerator lower than the highest power in the denominator. The expression

$$\frac{as^3 + bs^2 + cs + d}{es^4 + fs^3 + gs^2 + hs + j}$$

usually describes a minimum phase system.

Some processes are non-minimum phase. The classic examples are shown on Fig. 5.1a and b, both from the power generation industry. The first example is level control of feedwater for a steam turbine. The boiler drum contains a mixture of steam and water, with the water containing a myriad of steam bubbles. When more (cold) water is added, the steam bubbles condense and the water level initially falls as Fig. 5.1b despite the introduction of water. The second example occurs in hydroelectric stations, where turbine speed is regulated by control of water flow from a reservoir. If the

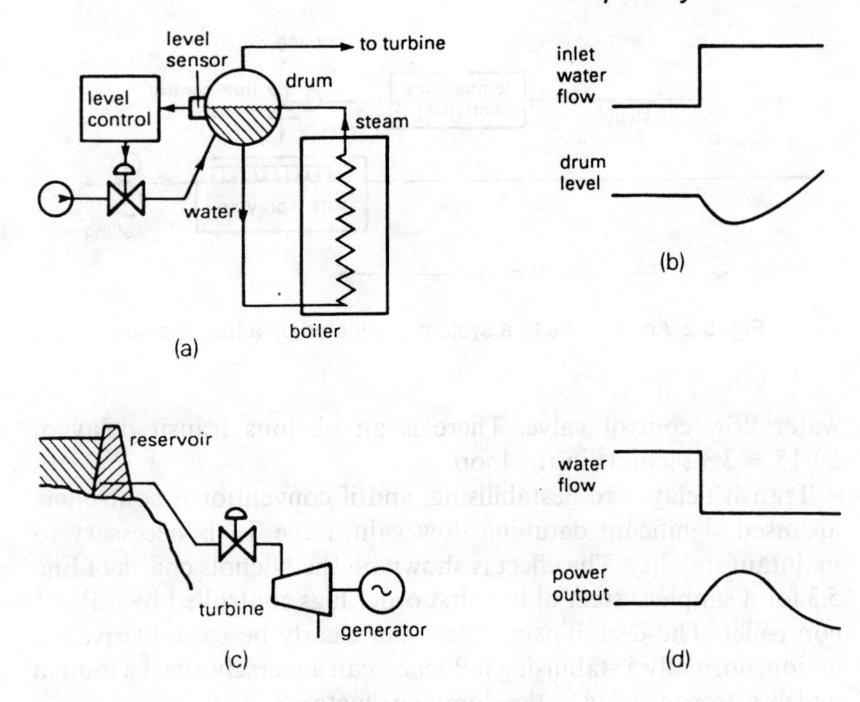

Fig. 5.1 Examples of non-minimum phase systems, (a) Drum water system; (b) Effect of change in inlet flow; (c) Hydroelectric power station; (d) Effect of change in water flow.

flow is reduced, the inertia of the water in the pipes maintains the flow as the valve closes, leading to an increase in flow velocity from the nozzle and a temporary increase in power. It is thought that this effect, occurring at the Niagara Falls hydroelectric station, was the initial trigger for the power failure which affected much of north-east America in 1965.

This chapter examines systems which contain more than one loop, systems which are difficult to control and systems which are non-linear.

5.2. Systems with transit delays

Transit delays are a function of speed, time and distance. A typical example from the steel industry is the tempering process of Fig. 5.2 where red hot rolled steel travelling at the order of 15 m/sec is quenched by passing beneath high pressure water sprays. The recovery temperature, some 50 metres downstream, is the controlled variable which is measured by a pyrometer and used to adjust the

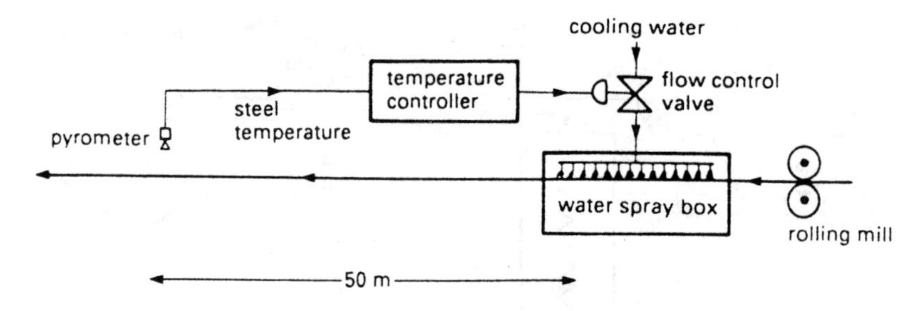

Fig. 5.2 An example of a system dominated by a transit delay.

water flow control valve. There is an obvious transit delay of 50/15 = 3·3 seconds in the loop.

Transit delays are destabilising, and if conventional controllers are used significant detuning (low gain, large T_i) is necessary to maintain stability. The effect is shown on the Nichols charts of Fig. 5.3 for a simple system of two first order lags controlled by a P + I controller. The destabilising effect can clearly be seen. Derivative action, normally a stabilising influence, can adversely affect a loop in which a transit delay is the dominant factor.

The effects of a transit delay can be reduced by the arrangement of Fig. 5.4 called a Smith predictor. The plant is considered to be an ideal plant followed by a transit delay (this may not be true, but the position of the transit delay, before or after the plant, makes no difference to the plant behaviour). The plant and its associated delay are modelled as accurately as possible in the controller.

The controller output OP is applied to the plant and to the internal controller model. Signal A should thus be the same as the notional (and unmeasurable) plant signal X, and the signal B should be the same as the measurable controlled variable signal Y.

The P + I + D controller, however, is primarily controlling the model, not the plant, via summing junction 1. There are no delays in this loop, so the controller can be tuned for tight operation. With the model being the only loop, however, the plant is being operated in open loop control, and compensation will not be applied for model inaccuracies or outside disturbances.

Signals Y and B are therefore compared by a subtractor to give an error signal which encompasses errors from both disturbances and the model. These are added to the signal A from the plant model to give the feedback signal to the P + I + D control block.

Discrepancies between the plant model and the real plant will be compensated for in the outer loop, so exact modelling is not

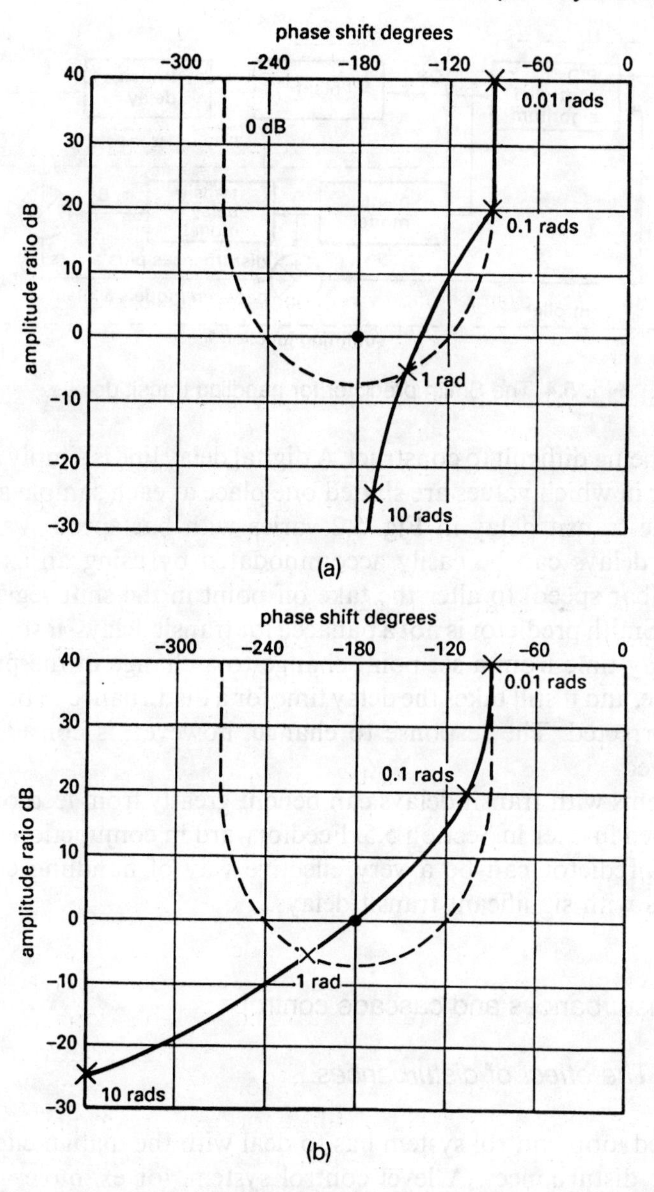

Fig. 5.3 The effect of a transit delay shown on a Nichols chart. The system comprises a P + I controller (k = 5, T_i = 5 s) and two lags of time constants 4 s and 2 s, (a) System without transit delay; (b) System with 1 s transit delay.

necessary. The poorer the model, however, the less tight the control that can be applied in the P + I + D block as the errors have to be compensated via the plant transit delay.

Smith predictors are usually implemented digitally, analog transit

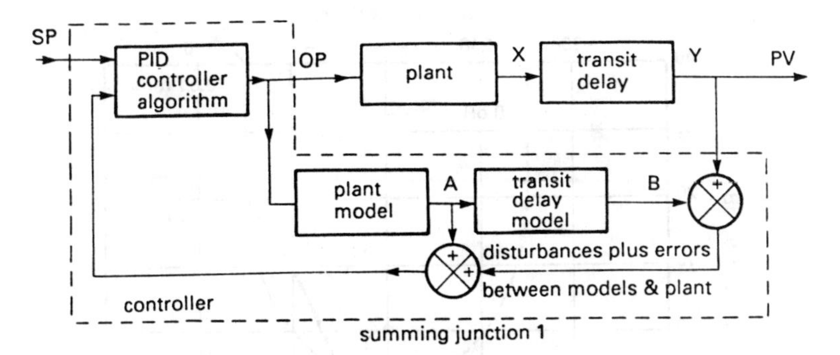

Fig. 5.4 The Smith predictor for handling transit delays.

delays being difficult to construct. A digital delay line is simply a shift register in which values are shifted one place at each sample as Fig. 5.5. The transit delay in Fig. 5.2 varies with bar speed. Variable transit delays can be easily accommodated by using an external signal (bar speed) to alter the take off point in the shift register.

The Smith predictor is not a panacea for transit delays, it still takes the delay time from a set point change to a change in the process variable, and it still takes the delay time for a disturbance to be noted and corrected. The response to change, however, is considerably improved.

Systems with transit delays can benefit greatly from feedforward described further in Section 5.5. Feedforward in conjunction with a Smith predictor can be a very effective way of handling control systems with significant transit delays.

5.3. Disturbances and cascade control

5.3.1. The effect of disturbances

A closed loop control system has to deal with the malign effects of outside disturbances. A level control system, for example, has to handle varying throughput, or a gas fired furnace may have to cope with changes in gas supply pressure. Although disturbances can enter a plant at any point, it is usual to consider disturbances at two points; supply disturbances at the input to the plant and load/demand disturbances at the point of measurement.

The closed loop block diagram can be modified to include disturbances as shown in Fig. 5.6b. A similar block diagram could be drawn for load disturbances or disturbances entering at any point by

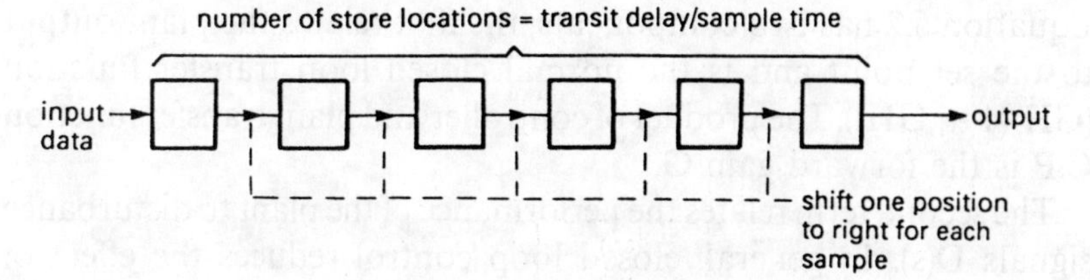

Fig. 5.5 Digital simulation of a transit delay.

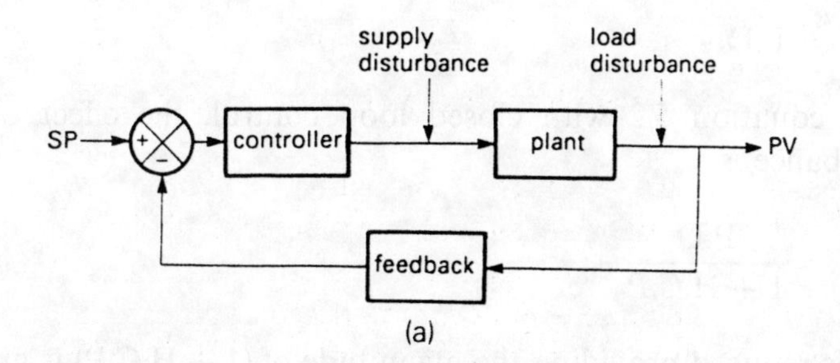

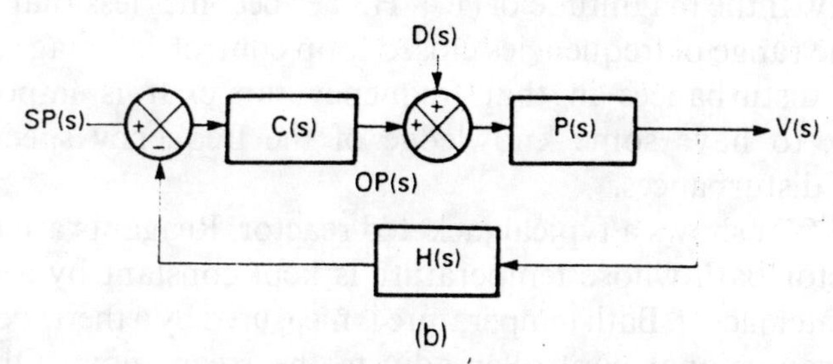

Fig. 5.6 The effect of supply and load disturbances, (a) Points of entry of disturbances; (b) Block diagram of system disturbances.

subdividing the plant block. By normal closed loop theory we have (omitting s for simplicity)

$$V = P(OP + D) \tag{5.1}$$

$$\text{and } OP = C(SP - H.V)$$

Substituting into equation 5.1 gives

$$V = P(C(SP - H.V) + D)$$

$$\text{or } V(1 + HCP) = C.P.SP + P.D$$

$$\text{or } V = \frac{C.P.SP}{1 + H.C.P} + \frac{P.D.}{1 + H.C.D} \tag{5.2}$$

Equation 5.2 has two components; the first relates the plant output to the set point and is the normal closed loop transfer function (GH/(1 + GH)). The product of controller and plant transfer function C.P is the forward gain G.

The second term relates the performance of the plant to disturbance signals D(s). In general, closed loop control reduces the effect of disturbances. If the plant was run open loop, the effect of the disturbances on the output would be simply

$$V = P.D$$

From equation 5.2 with closed loop control, the effect of the disturbance is

$$V = \frac{P.D}{1 + H.C.P}$$

i.e. it is reduced providing the magnitude of (1 + H.C.P) is greater than unity. If the magnitude of (1 + H.C.P) becomes less than unity over some range of frequencies closed loop control will magnify the effect of disturbances in that frequency range. It is important, therefore, to have some knowledge of the frequency spectra of expected disturbances.

Figure 5.7a shows a typical jacketed reactor. Reagants are mixed in a reactor bath whose temperature is kept constant by a steam heated water jacket. Bath temperature is measured by a thermocouple and a proportional controller adjusts the steam flow. Obvious disturbances are the steam temperature (a supply disturbance) and the cold reactants (a load/demand disturbance which will vary with throughput).

The system can be modelled by Fig. 5.7b. The jacket has a fairly long time constant of one minute, the stirred bath a shorter time constant of 10 seconds. The thermocouple and its protective sheath have a time constant of 4 seconds. The proportional control gain is set for 8.

The open loop response to a disturbance D(s) is

$$P.D(s) = \frac{0{\cdot}24}{(1 + 60s)(1 + 10s)} \cdot D(s) \tag{5.3}$$

To find the low frequency DC response, we let terms in s go to zero (s^n represents d^n/dt^n remember) giving 0·24D where D is the size of the disturbance.

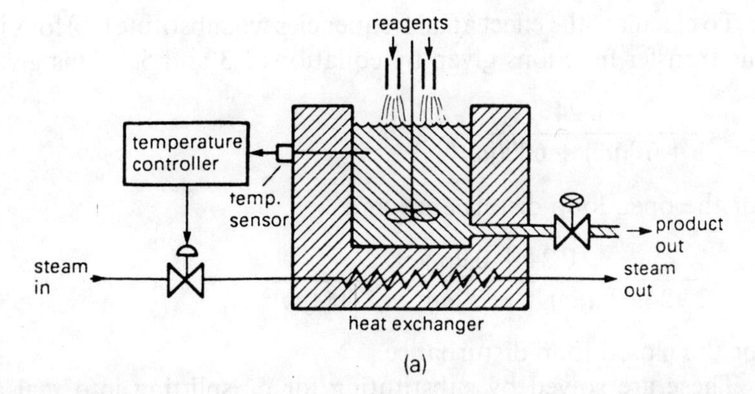

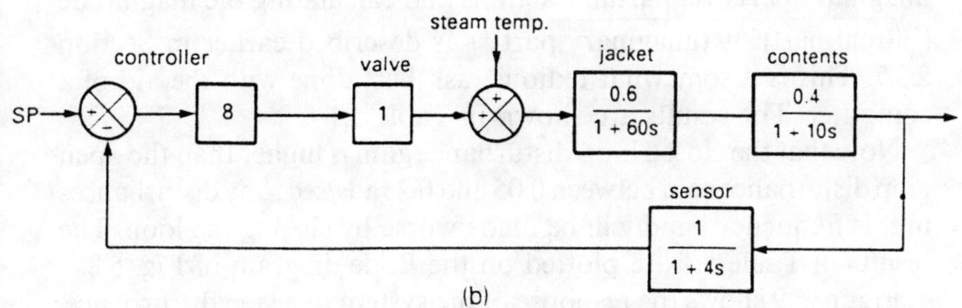

Fig. 5.7 System with supply disturbances, (a) A jacketed reactor; (b) Block diagram of jacketed reactor.

The closed loop response is

$$\frac{P}{1 + H.C.P} \cdot D(s)$$

$$= \frac{0{\cdot}24/(1 + 60s)(1 + 10s)}{1 + 1{\cdot}92/(1 + 60s)(1 + 10s)(1 + 4s)} \cdot D(s)$$

$$= \frac{0{\cdot}24(1 + 4s)}{(1 + 60s)(1 + 10s)(1 + 4s) + 1{\cdot}92} \cdot D(s) \qquad (5.4)$$

As before, we can find the low frequency gain by letting s go to zero giving

$$\frac{0{\cdot}24D}{1 + 1{\cdot}92} = \frac{0{\cdot}24D}{2{\cdot}92}$$

$$= 0{\cdot}08D$$

Closed loop control has reduced the effect of the disturbance by a factor of 3 at low frequencies.

To calculate the effect at all frequencies we substitute (jω) for s into the transfer functions given by equations 5.3 and 5.4. This gives

$$\frac{0{\cdot}24}{1 + 70(j\omega) + 600(j\omega)^2}$$

for the open loop disturbance and

$$\frac{0{\cdot}24 + 0{\cdot}96(j\omega)}{2{\cdot}92 + 74(j\omega) + 880(j\omega)^2 + 2400(j\omega)^3}$$

for the closed loop disturbance.

These are solved by substituting for ω, splitting into real and imaginary parts via partial fractions and calculating the magnitude ($\sqrt{(\text{real part})^2 + (\text{imaginary part})^2}$) as described earlier in Section 2.3.7. This is a somewhat tedious task best done with the aid of a computer. The results are shown in Table 5.1

Note that the closed loop disturbance gain is higher than the open loop disturbance gain between 0·05 and 0·2 rads/sec. Any disturbances in this frequency band will be made worse by closing the loop. The results of Table 5.1 are plotted on the Bode diagram of Fig. 5.8.

Figure 5.9 shows the response of the system to a step disturbance. It can be seen that the closed loop system settles more quickly to a lower error with a barely detectable overshoot. The error is not removed totally because the controller has no integral action. The

Table 5.1

Omega (rad/sec)	*Open loop disturbance gain dB*	*Closed loop disturbance gain*	
		K = 8 dB	*K = 50 dB*
0·02	−16	−22	−34
0·03	−19	−22	−34
0·04	−21	−22	−34
0·05	−23	−23	−33
0·06	−25	−24	−33
0·07	−27	−25	−32
0·08	−28	−27	−31
0·09	−30	−28	−30
0·1	−31	−30	−28
0·2	−41	−41	−37
0·3	−48	−47	−46
0·4	−52	−52	−52
0·5	−56	−56	−56
0·6	−59	−59	−59

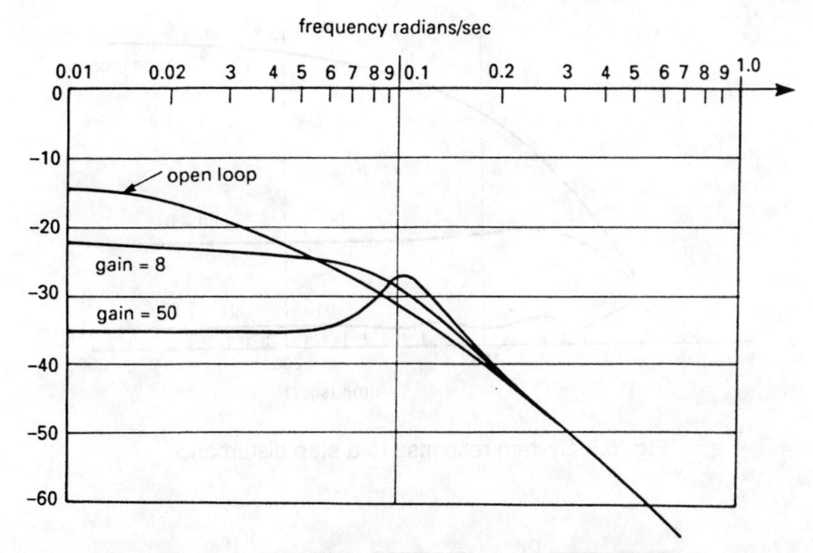

Fig. 5.8 Bode diagram of system response to a disturbance. Note that closed loop response is worse than open loop response for some frequencies.

disturbance effect can also be reduced by increasing the controller gain, but this, of course, has stability considerations and can accentuate the frequency band where the effect of disturbances are made worse by closed loop control.

5.3.2. Cascade control

Closed loop control gives increased performance over open loop control, so it would seem logical to expect benefits from adding an inner control loop around plant items that are degrading overall performance. Figure 5.10 shows a typical example, here the output of the outer loop controller becomes the set point for the inner controller. Any problems in the inner loop (disturbances, non-linearities, phase lag, etc.) will be handled by the inner controller, thereby improving the overall performance of the outer loop. This arrangement is known as cascade control. The simplest cascade controller is a valve positioner.

To apply cascade control, there must obviously be some intermediate variable that can be measured (PV_i on Fig. 5.10) and some actuation point that can be used to control it.

A more detailed block diagram is shown on Fig. 5.11. The inner loop has a transfer function, G(s) given by

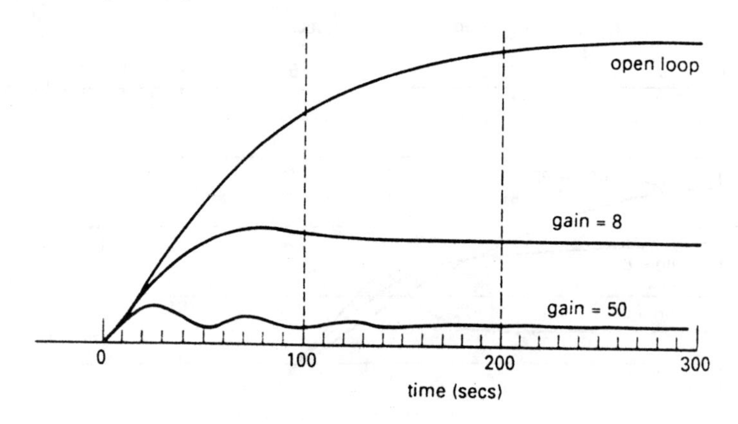

Fig. 5.9 System response to a step disturbance.

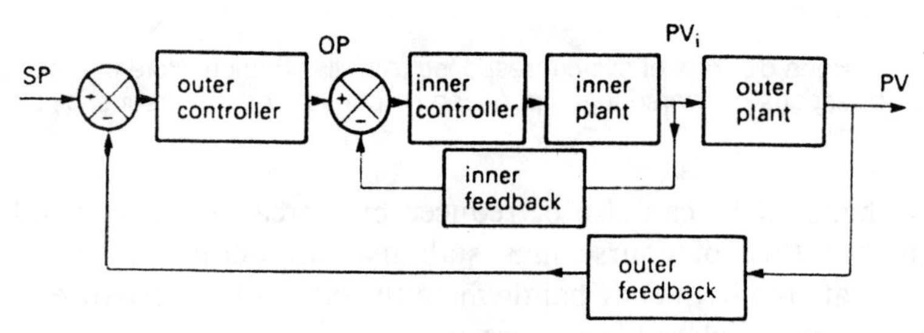

Fig. 5.10 A system with cascade control.

$$G = \frac{V_i}{OP} = \frac{C_2P_2}{1 + C_2P_2H_2}$$

The transfer function of the outer loop is

$$\frac{V_o}{SP} = \frac{C_1GP_1}{1 + C_1GP_1H_1}$$

Substituting for G gives

$$V_o = \frac{\dfrac{C_1C_2P_2P_1}{(1 + C_2P_2H_2)}}{1 + \dfrac{C_1C_2P_2P_1H_1}{(1 + C_2P_2H_2)}}$$

$$= \frac{C_1C_2P_1P_2}{1 + C_2P_2H_2 + C_1C_2P_1P_2H_1} \tag{5.5}$$

Applying a similar analysis to the disturbance gain gives

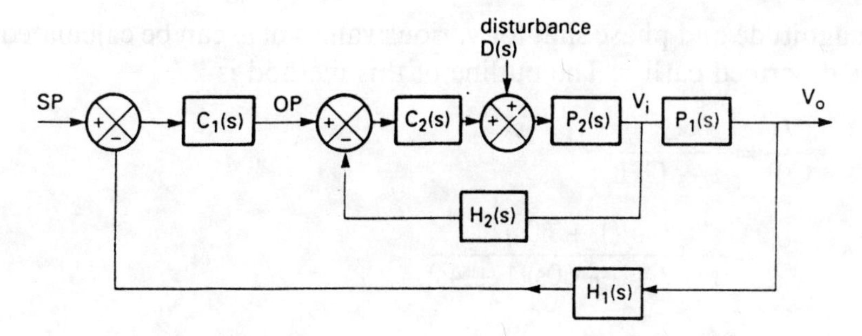

Fig. 5.11 Block diagram of cascade system affected by a disturbance.

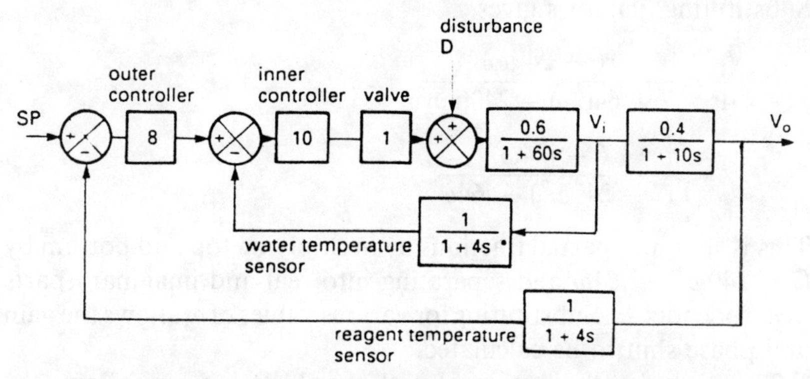

Fig. 5.12 Cascade control applied to jacketed reactor.

$$\frac{V_o}{D} = \frac{P_2P_1}{1 + C_2P_2H_2 + C_1C_2P_1P_2H_2} \tag{5.6}$$

Evaluating equations 5.5 and 5.6 is difficult and tedious (a computer program helps tremendously).

Figure 5.12 shows cascade control added to the jacketed reactor examined in the previous section. The inner loop is placed around the water jacket and aims to reduce the 60 second time constant which dominates the process.

Analysing the inner loop is complicated by the time constant of the thermocouple in the feedback loop. This precludes the direct use of Nichols charts and similar techniques which can assume unity feedback. The problem can be handled in two ways. Both use frequency response methods.

The first technique is to build up the transfer function G/(1 + GH) and substitute ($j\omega$) for s. The transfer function can then be split into real and imaginary parts via partial fractions from which the magnitude and phase shift for various values of ω can be calculated as described earlier. The outline of this method is

$$\frac{V_i}{OP} = \frac{G}{1 + GH}$$

$$= \frac{6/(1 + 60s)}{1 + 6/(1 + 60s)(1 + 4s)}$$

$$= \frac{6 + 24s}{7 + 64s + 240s^2}$$

Substituting (jω) for s gives

$$\frac{V_i}{OP} = \frac{6 + 24(j\omega)}{7 + 64(j\omega) + 240(j\omega)^2}$$

$$= \frac{6 + j24\omega}{(7 - 240\omega^2) + j64\omega}$$

This is split into partial fractions by multiplying top and bottom by $(7 - 240\omega^2) - j64\omega$ and separating into real and imaginary parts (see Appendix A). Substituting for various values of ω allows the gain and phase shift to be calculated.

The second technique involves less calculation and allows the Nichols chart to be used. The transfer function is

$$\frac{V_i}{OP} = \frac{G}{1 + GH}$$

where G represents the forward elements and H the feedback elements. This can be rearranged as

$$\frac{GH}{1 + GH} \cdot \frac{1}{H}$$

with the feedback elements combined with the forward elements and an extra block 1/H added outside the loop as Fig. 5.13.

The closed loop frequency response can now be formed in a few stages. First the frequency response of GH is plotted onto a Nichols chart and the closed loop frequency response plotted onto a Bode diagram. Next the frequency response for (1/H) is added to the closed loop Bode diagram to give the complete closed loop response.

Whichever method is used, the result is the closed loop frequency response on a Bode diagram which can now be compared with standard blocks. The Bode diagram for the inner loop of Fig. 5.12 is

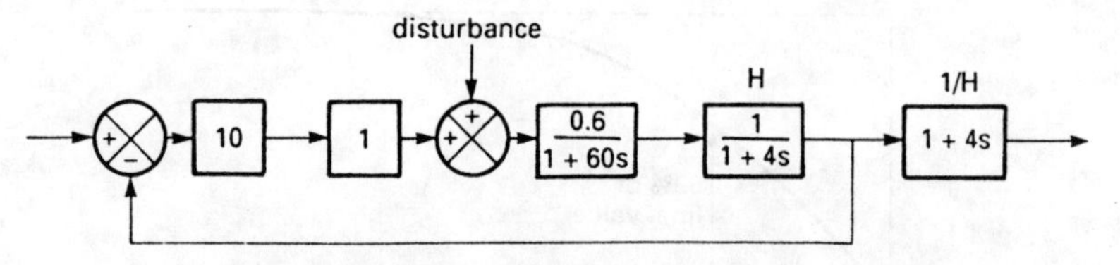

Fig. 5.13 Inner loop rearranged to have unity feedback.

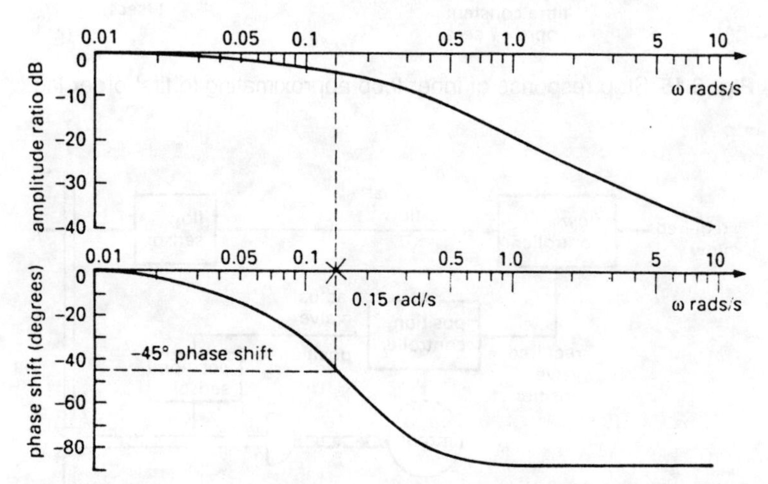

Fig. 5.14 Closed loop response of inner loop plotted on Bode diagram approximates to first order lag.

drawn on Fig. 5.14. It can be seen that it has a strong resemblance to a first order lag (phase shift goes from 0 to −90°, gain is roughly −3 dB when phase shift is 45°, high frequency asymptote 6 dB/octave). The corner frequency is about 0·15 rads, giving an apparent time constant of 1/0·15 = 6·7 sec. Applying cascade control around the inner loop has reduced the time constant from 60 sec to just under 7 sec. This is confirmed by the response to a step change in set point shown in Fig. 5.15. This reduction in response time leads to a similar improvement in the performance of the outer loop.

Calculation of the effect of disturbances can be done in a similar manner, and it is found that these are reduced by a similar amount.

5.3.3. General comments

Cascade control, therefore, brings several benefits. The secondary controller will deal with disturbances before they can affect the outer loop. Phase shift within the inner loop is reduced, leading to

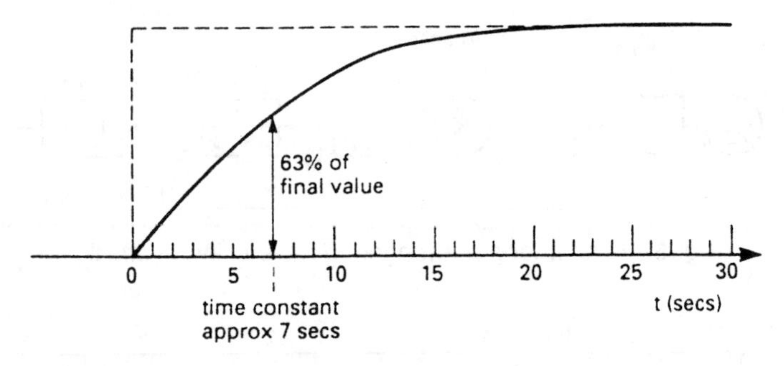

Fig. 5.15 Step response of inner loop approximating to first order lag.

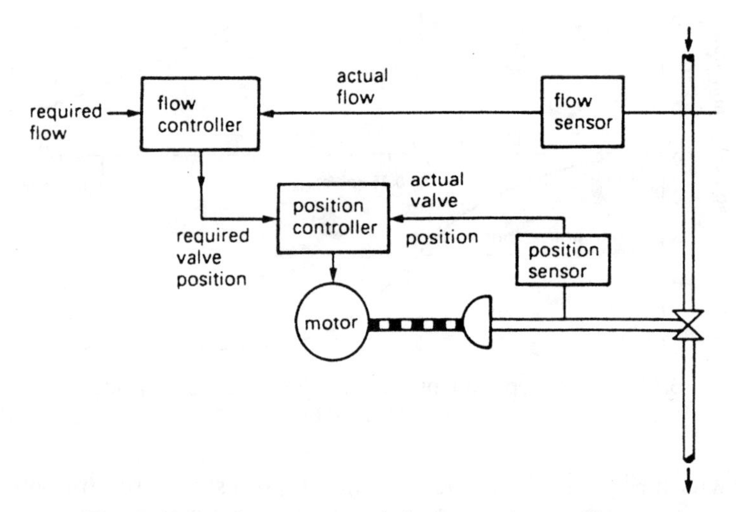

Fig. 5.16 A common cascade loop, a valve positioner.

increased stability and speed of response in the outer loop. Devices with inherent integral action (such as a fail in current position motorised valve) introduce the inherent $-90°$ integrator phase lag. This can be removed by adding a valve positioner in cascade as shown for the flow control loop of Fig. 5.16. Cascade control will also reduce the effect of non-linearisation (e.g. non-linear gain, backlash, etc.) in the inner loop.

There are a few precautions that need to be taken, however, when applying cascade control. The analysis so far ignores the fact that components saturate. The analysis of the cascade control for the jacketed reactor of Fig. 5.12 assumes an infinite supply of steam, which is obviously impossible. Problems can arise when the inner loop saturates. This can be overcome by limiting the demands that the outer loop controller can place on the inner loop (i.e. ensuring the outer loop controller saturates first) or by providing a signal from the inner to the outer controller which inhibits the outer integral term when the inner loop is saturated.

The application of cascade control requires an intermediate variable and control action point, and should include, if possible, the plant item with the shortest time constant. In general, high gain proportional only control will suffice for the inner loop, any offset is of little concern as it will be removed by the outer controller. For stability, the inner loop should obviously be faster than the outer loop.

Tuning a system with cascade control requires a methodical approach. The inner loop must be tuned first with the outer loop steady in manual control. Once the inner loop is tuned satisfactorily, the outer loop can be tuned as normal. A cascade system, once tuned, should be observed to ensure that the inner loop does not saturate, which can lead to instability or excessive overshoot on the outer loop. If saturation is observed limits must be placed on the output of the outer loop controller, or a signal provided to prevent integral windup as described earlier.

5.4. Feedforward control

Cascade control can reduce the effect of disturbances occurring early in the forward loop, but generally cannot deal with load/demand disturbances which occur close to, or affect directly, the process variable as there is no intermediate variable or accessible control point.

Disturbances directly affecting the process variable must produce an error before the controller can react. Inevitably, therefore, the output signal will suffer, with the speed of recovery being determined by the loop response. Plants which are difficult to control tend to have low gains and long integral times for stability and hence have a slow response. Such plants are prone to error from disturbances.

It was shown earlier that a closed loop system can be considered to behave as a second order system, with a natural frequency ω_n and a damping factor. At frequencies above ω_n the closed loop gain falls off rapidly (at 20 dB/decade). Disturbances occurring at a frequency much above $2\omega_n$ will be uncorrected. If the closed loop damping factor is less than unity (representing an underdamped system), the effect of disturbances with frequency components around ω_n can be magnified.

A system being affected by a disturbance is shown in Fig. 5.17a. Cascade control cannot be applied because there is no intermediate

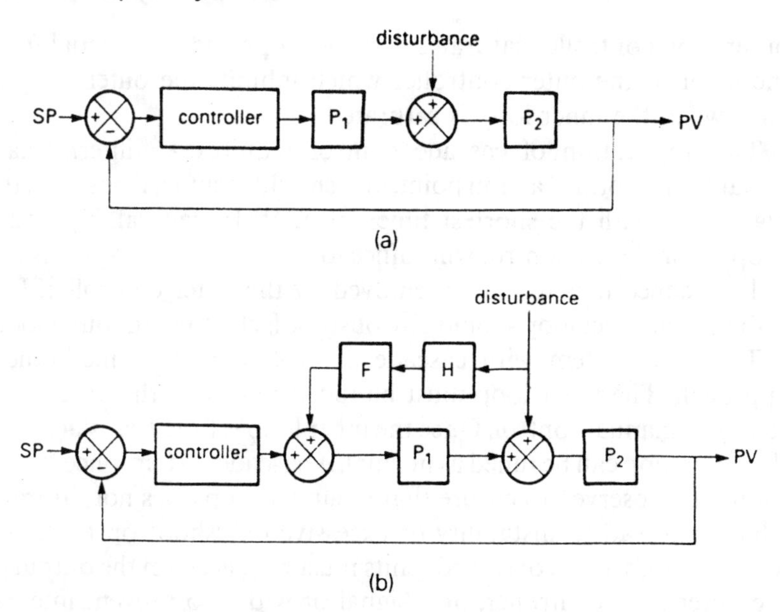

Fig. 5.17 Disturbance effect reduced by feedforward, (a) System with disturbance to which cascade control cannot be applied; (b) Correcting signal derived by measuring the disturbance.

variable between the point of entry and the process variable. If the disturbance can be measured, and its effect known (even approximately), a correcting signal can be added to the controller output signal to compensate for the disturbance. This is known as feedforward control.

This correcting signal, arriving by blocks H, F, and P_1, must exactly cancel the original disturbance, both in the steady state and dynamically under changing conditions. The transfer functions of the transducer H and plant P_1 are fixed, with F a compensator block designed to match H and P_1.

Considering only the effects of the disturbance, for cancellation

$$FHP_1d = -d$$

or

$$F = -\frac{1}{HP_1}$$

If the transfer function of the transducer is unity, i.e. it has negligible lag and unity gain (in terms of the internal control signals) then

$$F = -\frac{1}{P_1}$$

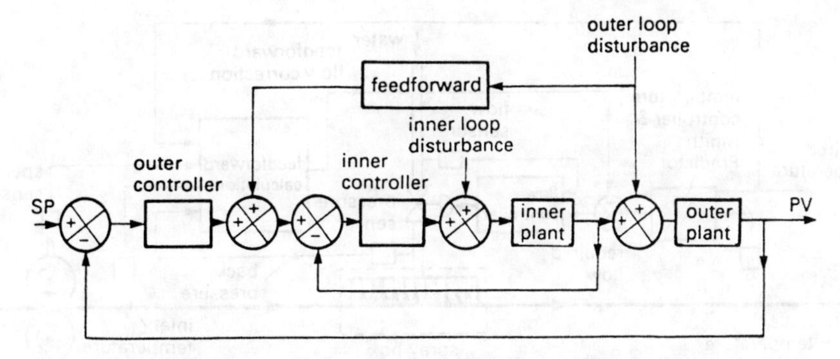

Fig. 5.18 A common way of reducing the effects of disturbances. The inner loop uses cascade control and the outer loop uses feedforward.

If the plant transfer function behaves as a lag $1/(1 + sT)$ with time constant T sec, the required transfer function will be a lead circuit $(1 + sT)$. This is a P + D controller with derivative action time T.

Cascade control can usually deal with supply disturbances, and feedforward with load/demand disturbances. These neatly complement each other so it is very common to find a system similar to Fig. 5.18 where the feedforward component modifies directly the input to the cascade loop.

If the cascade loop is dominated by a P + I controller, the cascade loop gain G is

$$G = \frac{AK(1 + sT_i)}{sT_i}$$

Giving a closed loop transfer function $G/(1 + G)$ or

$$\frac{AK(1 + sT_i)/sT_i}{1 + AK(1 + sT_i)/sT_i}$$

$$= \frac{AK(1 + sT_i)}{AK(1 + sT_i(1 + 1/AK))}$$

$$= \frac{1 + sT_i}{1 + sT_b}$$

where $T_b = T_i(1 + AK)/AK$.

The required feedforward compensation is thus

$$F(s) = \frac{1 + sT_b}{1 + sT_i}$$

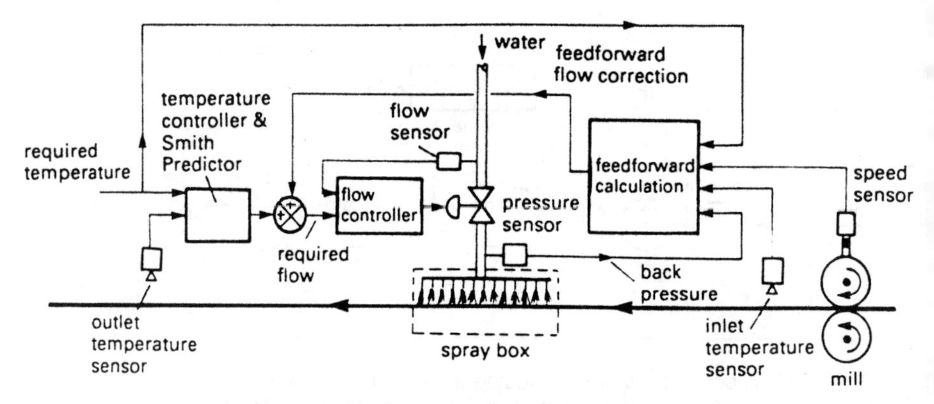

Fig. 5.19 Feedforward applied to tempering process.

where $T_b > T_1$. This is the transfer function of a lead compensator (see Section 4.3.2).

The feedforward compensation does not have to match exactly the plant characteristics; even a rough model will give a significant improvement (although a perfect model will give perfect control). In most cases, one of the compensators of Section 4.3.2 will suffice. It should be noted that it is not possible to compensate for a transit delay as this requires extra sensory perception.

A process can be affected by many disturbances. The tempering process given earlier in Fig. 5.2 is affected by manifold back pressure (a function of spray nozzle blockage and wear), bar speed and ingoing bar temperature. These can all become part of a feedforward system as Fig. 5.19, with separate compensators whose outputs are summed to correct the temperature controller output to the cascade flow controller. The temperature controller is set up as a relatively slow acting trim control subject to a transit delay so the set point is also used as input to the feedforward compensators to give an immediate (and hopefully correct) change in flow for a change in set point temperature.

5.5. Ratio control

It is a common requirement for two flows to be kept in precise ratio to each other; gas/oil and air in combustion control, or reagants being fed to a chemical reactor are typical examples. In ratio control, one flow is declared to be the master. This flow is set to meet higher level requirements such as plant throughput or furnace temperature.

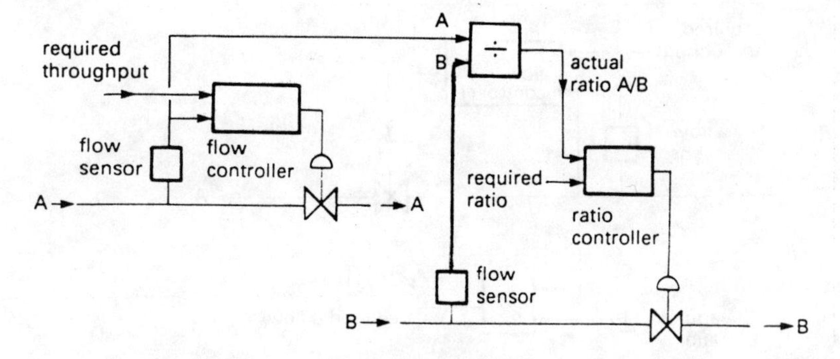

Fig. 5.20 An intuitive, but incorrect, method of ratio control. The loop gain varies with throughput.

The second flow is a slave and is manipulated to maintain the set flow ratio.

The controlled variable here is ratio, not flow, so an intuitive solution might look similar to Fig. 5.20 where the actual ratio A/B is calculated by a divider module (or routine in a computer) and used as the process variable for a controller which manipulated the slave control valve.

This scheme has a hidden problem. The slave loop includes the divider module and hence the term A. The loop gain varies directly with the flow A, leading to a sluggish response at low flows and possible instability at high flow. If the inverse ratio B/A is used as the controller variable the saturation becomes worse as the term 1/A now appears in the slave loop giving a loop gain which varies inversely with A, becoming very high at low flows. Any system based on Fig. 5.20 would be impossible to tune for anything other than constant flow rates.

Ratio control systems are generally based on Fig. 5.21. The master flow is multiplied by the ratio to produce the set point for the slave flow controller. The slave flow thus follows the master flow. Note that in the event of failure in the master loop (a jammed valve, for example) the slave controller will still maintain the correct ratio.

The slave flow will tend to lag behind the master flow. On a gas/air burner, the air flow could be master and the gas loop the slave. Such a system would run lean on increasing heat and run rich on decreasing heat. To some extent this can be overcome by making the master loop slower acting than the slave loop, possibly by tuning.

In a ratio system, a choice has to be made for master and slave loops. The first consideration is usually safety. In a gas/air burner, for example, air master/gas slave (called gas follow air) is usually

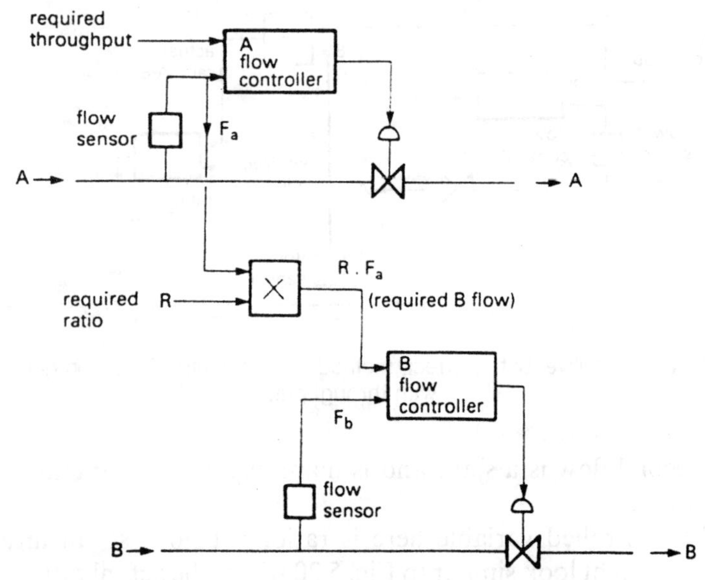

Fig. 5.21 Acceptable ratio control with stable loop gains.

chosen as most failures in the air loop cause the gas to shut down. If there are no safety considerations, the slowest loop should be the master and the fastest loop the slave to overcome the lag described above. Since 'fuel' (in both combustion and chemical terms) is usually the smallest flow in a ratio system and consequently has smaller valves/actuators, the safety and speed requirements are often the same.

The ratio block is a simple multiplier. If the ratio is simply set by an operator this can be a simple potentiometer acting as a voltage divider (for ratios less the unity) or an amplifier with variable gain (for ratios greater than unity). In digital control systems, of course, it is a simple multiply instruction.

If the ratio is to be changed remotely (a trim control from an automatic sampler on a chemical blending system, for example) a single quadrant analog multiplier is required.

Ratio blocks are generally easier to deal with in digital systems working in real engineering units. True ratios (an air/gas ratio of 8/1, for example) can then be used. In analog systems the range of the flow meters needs to be considered. Suppose we have a master flow with FSD of 12 000 litres/minute, a slave flow of FSD 2000 litres/minute and a required ratio (master/slave) of 10/1. The required setting of R on Fig. 5.21 would be 0·6. In a well-designed plant with correctly sized pipes, control valves and flow meters, analog ratios are usually close to unity. If not, the plant design should be examined.

Problems can arise with ratio systems if the slave loop saturates before the master. A typical scenario on a gas follow air burner control could go: (i) the temperature loop calls for a large increase in heat (because of some outside influence). (ii) The air valve (master) opens fully, and the gas valve follows correctly but cannot match the requested flow. (iii) The resulting flame is lean and cold (flame temperature falls off rapidly with too lean a ratio) and the temperature does not rise. The system is now locked with the temperature loop demanding more heat and the air/gas loops saturated, delivering full flow but no temperature rise. The moral is: the master loop must saturate before the slave. If this is not achieved by pipe sizing (as in the example calculation above) the output of the master controller should be limited.

Figure 5.22 shows a variation on ratio control often used on large combustion control systems where it is dangerous to allow an excess of fuel to accumulate, even transiently. The fuel controller gets its set point from a low selector which passes the lowest of the heat demand

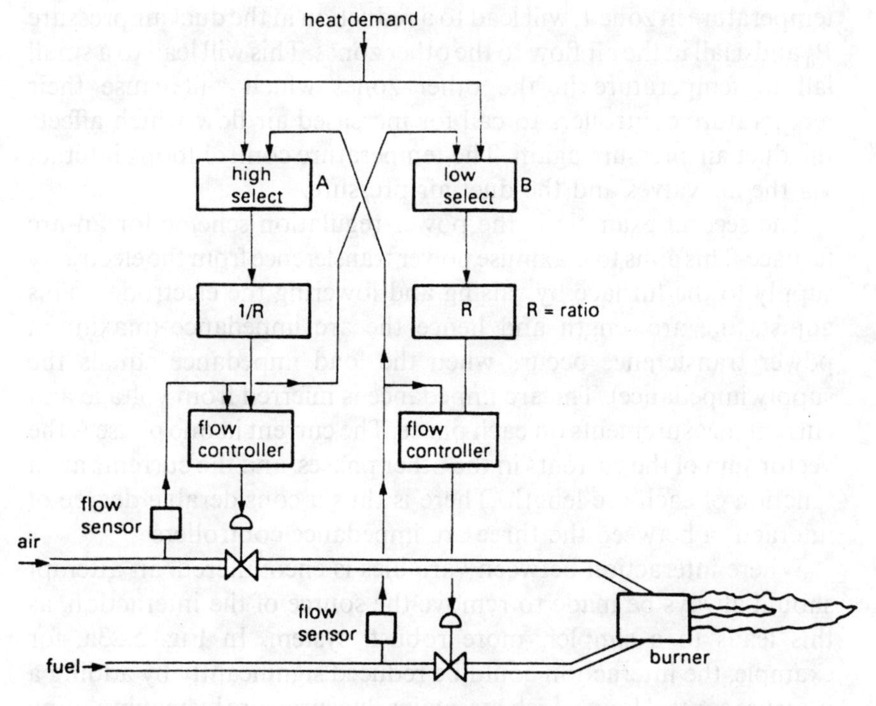

Fig. 5.22 Lead/lag combustion control. The air leads for increasing heat and lags for decreasing heat.

signal or the ratioed air flow. The air flow controller gets its set point from the highest of the heat demand or ratioed gas flow signals. The air valve thus follows the gas valve on reduction of heat, and the gas valve follows the air valve on increase of heat. Transients therefore always lead to a gas lean flame. The system is also inherently safer as faults will result in full air and no fuel. The arrangement is called lead–lag combustion control (since the air leads on opening and lags on closing). The term should not be confused with lead–lag compensators.

5.6. Multivariable control

Classical process control analysis is concerned with single loops having a single set point, single actuator and a single controlled variable. Unfortunately, in practice, plant variables often interact, leading to interaction between control loops.

Typical interactions are shown in Fig. 5.23a and b. In the first example, a single combustion air fan feeds several burners in the multizone furnace. An increase in air flow, via V_1 say to raise the temperature in zone 1, will lead to a reduction in the duct air pressure P_d and a fall in the air flow to the other zones. This will lead to a small fall in temperature in the other zones which will cause their temperature controllers to call for increased air flow which affects the duct air pressure again. The temperature control loops interact via the air valves and the duct air pressure.

The second example is the power regulation scheme for an arc furnace. This aims to maximise power transference from the electricity supply to the furnace by raising and lowering the electrodes. This adjusts the arc length and hence the arc impedance (maximum power transference occurs when the load impedance equals the supply impedance). The arc impedance is inferred from voltage and current measurements on each phase. The current in one phase is the vector sum of the currents in the other phases, and the currents are a function of each arc length. There is thus a considerable degree of interaction between the three arc impedance controllers.

Where interaction between variables is encountered, an attempt should always be made to remove the source of the interaction, as this leads to a simpler, more robust, system. In Fig. 5.23a, for example, the interaction could be reduced significantly by adding a pressure control loop which maintains duct pressure by manipulating a vane on the combustion fan output. Often, however, the interaction is inherent and cannot be removed.

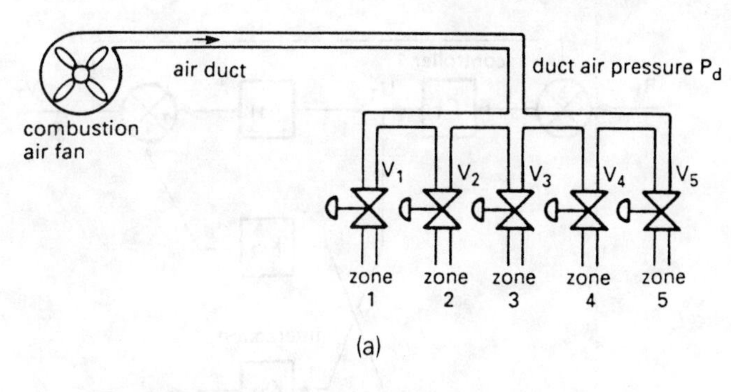

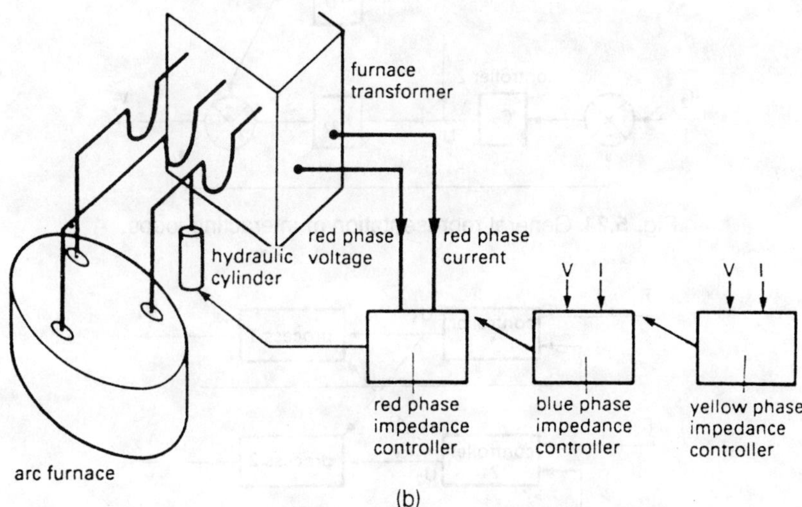

Fig. 5.23 Examples of multivariable control, (a) Interaction between air flows via changes in duct pressure; (b) Interactions between impedance controllers on an arc furnace.

Figure 5.24 is a general representation of two interacting control loops. The blocks C_1 and C_2 represent the controllers comparing set point R with process variable V to give a controller output U. The blocks K_{ab} represent the transfer function relating variable a to controller output b. Blocks K_{11} and K_{22} are the normal forward control path, with blocks K_{21} and K_{12} representing the interaction between the loops.

An alternative way of representing control loops is shown on Fig. 5.25. The process gain of process 1 can be defined as $\Delta V_1/\Delta U_1$ where Δ denotes small change. This process gain can be measured with loop 2 in open loop (i.e. U_2 fixed) or loop 2 in closed loop control (i.e. V_2 fixed) we can thus observe two gains

$$K_{2OL} = \left(\frac{\Delta V_1}{\Delta U_1}\right) \text{ for loop 2 open loop}$$

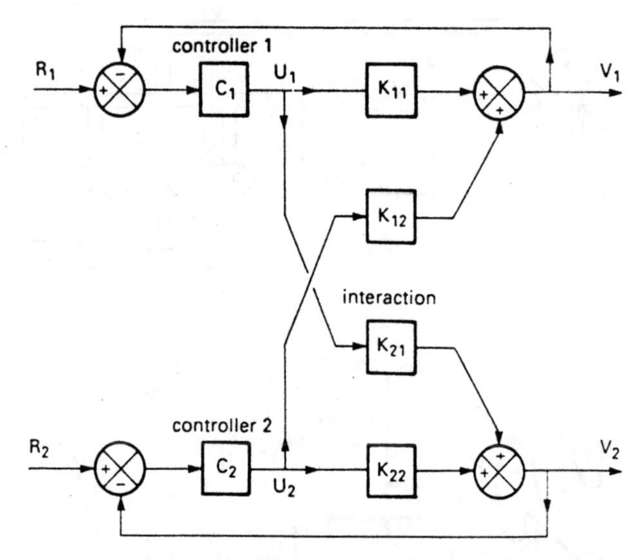

Fig. 5.24 General representation of interacting loops.

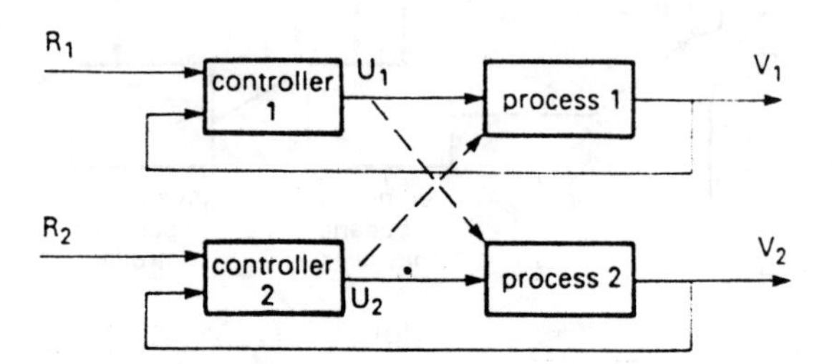

Fig. 5.25 Open and closed loop process gains.

and

$$K_{2CL} = \left(\frac{\Delta V_1}{\Delta U_1}\right) \text{ for loop 2 closed loop}$$

The gains will, of course, vary with frequency and have magnitude and phase shift components. We can now define a relative gain for loop 1

$$\lambda = \frac{K_{2OL}}{K_{2CL}}$$

If λ is unity, changing from manual to auto in loop 2 does not affect loop 1, and there is no interaction between the loops.

If $\lambda < 1$ the interaction will apparently increase process 1 gain

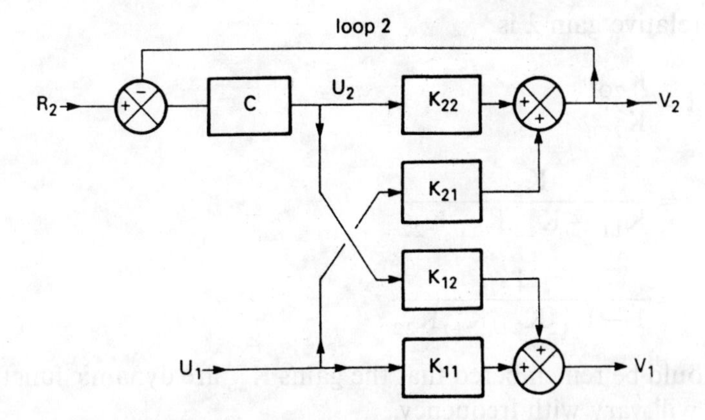

Fig. 5.26 Redrawn version of the Fig. 5.24 to evaluate process gains of loop 1.

when loop 2 is switched to automatic. If $\lambda > 1$ process 1 gain will apparently be decreased when loop 2 is in automatic.

This apparent change in gain can be shown by redrawing Fig. 5.24 as Fig. 5.26. With loop 2 in manual, U_2 is fixed, so K_{2OL} is simply K_{11}. To find K_{2CL} we must consider what happens when loop 2 effectively shunts K_{11}. We have

$$V_1 = K_{11}U_1 + K_{12}U_2 \tag{5.7}$$

$$V_2 = K_{22}U_2 + K_{21}U_1 \tag{5.8}$$

Rearranging equation 5.8 gives

$$U_2 = \frac{V_2 - K_{21}U_1}{K_{22}}$$

which can be substituted in equation 5.7 giving

$$V_1 = K_{11}U_1 + \frac{K_{12}}{K_{22}}(V_2 - K_{21}U_1)$$

The process 1 gain with loop 2 in auto is

$$K_{2CL} = \frac{dV_1}{dU_1}$$

$$= K_{11} - \frac{K_{12}.K_{21}}{K_{22}}$$

$$= \frac{K_{11}K_{22} - K_{12}.K_{21}}{K_{22}}$$

The relative gain λ is

$$\lambda = \frac{K_{2OL}}{K_{2CL}}$$

$$= \frac{K_{11}}{K_{11} - K_{12}.K_{21}/K_{22}}$$

$$= \frac{1}{1 - K_{12}K_{21}/K_{11}K_{22}} \quad (5.9)$$

It should be remembered that the gains K_{ab} are dynamic functions, so λ will vary with frequency.

The term $(K_{12}K_{21}/K_{11}K_{22})$ is the ratio between the interaction and forward gains. This should be in the range 0 to 1. If the term is greater than unity the interactions have more effect than the supposed process and the process variables are being manipulated by the wrong actuators!

It is possible to determine the range of λ from the relationship $(K_{12}K_{21}/K_{11}K_{12})$. If this is positive λ will be greater than unity, and loop 1 process gain will decrease when loop 2 is switched to auto. This will occur if there is an even number of K_{ab} blocks with negative sign (0, 2 or 4). If the relationship is negative λ will be less than unity and loop 1 process gain will increase when loop 2 is closed. This occurs if there is an odd number of blocks with negative sign (1 or 3).

The combustion air flow system of Fig. 5.23a is redrawn on Fig. 5.27a. Increasing U_1 obviously decreases V_2 and increasing U_2 similarly decreases V_1. The interaction block diagram thus has the signs of Fig. 5.27b. There are two negative blocks, so λ is greater than unity.

In Fig. 5.28a air is required by a process, with both flow and source pressure required to be controlled. There are two vanes, the upstream vane being used to regulate the pressure and the downstream vane the flow. This has the block diagram of Fig. 5.28b. There is one block with a negative sign, so the apparent process gains will increase in automatic control.

If λ is greater than unity, the interaction can be considered benign as the reduced process gain will tend to increase the loop stability (albeit at the expense of response time). The loops can be tuned individually in the knowledge that they will remain stable with all loops in automatic control.

If λ is in the range $0 < \lambda < 1$ care must be taken as it is possible for loops to be individually stable but collectively unstable requiring a

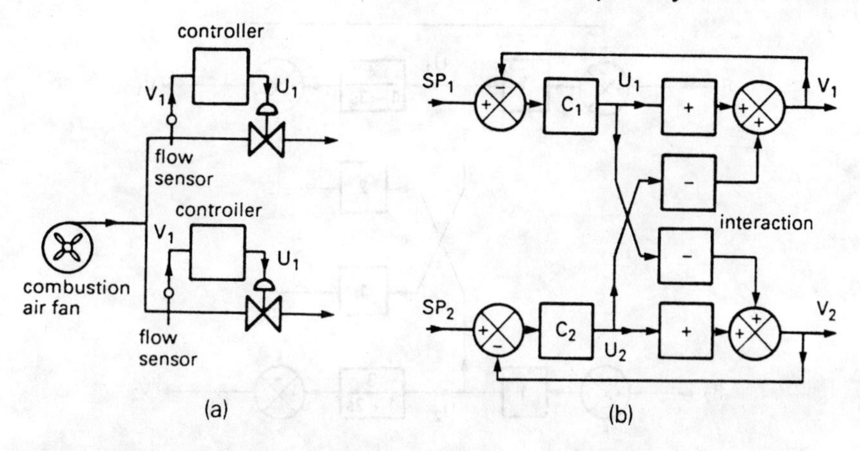

Fig. 5.27 A system with benign interaction, (a) Combustion air flow system; (b) Interaction block diagram. With two negative blocks the interaction decreases the apparent process gain.

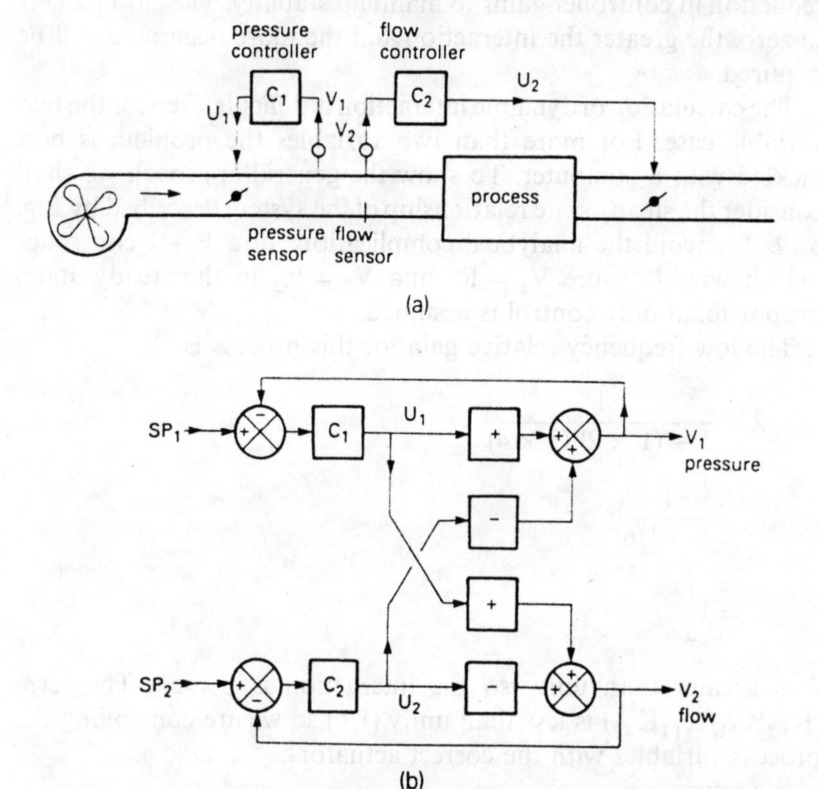

Fig. 5.28 A system with non-benign interaction, (a) Process with pressure controlled by upstream vane and flow by downstream vane; (b) Block diagram of interactions. There is one negative block so the process gain is apparently increased.

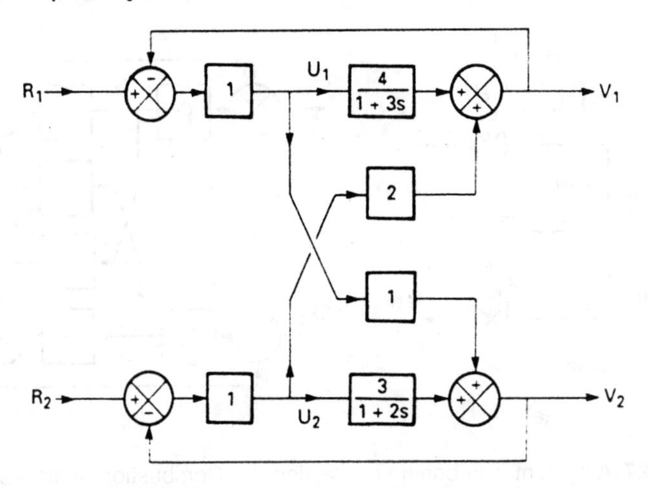

Fig. 5.29 System used to analyse interactions.

reduction in controller gains to maintain stability. The closer λ gets to zero, the greater the interaction and the more degaining will be required.

The calculation of dynamic interaction is difficult, even for the two variable case. For more than two variables the problem is best tackled with a computer. To show the general approach we shall consider the steady state relationship of the system described by Fig. 5.29. To avoid the analytical complications of a P + I controller (which would ensure $V_1 = R_1$ and $V_2 = R_2$ in the steady state) proportional only control is assumed.

The low frequency relative gain for this process is

$$\lambda = \frac{1}{1 - (1 \times 2)/(3 \times 4)}$$

$$= \frac{1}{1 - 1/6}$$

$$= 1\frac{1}{5}$$

λ is greater than unity so the interaction is benign. The term $(K_{12}K_{21}/K_{11}K_{22})$ is less than unity (1/6) so we are controlling the process variables with the correct actuators.

We have

$$U_1 = R_1 - V_1 \tag{5.10}$$

$$U_2 = R_2 - V_2 \tag{5.11}$$

$$V_1 = 4U_1 + 2U_2 \tag{5.12}$$

$$V_2 = 3U_2 + U_1 \tag{5.13}$$

Substituting equation 5.10 into equation 5.12 gives

$$V_1 = 4R_1 + 2R_2 - 2V_2$$

$$\text{or } 5V_1 = 4R_1 + 2R_2 - 2V_2 \tag{5.14}$$

Similarly, combining equations 5.11 and 5.13 gives

$$4V_2 = 3R_2 + R_1 - V_1 \tag{5.15}$$

Substituting for V_2 from equation 5.14 into equation 5.15 gives V_1 in terms of R_1 and R_2

$$V_1 = \frac{7}{9}R_1 + \frac{1}{9}R_2$$

Similarly substituting for V_1 from equation 5.14 into equation 5.15 gives V_2 also in terms of the set points.

$$V_2 = \frac{13}{18}R_2 + \frac{1}{18}R_1$$

Similar, if rather tedious, calculations could be performed to the dynamic behaviour with a more realistic controller.

With more interacting variables, the analysis becomes exceedingly complex and computer solutions are best used. Ideally, however, interactions once identified, should be engineered out wherever possible. A typical example would be adding an air pressure control loop to the system of Fig. 5.23. This would keep the air pressure at the manifold constant despite flow changes, and eliminate the interaction between the zones.

5.7. Dealing with non-linear elements

5.7.1. Introduction

All systems are non-linear to some degree. Valves have non-linear transfer functions, actuators often have a limited velocity of travel and saturation is possible in every component. A controller output is limited to the range 4–20 mA, say, and a transducer has only a restricted measurement range.

One of the beneficial effects of closed loop control is the reduced effect of non-linearities. The majority of non-linearities are therefore simply lived with, and their effect on system performance is negligible. Occasionally, however, a non-linear element can dominate a system and in these cases its effect must be studied.

Some non-linear elements can be linearised with a suitable compensation circuit. Differential pressure flow meters have an output which is proportional to the square of flow. Following a non-linear differential pressure flow transducer with a non-linear square root extractor gives a linear flow measurement system.

Cascade control can also be used around a non-linear element to linearise its performance as seen by the outer loop. Butterfly valves are notoriously non-linear. They have an s-shaped flow/position characteristic, suffer from backlash in the linkages and are often severely velocity limited. Enclosing a butterfly valve within a cascade flow loop as Fig. 5.30 will make the severely non-linear flow control valve appear as a simple linear first order lag to the rest of the system.

There are two basic methods of analysing the behaviour of systems with non-linear elements. It is also possible, of course, to write computer simulation programs (or use analog computers). Such techniques are the only practical way of analysing complex non-linearities.

5.7.2. The describing function

If a non-linear element is driven by a sine wave, its output will probably not be sinusoidal, but it will be periodic with the same frequency as the input, with differing shape and possibly shifted in phase as shown on Fig. 5.31. Often the shape and phase shift are related to the amplitude of the driving signal.

Fourier analysis is a technique that allows the frequency spectrum of any periodic waveform to be calculated. A simple pulse can be considered to be composed of an infinite number of sine waves as Fig. 5.32.

The non-linear output signals of Fig. 5.31 could therefore be represented as a frequency spectrum, obtained from Fourier analysis. This is, however, unnecessarily complicated. Process control is generally concerned with only dominant effects, and as such it is only necessary to consider the fundamental of the spectrum. We can therefore represent a non-linear function by its gain and phase shift at the fundamental frequency. This is known as the describing function, and will probably be frequency and amplitude dependent.

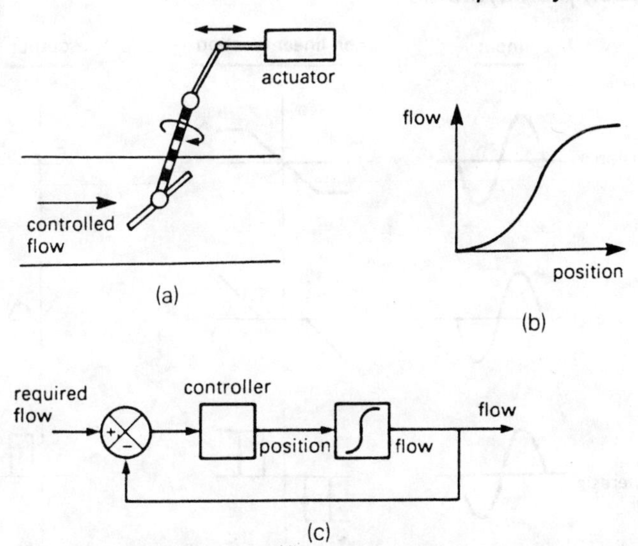

Fig. 5.30 Linearising a system with a cascade loop, (a) Butterfly flow control valve; (b) System characteristics; (c) The use of a cascade loop linearises the system.

Figure 5.33 shows a very crude bang/bang servo system used to control level in a header tank. The level is sensed by a capacitive probe which energises a relay when a nominal depth of probe is submerged. The relay energises a solenoid which applies pneumatic pressure to open a flow valve. This system is represented by Fig. 5.34.

The level sensor can be considered to be a level transducer giving a 0–10 V signal. The signal is filtered with a 2 second time constant to overcome noise from splashing, ripples, etc. The level transducer output is compared with the voltage from a set point control and the error signal energises or de-energises the relay. We shall assume no hysteresis for simplicity although this obviously would be desirable in a real system.

The relay drives a solenoid assumed to have a small delay in operation which applies 15 psi to an instrument air pipe to open the valve. The pneumatic signal takes a finite time to travel down the pipe, so the solenoid valve and piping are considered as a 0·5 sec transit delay. The valve actuator turns on a flow of 150 cubic metres per minute for an applied pressure of 15 psi. We shall assume it is linear for other applied pressures. The actuator/valve along with the inertia of the water in the pipe appear as a first order lag of 4 seconds time constant. The tank itself appears as an integrator from flow to level.

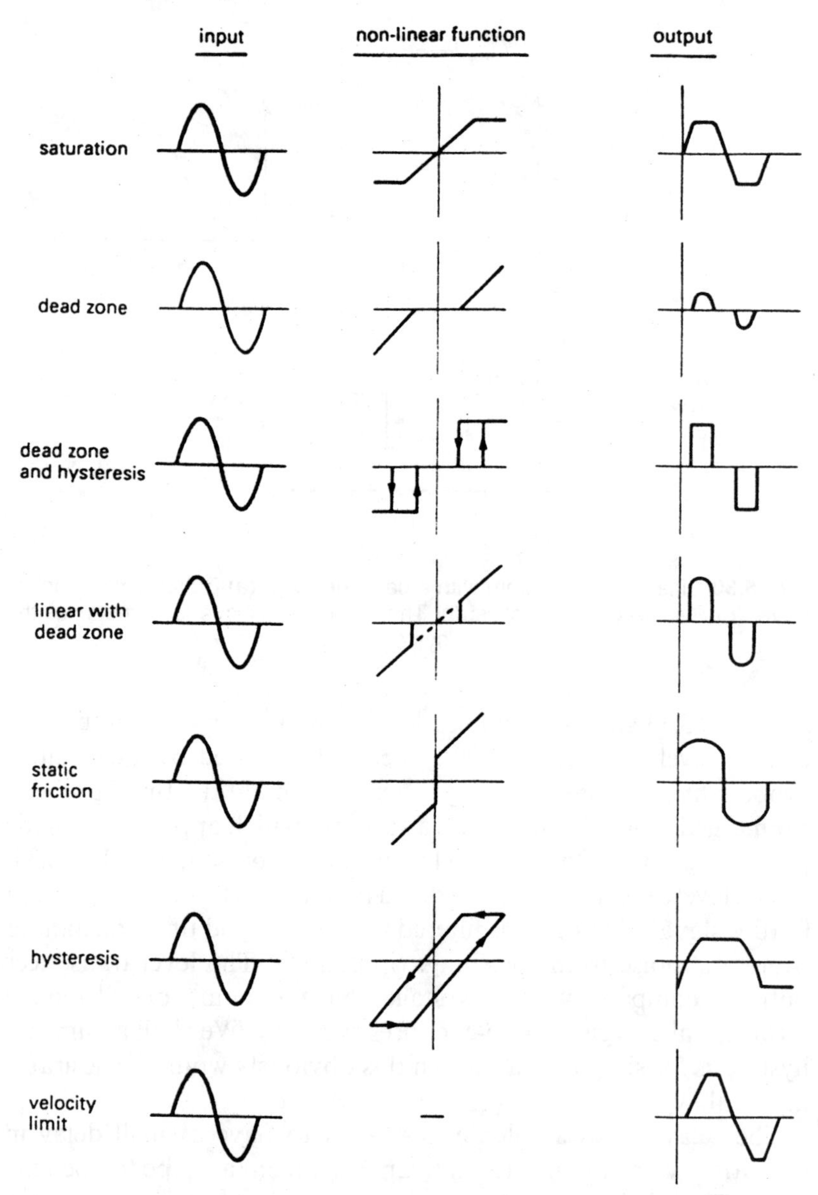

Fig. 5.31 Common non-linearities.

This system is dominated by the non-linear nature of the level probe and the solenoid. The rest of the system can be considered linear if we combine the level comparator, relay and solenoid into a single element which switches 0 to 15 psi according to the sign of the error signal (15 psi for negative error, i.e. low level).

This non-linear element will therefore have the response of

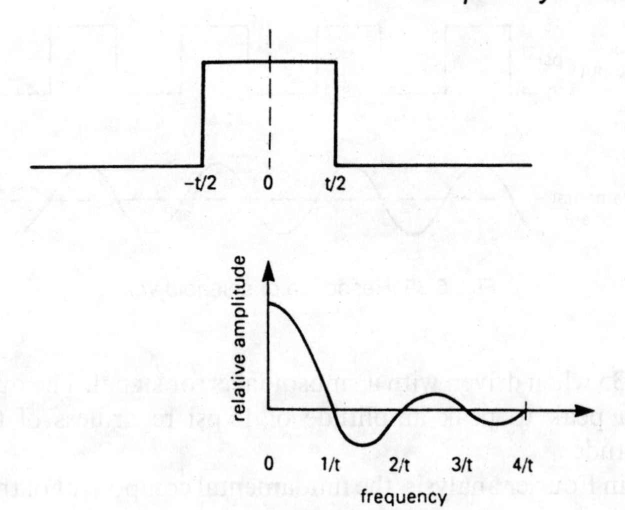

Fig. 5.32 Fourier transform of a single pulse.

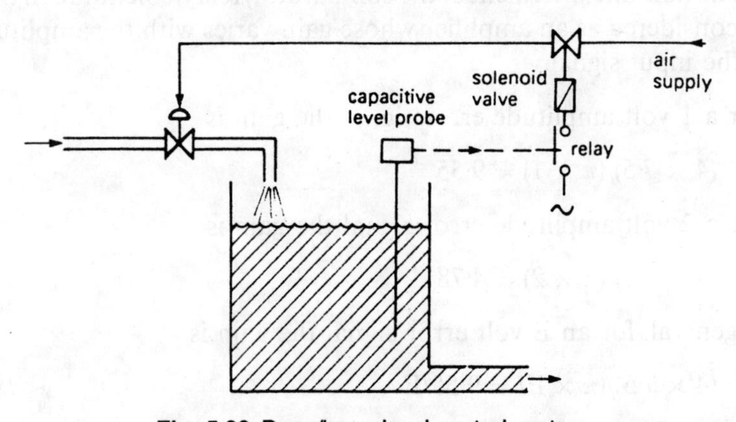

Fig. 5.33 Bang/bang level control system.

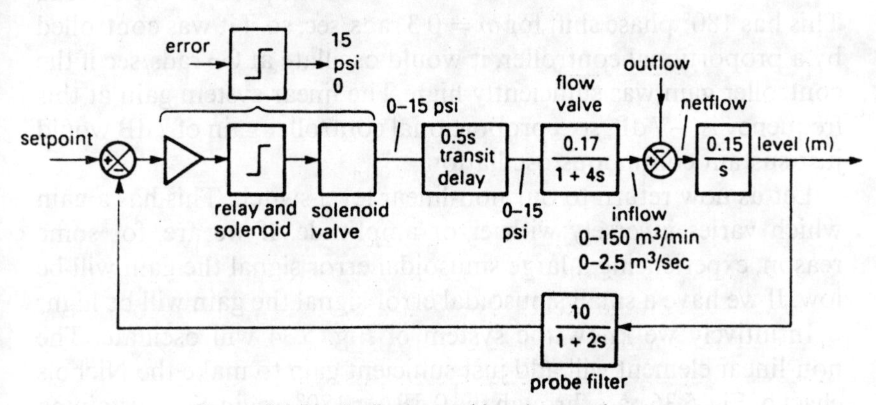

Fig. 5.34 Block diagram of level control system.

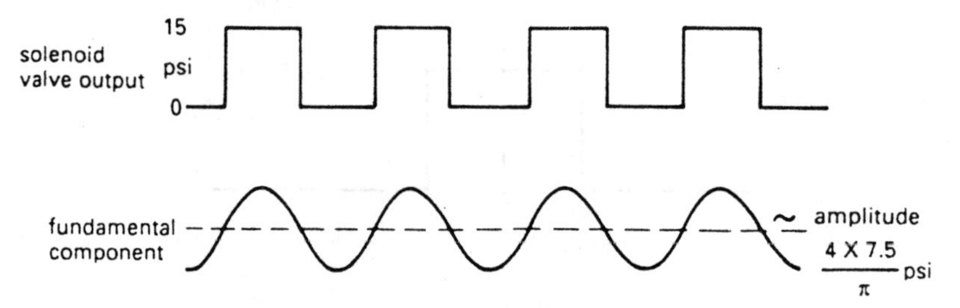

Fig. 5.35 Response of solenoid valve.

Fig. 5.35 when driven with a sinusoidal error signal. The output will have a peak to peak amplitude of 15 psi regardless of the error magnitude.

From Fourier analysis, the fundamental component of the output signal is a sine wave with amplitude $4 \times 7{\cdot}5/\pi$ psi as shown. The phase shift is zero at all frequencies.

The non-linear element of the comparator/relay/solenoid can thus be considered as an amplifier whose gain varies with the amplitude of the input signal.

For a 1 volt amplitude error signal the gain is

$$(4 \times 7{\cdot}5)/(\pi \times 1) = 9{\cdot}55$$

For a 2 volt amplitude error signal the gain is

$$(4 \times 7{\cdot}5)/(\pi \times 2) = 4{\cdot}78$$

In general, for an E volt error signal the gain is

$$(4 \times 7{\cdot}5)/(\pi \times E) = 9{\cdot}55/E$$

Figure 5.36 is a Nichols chart for the linear parts of the system. This has 180° phase shift for $\omega = 0{\cdot}3$ rads/sec, so if it was controlled by a proportional controller, it would oscillate at 0·3 rads/sec if the controller gain was sufficiently high. The linear system gain at this frequency is -7 dB, so a proportional controller gain of 7 dB would just sustain continuous oscillation.

Let us now return to our non-linear level switch. This has a gain which varies inversely with error amplitude. If we are, for some reason, experiencing a large sinusoidal error signal the gain will be low. If we have a small sinusoidal error signal the gain will be high.

Intuitively we know the system of Fig. 5.34 will oscillate. The non-linear element will add just sufficient gain to make the Nichols chart of Fig. 5.36 pass through the 0 dB/ $-$ 180° origin. Self-sustaining

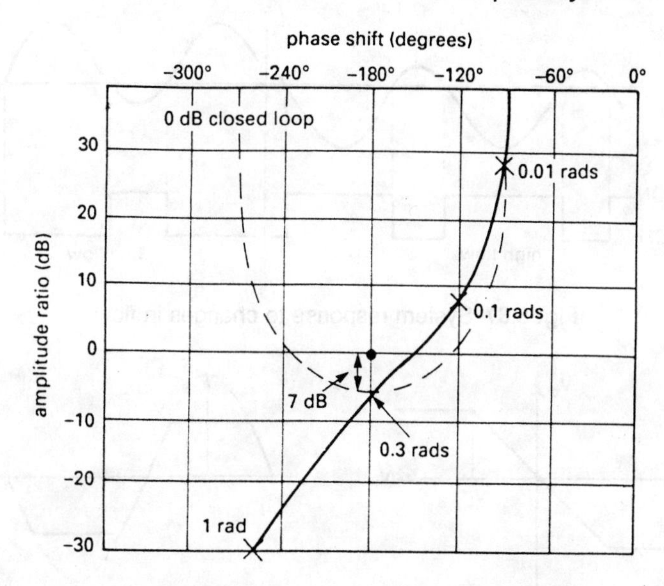

Fig. 5.36 Nichols chart for linear parts of system.

oscillations will result at 0·3 rads/sec. If these increase in amplitude for some reason, the gain will decrease causing them to decay again. If they cease, the gain will increase until oscillations recommence. The system stabilises with continuous constant amplitude oscillation.

To achieve this the non-linear element must contribute 7 dB gain, or a linear gain of 2·24. From above, the gain is 9·55/E where E is the error amplitude. The required gain is thus given by an error amplitude of $9{\cdot}55/2{\cdot}24 = 4{\cdot}26$ V. This corresponds to an oscillation in level of 0·426 metres.

The system will thus oscillate about the set level with an amplitude of 0·4 metres (the assumptions and approximations give more significant figures a relevance they do not merit) and an angular frequency of 0·3 rads/sec (period fractionally over 20 seconds). There is a hidden assumption in the above analysis that the outgoing flow is exactly half the available ingoing flow to give equal mark/space ratio at the valve. Other flow rates will give responses similar to Fig. 5.37, exhibiting a form of pulse width modulation. The relatively simple analysis however has told us that our level control system will sustain constant oscillation with an amplitude of around half a metre and a period of about 20 seconds at nominal flow.

Similar techniques can be applied to other non-linearities; saturation (shown in Fig. 5.38) and limiters, for example, have unity gain for input amplitudes less than the limiting level. For increasing amplitude

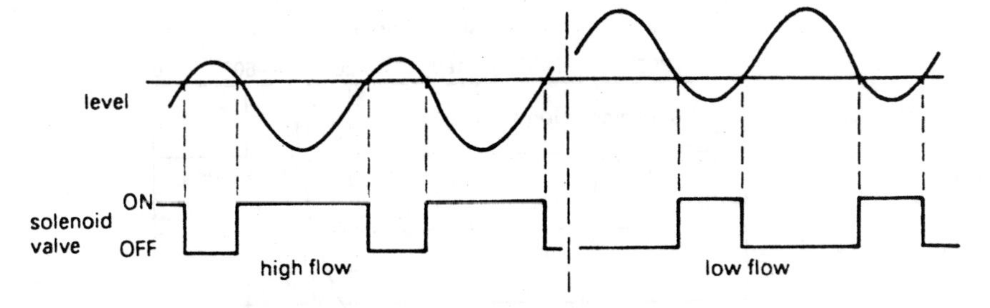

Fig. 5.37 System response to changes in flow.

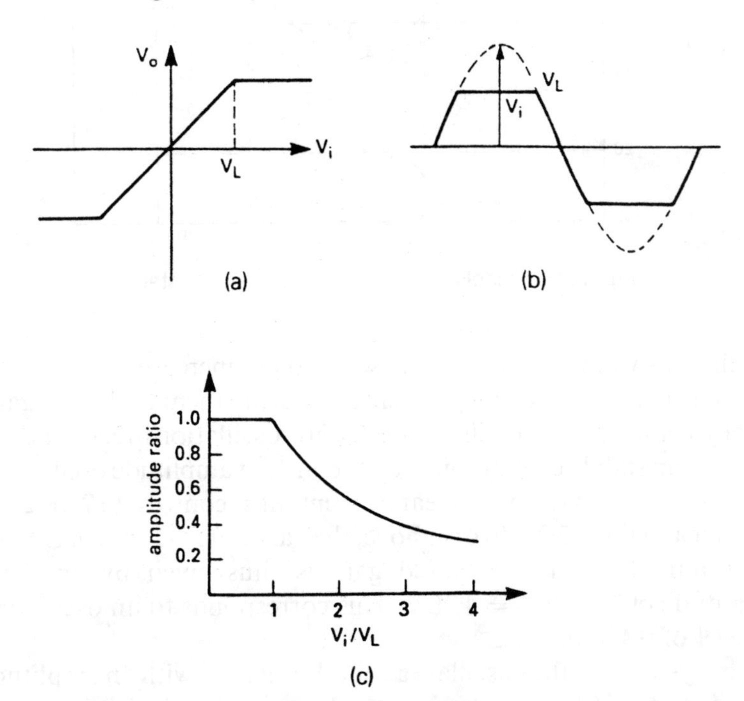

Fig. 5.38 The effect of a limiter circuit, (a) System response; (b) Effect of limiting; (c) Relationship between input and output signals for various levels of input.

the apparent gain will decrease. The describing function when limiting occurs is a gain of

$$G = \frac{2}{\pi}\left(\sin^{-1}\left(\frac{1}{R}\right) + \frac{1}{R}\sqrt{1 - \frac{1}{R^2}}\right)$$

where R is the ratio between the input signal amplitude and the limiting level (V_i/V_L). This is plotted on Fig. 5.38c. There is no phase shift between input and output.

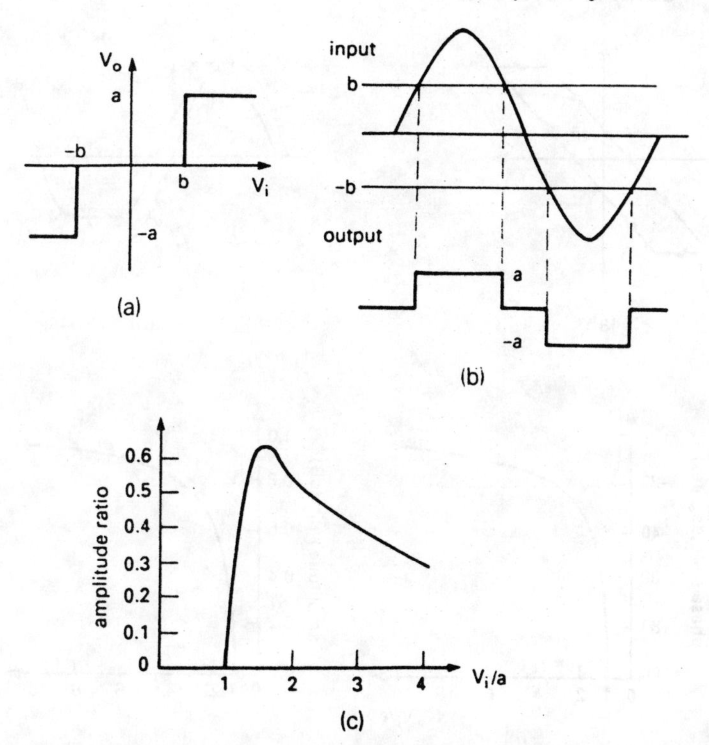

Fig. 5.39 The effect of a relay with dead zone, (a) System response; (b) Effect of relay with dead zone; (c) Relationship between input and output signals.

A relay with dead zone is often used to prevent chatter from noise, this has the response of Fig. 5.39. The describing function is zero for input amplitudes less than the deadband width. For higher amplitudes the gain is

$$G = \frac{4}{\pi R}\sqrt{1 - \frac{1}{R^2}}$$

where R is the ratio between the input signal V_i and the output switched level V_o. This is plotted in Fig. 5.39.

Hysteresis, shown in Fig. 5.40, introduces a phase shift, and a flat top to the output waveform. This is not the same waveform as the limiter; the top is simply levelled off at 2a below the peak where a is half the dead zone width. If the input amplitude is large compared to the dead zone, the gain is unity and the phase shift can be approximated by:

$$\phi = \sin^{-1}(a/V_i)$$

As the input amplitude decreases, the gain decreases and becomes

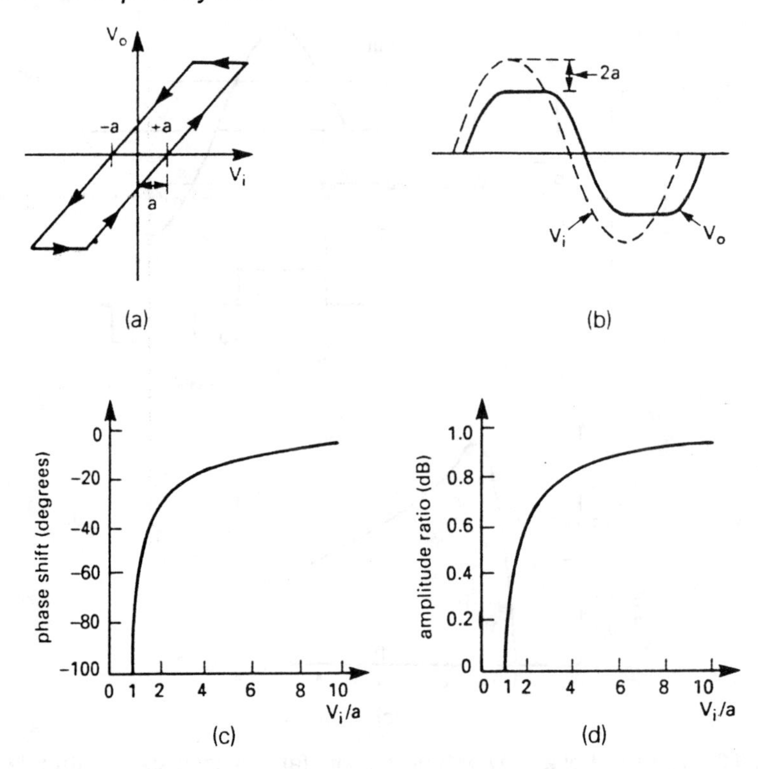

Fig. 5.40 The effect of hysteresis, (a) System response; (b) Effect of hysteresis; (c) Phase shift; (d) Amplitude ratio.

zero when the input peak to peak amplitude is less than the dead zone width. The exact relationship is complex, but is shown on Fig. 5.40c, d.

Non-linear elements generally have gains and phase shift which increase or decrease with input amplitude (usually a representation of the error signal). Figure 5.41 illustrates the two gain cases. For a loop gain of unity, constant oscillations will result. For loop gains greater than unity, oscillations will increase in amplitude, for loop gain less than unity oscillations will decay.

In Fig. 5.41a the gain falls off with increasing amplitude. The system thus tends to approach point X as large oscillations will decay and small oscillations increase. The system will oscillate at whatever gain gives unity loop gain. This is called limit cycling. Most non-linearities (bang/bang servo, saturation, etc.) are of this form.

Where loop gain increases with amplitude as Fig. 5.41b, decreasing gain gives increasing damping as amplitude decreases, so oscillations will quickly die away. This response (described in Section 4.8) is

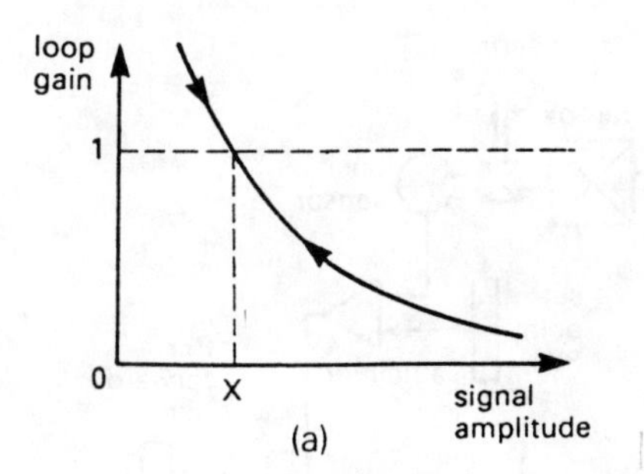

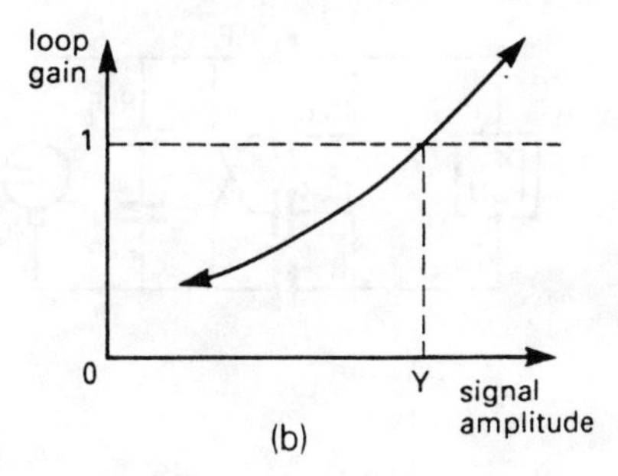

Fig. 5.41 Possible relationships between gain and signal amplitude, (a) Gain decreases with increasing amplitude; (b) Gain increases with increasing amplitude.

sometimes deliberately introduced into level controls. If, however, the system is provoked beyond Y by a disturbance, the oscillation will rapidly increase in amplitude and control will be lost.

5.7.3. State space and the phase plane

Figure 5.42a shows a simple position control that could be used to position, say, an aerial rotator. The position is sensed by a potentiometer, and compared with the position set on potentiometer RV1. The resulting error signal is compared with an error 'window' by comparators C_1, C_2. Preset RV2 sets the deadband, i.e. the width of the window. The comparators energise relays RLF and RLR which drive the aerial to the left and right respectively.

Initially, we shall analyse the system with RV2 set to zero, i.e. no deadband. This has the block diagram of Fig. 5.42b, with a first order lag of time constant T arising from the inertia of the system, and the integral action converting motor velocity to aerial position.

The system is thus represented by

$$x = \frac{\pm K}{s(1 + sT)} \tag{5.16}$$

where K represents the motor torque and the sign of K indicates the sign of the error. Representing equation 5.16 as a differential equation for positive error gives

$$T\frac{d^2x}{dt^2} + \frac{dx}{dt} = K$$

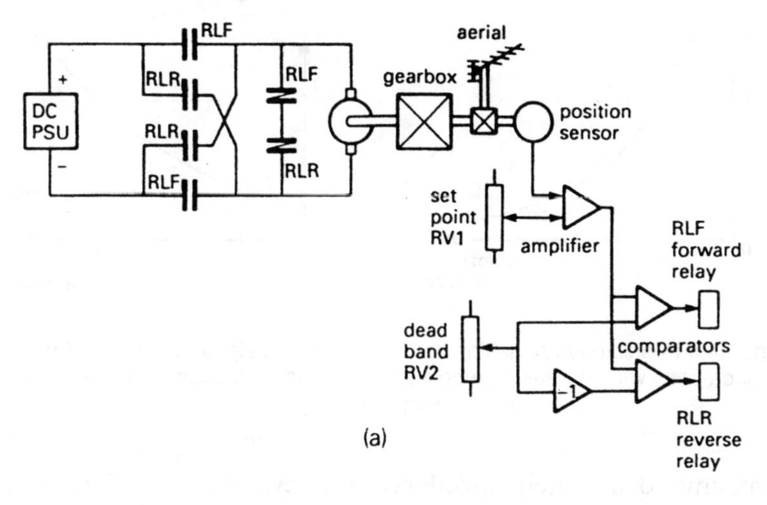

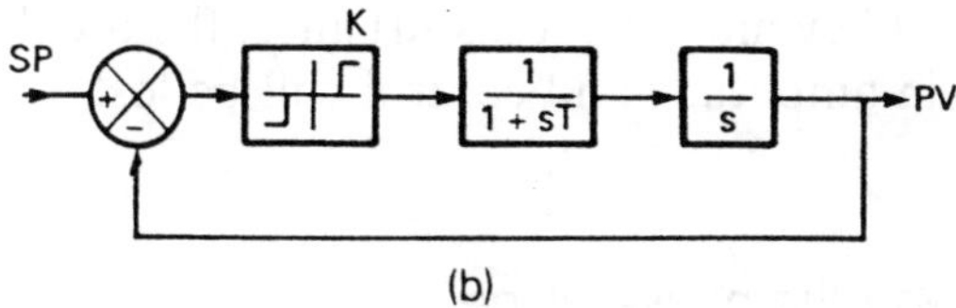

Fig. 5.42 An aerial position control system, (a) System diagram; (b) Block representation of system.

Using the techniques of Section 2.2.3 gives a solution

$$x = A + Kt + Be^{-t/T} \tag{5.17}$$

where A and B are constants depending on the initial position x_0 and initial velocity v_0

$$x = x_0 - TK + Tv_0 + Kt + T(K - v_0)e^{-t/T} \tag{5.18}$$

Differentiating gives the velocity v

$$v = \frac{dx}{dt}$$

$$= K - (K - v_0)e^{-t/T} \tag{5.19}$$

Equations 5.18 and 5.19 describe the behaviour of the system. These can be plotted graphically as Fig. 5.43 with velocity plotted against position for positive K.

Curve A represents a starting condition $X_0 = 0$, $Y_0 = 0$; curve B $X_0 = -12$, $V_0 = 0$; curve C $X_0 = 5$, $Y_0 = -1$; curve D $X_0 = -5$, $V_0 = -2$. These are plotted by calculating X and V at various times from $t = 0$.

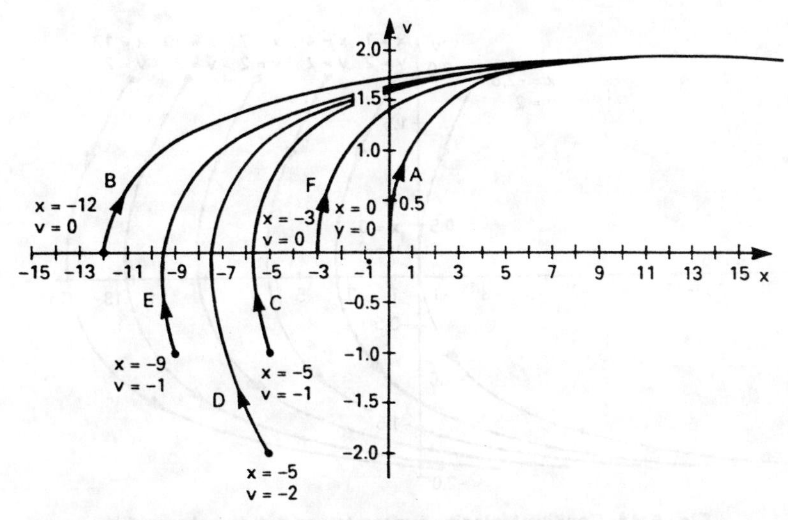

Fig. 5.43 Relationship between position and velocity for various initial conditions.

In each case, the curve tends towards V = 2 units/sec as t gets large. It can be seen from inspection that the family of curves have an identical shape, and the different starting conditions simply represent a horizontal shift of the curve.

A similar family of curves can be drawn for negative values of K. These are sketched on Fig. 5.44. In this case, the velocity tends towards V = − 2 units/sec.

Given these curves, we can plot the response of the system described by Fig. 5.42. Let us assume that the system is stationary at X = − 5, and the set point is switched to + 5. The subsequent behaviour is shown on Fig. 5.45. The system follows the curve passing through X = − 5, V = 0 for positive K, crossing X = 0 with a velocity 1·5 units/sec, reaching the set point at point X with a velocity of 1·76 units/sec. It cannot stop instantly, however, so it overshoots. At the instant the overshoot occurs K switches sign. The system now has a velocity of + 1·76, with K negative, so it follows the corresponding curve of Fig. 5.44 from point X to point Y. It can be seen that an overshoot to X = 7 occurs. At point Y, another overshoot occurs and K switches back positive. The system now follows the curve to Z with an undershoot of X = 4·1. At Z another overshoot occurs and the system spirals inwards as shown. The predicted step response is shown in Fig. 5.46.

In Fig. 5.47a, the deadband control (RV2 in Fig. 5.42) has been adjusted to energise RLF for error voltages more negative than − 1 unit and energise RLR for error voltages above + 1 unit. There is

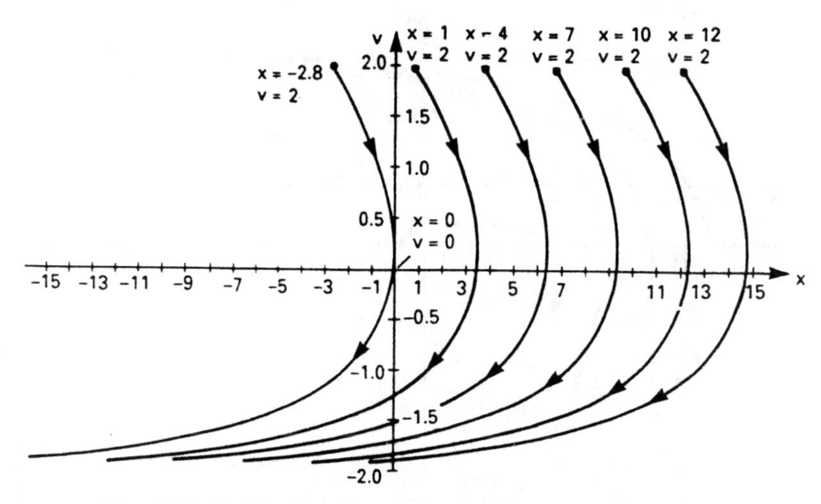

Fig. 5.44 Position/velocity curves for negative values of K.

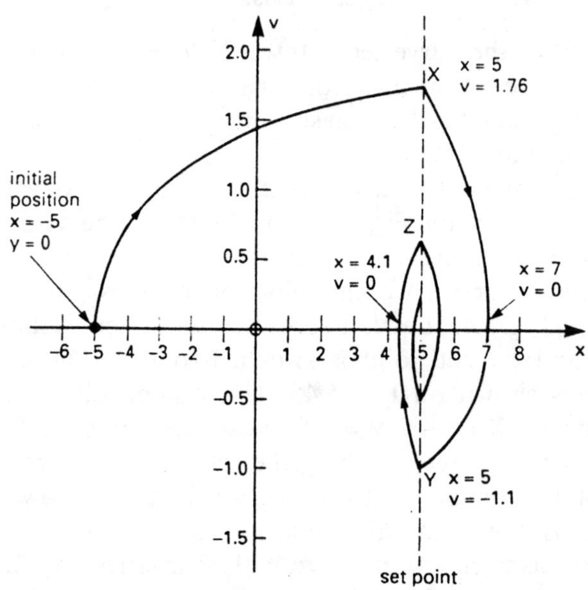

Fig. 5.45 System behaviour with change of set point from x=–5 to x= +–5.

thus a deadband 2 units wide around the set point.

Figure 5.47b shows the effect of this deadband. We will assume initial values of $X_0 = 0, V_0 = 0$ and we switch the set point to $X = 5$. The system accelerates to point U ($X = 4, V = 1{\cdot}40$) at which point RLF de-energises. The system loses speed ($K = 0$) until point V, where the position passes out of the deadband and RLR energises.

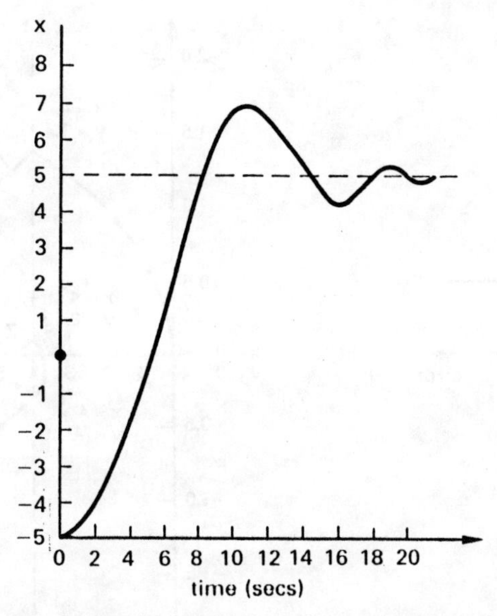

Fig. 5.46 Predicted step response for set point change.

The system reverses, and re-enters the deadband at point W, where RLR de-energises. An undershoot then occurs where the deadband is entered for the last time, coming to rest at point Z ($X = 4{\cdot}75$, $V = 0$). Note that the time taken to go to Z is, in fact, infinite as the speed theoretically falls exponentially to zero. This apparent atypical behaviour arises from the simplicity of our model. Figure 5.47c shows the predicted step response.

The position x and velocity dx/dt completely describe the system of equations 5.16 and 5.17 and are known as state variables. A linear system can be represented as a set of first order differential equations relating the various state variables. For a second order system there are two state variables, for higher order systems there will be more.

For the system described by equation 5.16, we can denote the state variables by x (position) and v (velocity). For a driving function K, we can represent the system by Fig. 5.48 which is called a state space model. This describes the position control system by the two first order differential equations.

$$T\frac{dv}{dt} = K - x \tag{5.20}$$

$$\text{and } v = \frac{dx}{dt} \tag{5.21}$$

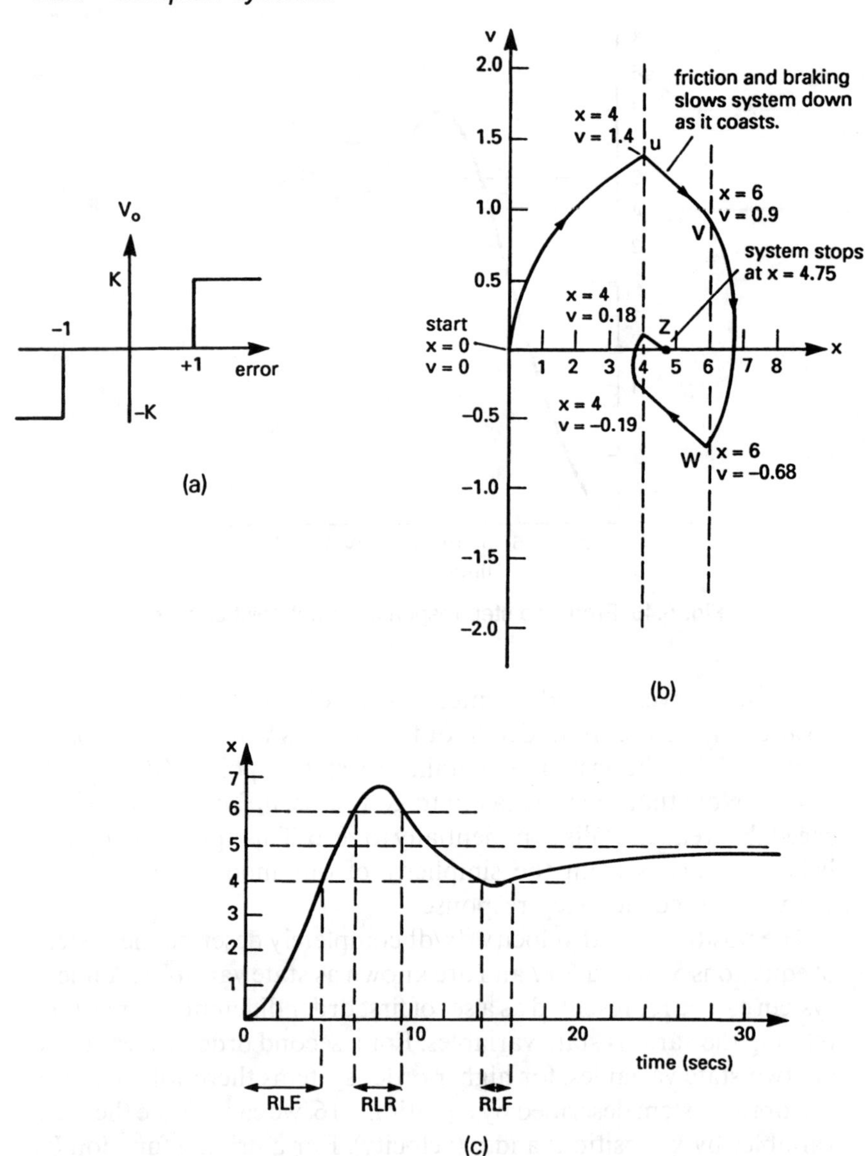

Fig. 5.47 System with deadband and friction, (a) Deadband response; (b) Position/velocity curve for set point change from x = 0 to x = 5; (c) Step response. Note that the system does not attain the set point.

Figures 5.45 and 5.47 plot velocity against position, and as such are a plot relating state variables. For two state variables (from a second order system) the plot is known as a phase plane. For higher order systems, a multi-dimensional plot, called state space, is required. Plots such as Fig. 5.43 and Fig. 5.44 which show a family of possible

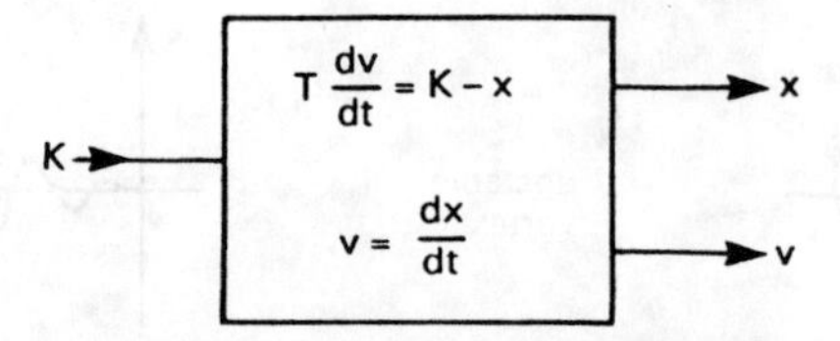

Fig. 5.48 State variables for position control system.

curves are called phase plane portraits.

Equations 5.20 and 5.21 involve a third variable; the time t. This can be eliminated to give the equations of the trajectories on the phase plane. For example, for $x_0 = v_0 = 0$, solving equation 5.19 for t yields

$$t = -T\log_e\left(1 - \frac{v}{K}\right)$$

which can be substituted back into equation 5.18 to yield the equation

$$x = -TK\left(\frac{v}{K} + \log\left(1 - \frac{v}{K}\right)\right)$$

which is the trajectory for RLF energised with initial conditions $x_0 = v_0 = 0$. This is curve A on Fig. 5.43 and the initial section of Fig. 5.47b.

In the deadband region of Fig. 5.47 $K = 0$. Substituting into equation 5.19 for initial conditions x_0, y_0 yields

$$t = -\log_e\left(\frac{v}{v_0}\right)$$

which substitutes into equation 5.18 giving

$$x = x_0 + T(v_0 - v)$$

which is the equation of a straight line, as can be seen from a visual inspection of Fig. 5.47b.

Similar phase planes can be drawn for other non-linearities such as saturation, hysteresis, etc. Various patterns emerge, which are summarised in Fig. 5.49. The system behaviour can be deduced from the shape of the phase trajectory.

In a linear closed loop system stability is generally increased by adding derivative action. In a position control system this is equivalent to adding velocity (dx/dt) feedback. The behaviour of a non-linear system can also be improved by velocity feedback.

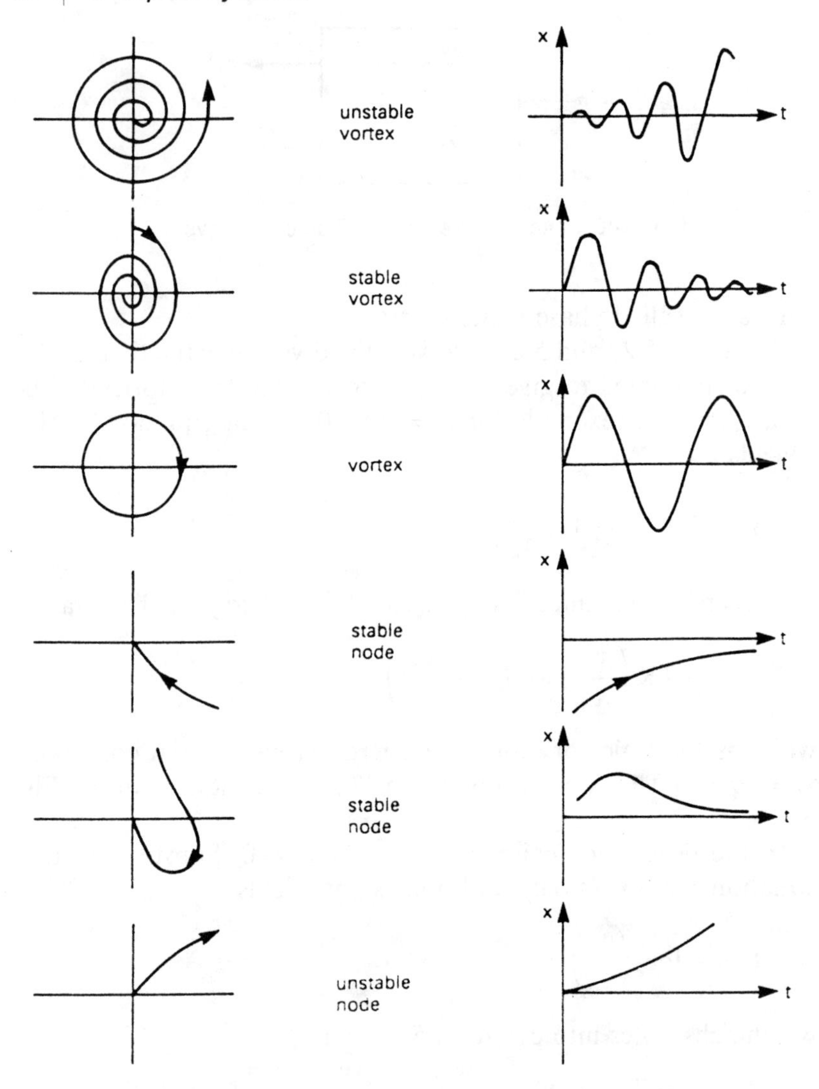

Fig. 5.49 Various possible trajectories and their response.

In Fig. 5.50 velocity feedback has been added to the simple bang/bang position servo of Fig. 5.42.

The switching point now occurs where

$$SP - x - Lv = 0$$

$$\text{or } v = \frac{1}{L}(SP - x)$$

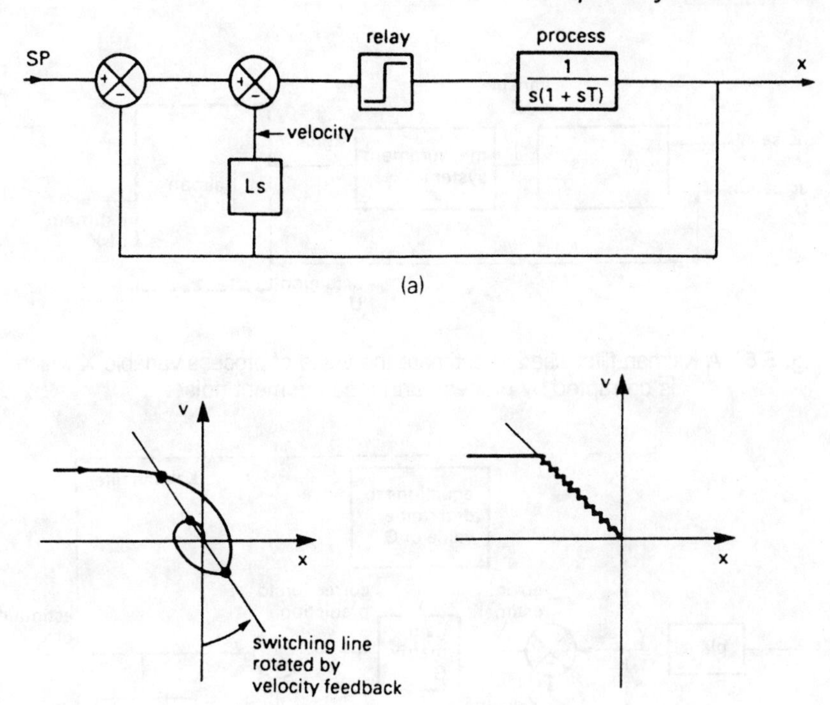

Fig. 5.50 The addition of velocity feedback to a non-linear system, (a) Non-linear system with velocity feedback; (b) System behaviour on position/velocity curve; (c) Overdamped system follows the switching line.

This is a straight line of slope $-1/L$, passing through $x = SP$, $v = 0$ on the phase plane. Note that L has the units of time. The line is called the switching line, and advances the changeover as shown in Fig. 5.50b, thereby reducing the overshoot. Too much velocity feedback as Fig. 5.50c simulates an overdamped system as the trajectory runs to the set point down the switching line.

5.8. Kalman filters

All filters are designed to remove noise from a signal. The Kalman filter is a recursive algorithm for a computer, which handles a signal corrupted by process and measurement noise to give a clean representation of a process variable and a prediction of its future behaviour. The filter can also estimate the values of unmeasurable variables within the process. The principles of the filter arose out of military work (Exocet missiles skim along the sea surface estimating

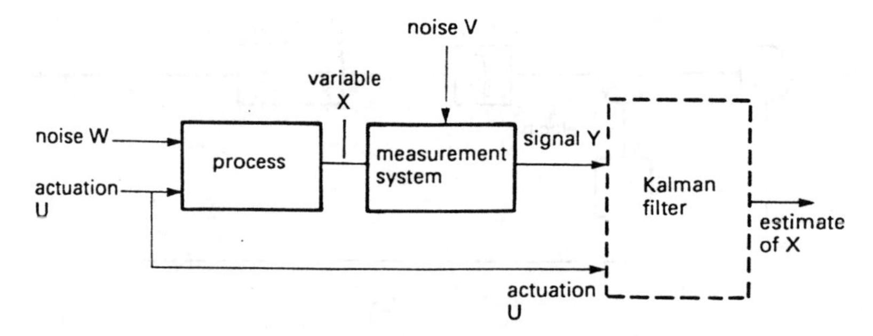

Fig. 5.51 A Kalman filter used to estimate the value of process variable X which is corrupted by process and measurement noise.

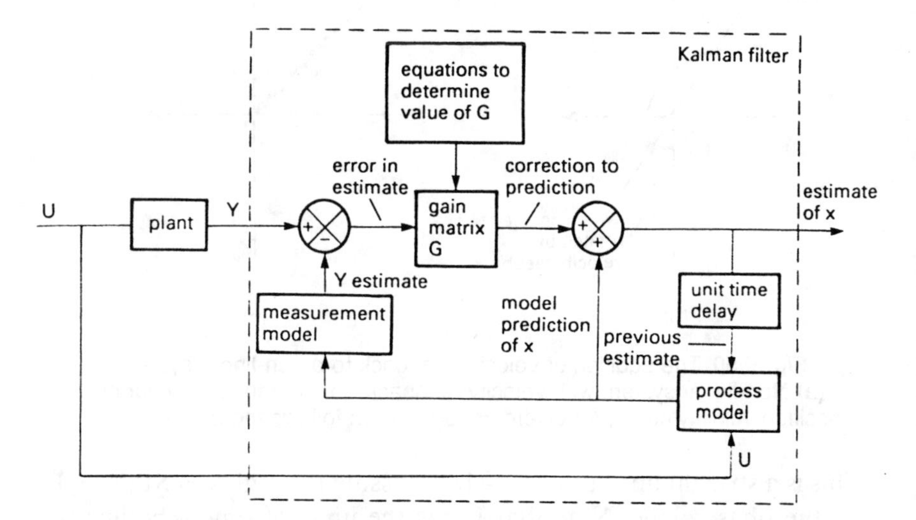

Fig. 5.52 The Kalman filter uses internal models to estimate the value of x.

their height above a rough sea with a Kalman filter) but the technique is becoming increasingly common in industrial controls. Typical applications include the load management of a power station, and measuring the internal steel temperature of an ingot inside a soaking pit (where all that is available is a pyrometer measurement of the steel surface temperature and data on the fuel and exhaust gas flows).

Figure 5.51 shows a typical process, with X being the variable we wish to measure. The nearest point of measurement is the signal Y, but this is corrupted by noise V. The process is driven by the actuation U, but the process is itself affected by noise W. The variables U–Y can be scalar or vector quantities. The Kalman filter requires the user to model the process and the measurement system, giving the filter block diagram of Fig. 5.52, and aims to minimise the variance (error) between the models and the observed process. The noise is assumed to be Gaussian white noise with zero mean value and to have variances provided by the user.

It can be seen that the filter evaluates the error between the model and the process and uses this to correct the signal estimate. The filter is thus a recursive feedback loop. The gain of this loop is determined by the matrix G, and is recalculated at each time step in the sampling.

Although the Kalman filter is an excellent tool for dealing with noisy processes, it does require a great deal of engineering effort to produce a working system. Fairly accurate models are needed for the measurement system and the process itself, plus knowledge of likely noise sources and their variance.

Chapter 6
Signals, noise and data transmission

6.1. Introduction

Much of the external aspect of process control is concerned with the conveyance of information from one location to another. A temperature transducer, for example, conveys temperature information from a chemical reactor to a control room, or a radio based telemetry system conveys control signals in a digital form to a remote unmanned valve station.

This information is conveyed over a transmission medium (or link) which has inherent limitations. The signal is invariably attenuated, and the link will have a limited bandwidth which will introduce errors from frequency dependent attenuation and phase shift. Noise, both from external sources and crosstalk from other signals, will be added to the information. Despite these limitations, the received information must still be useful to the recipient device.

There are basically two types of signal which need to be conveyed. Analog signals are continuous in time (e.g. measurement of process variables, control signals, audio and television signals). Digital signals convey information in the form of discrete logic states. These states are effectively a pulse train, and can represent digital data (e.g. text for display on a VDU or printer) or be a digital representation of an analog signal.

Once noise is added to a wanted signal, it becomes effectively part of the signal, and the original signal can only be reinstated if the noise has some characteristic that allows it to be removed. Differential amplifiers, as Fig. 6.1 for example, remove common mode noise because the interference affects both input lines equally, allowing the noise to be identified and subtracted from the (wanted signal + noise) signal. If the composition of the noise is known, it may be possible to design a filter which will pass the wanted signal but reject the noise component.

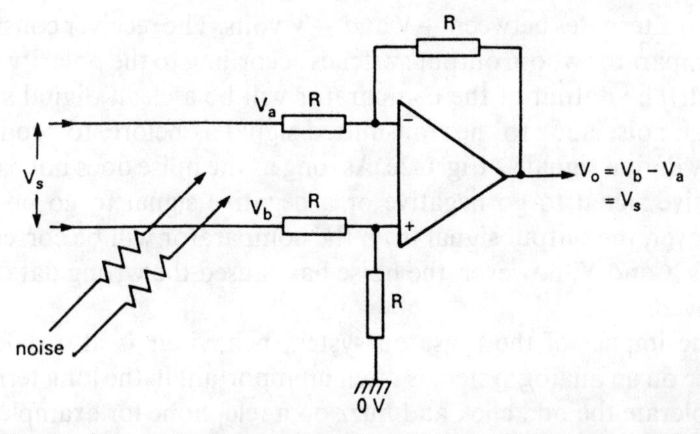

Fig. 6.1 Removal of common mode noise by a differential amplifier.

Noise affects analog and digital signals in different ways. An analog signal corrupted by noise as Fig. 6.2a is indistinguishable from an uncorrupted signal and the noise will be considered as signal unless it can be separated or removed by filtering. Noise will not, however, affect a digital system until a certain critical level is reached.

In Fig. 6.2b a digital transmission system sends data as a signal

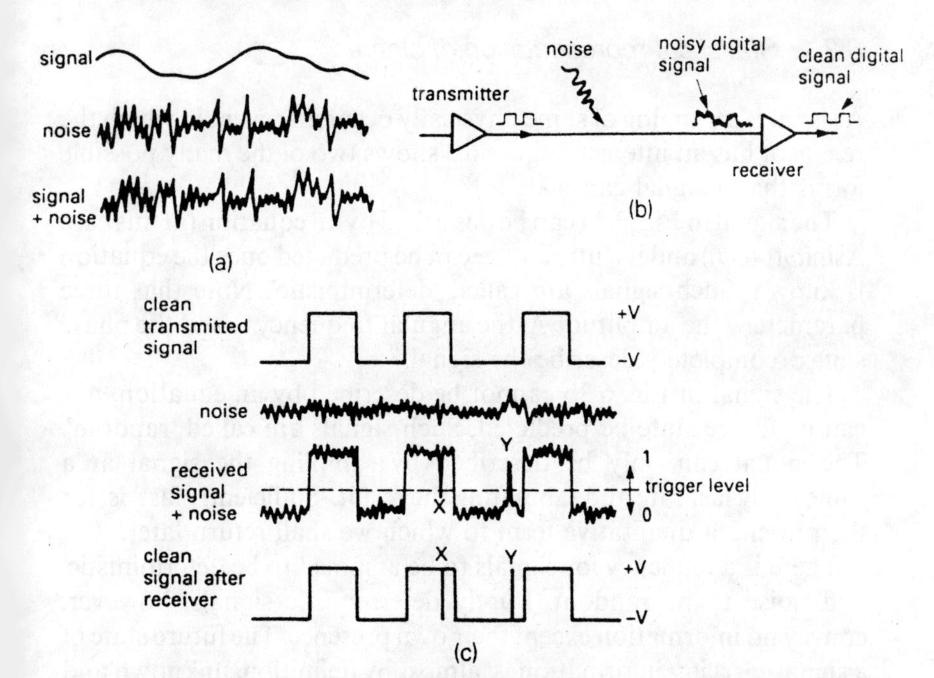

Fig. 6.2 The effect of noise on analog and digital signals, (a) Effect of noise on an analog signal, (b) Digital transmission system; (c) Effect of noise on digital signals.

which alternates between $+$V and $-$V volts. The receiver consists of a comparator whose output switches according to the polarity of the input. The output of the comparator will be a clean digital signal.

The noise adds to the transmitted signal as before, to produce a noisy digital signal as Fig. 6.2c. As long as the noise does not cause a positive signal to go negative or a negative signal to go positive, however, the output signal from the comparator will be correct. At times X and Y, however, the noise has caused the wrong data to be received.

The impact of the noise on system behaviour is also different. Noise on an analog system is often unimportant in the long term; we all tolerate the odd click and buzz on a telephone for example, and the odd kick of an indicator can be accepted in many applications. Although digital systems have an inherent noise rejection capability, a single error could have disastrous results in, say, the transmission of a computer data file or the digital control signals in a telemetry network.

6.2. Signals

6.2.1. Statistical representation of signals

A signal is an analog of some physically occurring event in which the recipient has an interest. Figure 6.3 shows two of the many possible forms that a signal can take.

The signal in Fig. 6.3a can be described by an equation (in this case $A\sin(\omega t + \phi)$) and its future state can be predicted once the equation is known. Such signals are called 'deterministic'. Note that three parameters, the amplitude A, the angular frequency ω and the phase shift ϕ completely describe the signal.

The signal in Fig. 6.3b cannot be described by an equation, nor can its future state be predicted. Such signals are called 'random'. The signal can only be described by sampling the signal at a sufficiently fast rate and tabulating the result. 'Sufficiently fast' is, for the present, a qualitative term to which we shall return later.

There is a tendency for signals to be assumed to be deterministic, and noise to be random. Purely deterministic signals, however, convey no information except their own presence. The future state of a signal carrying information is, almost by definition, unknown and as such must be considered random. Noise signals may be random but are often deterministic (mains induced 50/60 Hz hum, for example).

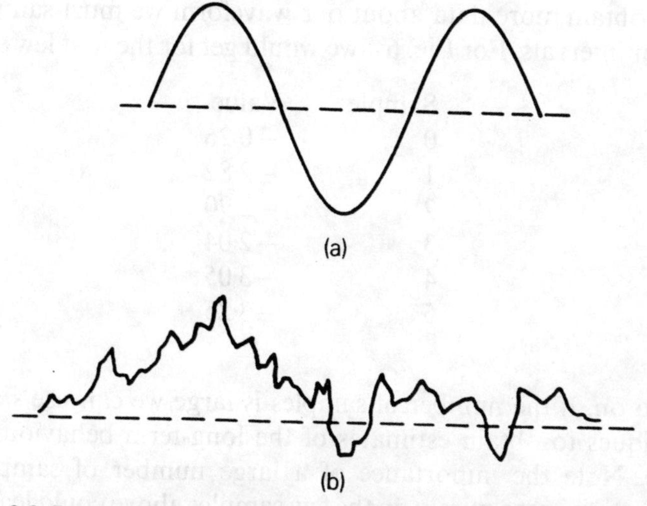

Fig. 6.3 Deterministic and random signals, (a) A deterministic signal; (b) A random signal.

Although a random signal can only be described by a table of samples, its characteristics can be described by statistical methods.

Figure 6.4 shows a random signal (generated by a computer program actually, but similar to many naturally occurring signals). This can be seen to vary over the period of observation, in the range -7 to $+7$. It can also be seen, intuitively, that the signal is varying around zero, and showing no tendency to increase or decrease in the long term.

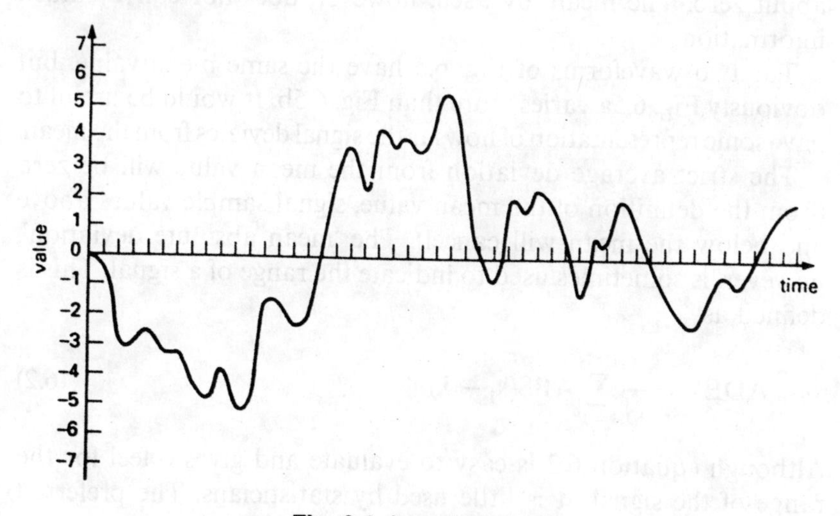

Fig. 6.4 A random signal.

To obtain more data about our waveform we must sample it at regular intervals. For Fig. 6.4 we would get for the first few samples

Sample	Value
0	−0·26
1	−2·82
2	−2·90
3	−2·04
4	−3·05
5	−3·25

and so on. If the number of samples is large we can use statistical techniques to obtain estimates of the long-term behaviour of the signal. Note the importance of a large number of samples. An observation based purely on the few samples above could lead to the erroneous observation that the signal was becoming increasingly negative.

The first statistical quantity of interest is the arithmetic mean. If we have N samples x_1 to x_n, the mean value $\bar{x}$ is defined as

$$\bar{x} = \frac{1}{N}\sum_{i=1}^{N} x_i \tag{6.1}$$

i.e. the sum of the sample divided by the number of samples.

The mean for Fig. 6.4 evaluates for the fifty samples shown to −0·05, confirming our intuitive impression that the signal is varying about zero. The mean, by itself, however, does not convey much information.

The two waveforms of Fig. 6.5 have the same mean value, but obviously Fig. 6.5a varies more than Fig. 6.5b. It would be useful to have some representation of how far the signal deviates from the mean.

The strict average deviation from the mean value will be zero (from the definition of the mean value, signal sample values above and below the mean will cancel). The 'mean absolute deviation', however, is sometimes used to indicate the range of a signal. This is defined as

$$\text{ADEV} = \frac{1}{N}\sum_{i=1}^{i=N} \text{ABS}(x_i - \bar{x}) \tag{6.2}$$

Although equation 6.2 is easy to evaluate and gives a feel for the range of the signal, it is little used by statisticians. The preferred measure of a signal variability is the 'variance', defined as

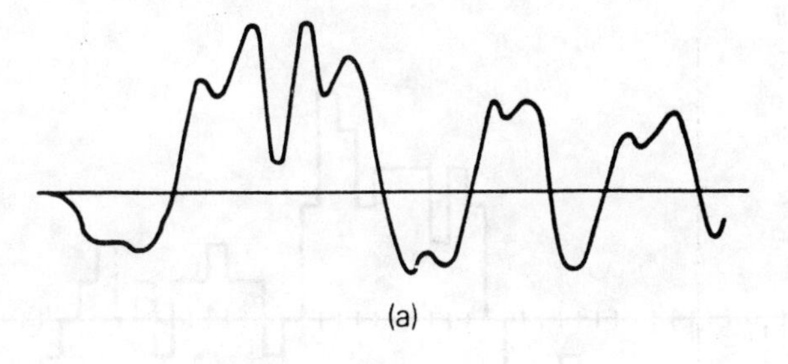

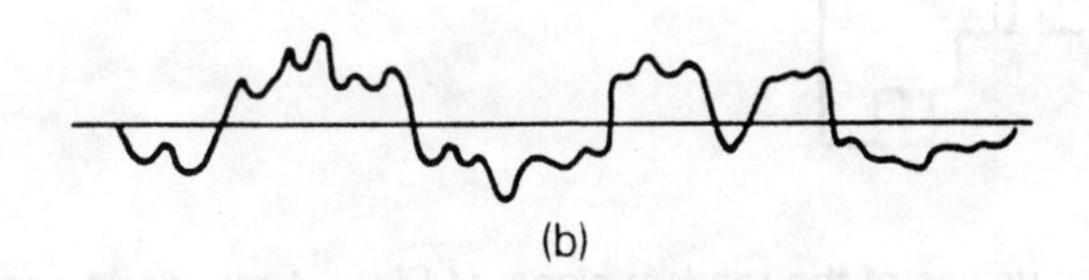

Fig. 6.5 The two traces have the same mean value, but trace (a) varies more than trace (b).

$$\mathrm{Var} = \frac{1}{N-1}\sum_{i=1}^{i=N}(x_i - \bar{x})^2 \tag{6.3}$$

and the 'standard deviation', denoted by σ, which is the square root of the variance, i.e.

$$\sigma = \sqrt{\mathrm{Var}} = \sqrt{\frac{1}{N-1}\sum_{i=1}^{i=N}(x_i - \bar{x})^2} \tag{6.4}$$

Often N is substituted for N − 1 in the denominators of equations 6.3 and 6.4. Strictly speaking, (N − 1) is used when the mean is derived from the same data as the samples, and N used when the mean is known from other data. If, however, so few samples are provided that the use of N or (N − 1) makes a significant difference to the result, the motives of the originator should be considered highly suspect.

For the signal of Fig. 6.4, the mean value is −0·05, the mean absolute deviation is 2·23, the variance is 7·15 and the standard deviation is 2·67.

Equations 6.1 and 6.4 can be applied to a deterministic signal. If the signal is described by x(t), and is observed for a time T_1 to T_2 we have

$$\text{Mean} \qquad \bar{x} = \frac{1}{T_2 - T_1}\int_{T_1}^{T_2} x(t)\,dt \tag{6.5}$$

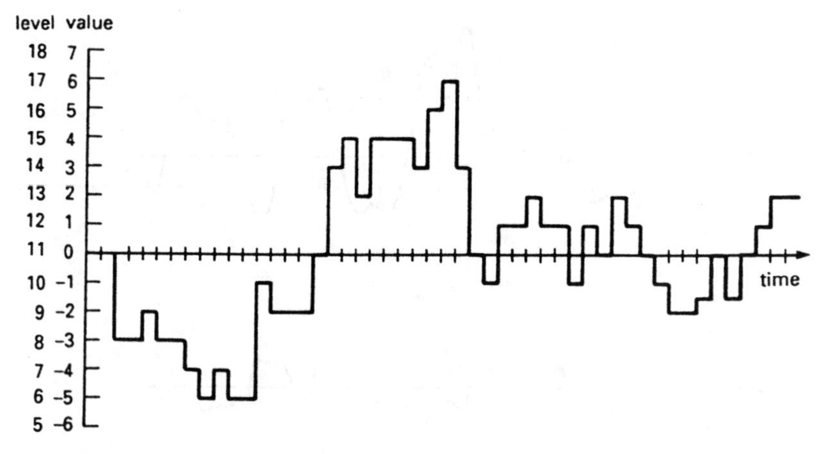

Fig. 6.6 The values of the random signal of Fig. 6.4 rounded to nearest integer value.

$$\text{Variance} \quad \text{Var} = \frac{1}{T_2 - T_1}\int_{T_1}^{T_2} (x(t) - \bar{x})^2 \, dt \tag{6.6}$$

$$\text{and standard deviation} \qquad \sigma = \sqrt{\text{Var}} \tag{6.7}$$

In Fig. 6.6 the signal has been sampled into twenty possible levels, each one unit wide. Level 1 corresponds to samples with values between −10·5 and −9·5, level 2 samples with values between −9·5 and −8·5 and so on. For Fig. 6.6, the number of samples occurring in each level are

Level	*Range*	*Number of samples*
5	−6·5 to −5·5	0
6	−5·5 to −4·5	3
7	−4·5 to −3·5	2
8	−3·5 to −2·5	4
9	−2·5 to −1·5	6
10	−1·5 to −0·5	6
11	−0·5 to 0.5	7
12	0·5 to 1·5	7
13	1·5 to 2·5	6
14	2·5 to 3·5	3
15	3·5 to 4·5	4
16	4·5 to 4·5	1
17	5·5 to 6·5	1
18	6·5 to 7·5	0

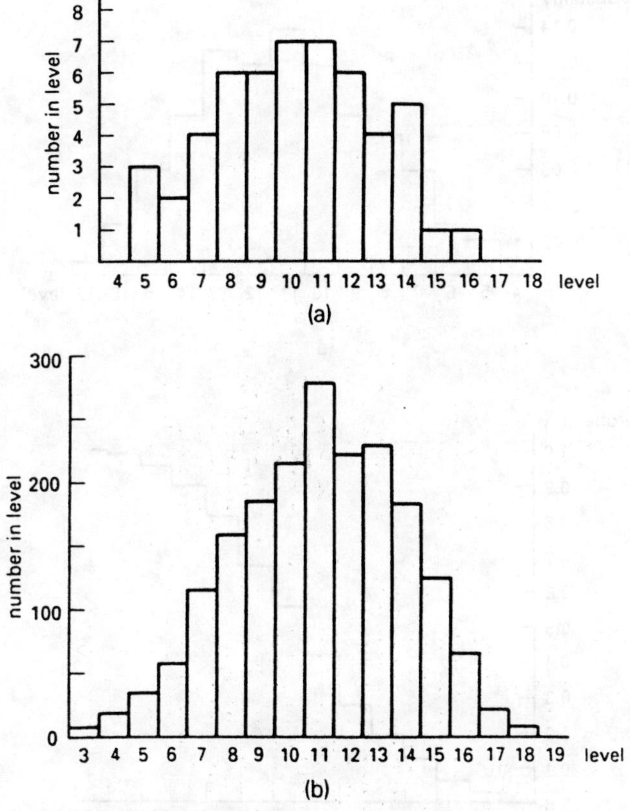

Fig. 6.7 Plot of random values in histogram form, (a) Histogram of values from Fig. 6.4 and 6.6; (b) Histogram of values from an extended run of 2000 samples.

This data is plotted in histogram form in Fig. 6.7a, along with data from an extended observation of 2000 samples in Fig. 6.7b. From these we can deduce the probability that the signal will be within any given range. The probability that the signal lies in the range −2·5 to −1·5 is

$$\text{probability} = \frac{\text{number of samples in range 9}}{\text{total number of samples}}$$

Based on Fig. 6.7b this is 184/2000 = 0·09.

In general, the probability that the signal occurs in range j is

$$P_j = \frac{\text{number of samples in range j}}{\text{total number of samples}} \tag{6.8}$$

Figure 6.8a is a redrawn version of Fig. 6.7b with probabilities

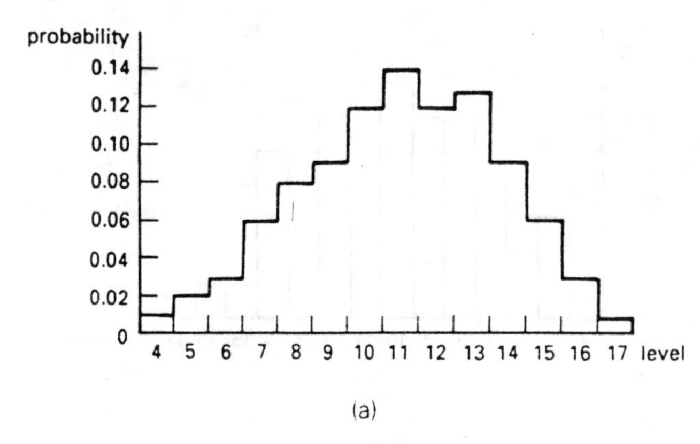

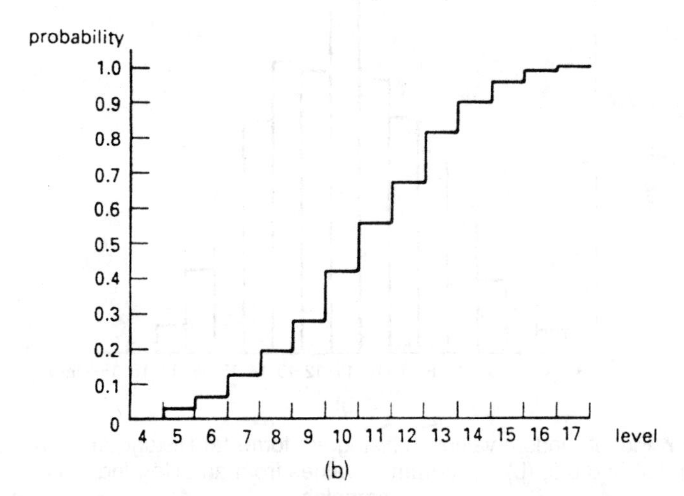

Fig. 6.8 Probability histograms derived from Fig. 6.7, (a) Probability histogram based on Fig. 6.7; (b) Cumulative probability.

rather than totals, shown for each range. Note that the total area under the curve is unity, since all samples are included. From this histogram we can see that the probability that the signal lies between $-2{\cdot}5$ and $+2{\cdot}5$ is simply the sum of the probabilities of ranges 9 to 13 inclusive, i.e.

$$
\begin{aligned}
P &= P_9 + P_{10} + P_{11} + P_{12} + P_{13} \\
&= 0{\cdot}09 + 0{\cdot}12 + 0{\cdot}14 + 0{\cdot}12 + 0{\cdot}13 \\
&= 0{\cdot}6
\end{aligned}
$$

The signal thus remains between $-2{\cdot}5$ and $+2{\cdot}5$ for about 60% of the time.

It is also useful to define the probability that the signal will lie in the first m ranges. This is known as the 'cumulative probability', and is defined as

$$C_m = \sum_{i=1}^{i=m} P_i \tag{6.9}$$

$$= \frac{1}{N} \sum_{i=1}^{i=m} x_i \tag{6.10}$$

where N is the total number of samples as before. Based on Fig. 6.8a, for example

$$C_8 = 0 + 0 + 0 + 0{\cdot}01 + 0{\cdot}02 + 0{\cdot}03 + 0{\cdot}06 + 0{\cdot}08$$
$$= 0{\cdot}20$$

i.e. there is a 20% probability that the signal will have a value more negative than $-2{\cdot}5$.

The cumulative probability is shown in Fig. 6.8b for the signal represented by Fig. 6.8a. Note that this approaches unity for higher ranges as all samples have been included.

The probability that a signal is more negative than a certain value can be read directly from its cumulative probability graph. From Fig. 6.8b, for example, the probability that the signal of Fig. 6.4 is more negative than $-1{\cdot}5$ units is 0·29.

Figure 6.8b is based on regions with a width of one unit. As the width of the units becomes less, the histogram sections become thinner, and ultimately a continuous curve results as the width tends to zero. This curve, shown in Fig. 6.9, is called the cumulative probability function. The probability that the signal is more negative than a specified value can be read from Fig. 6.9 in the same way as from Fig. 6.8b. Note that the probability that the signal is negative, for our example, is 0·5. As before, the function approaches unity for large values of x.

If the cumulative probability function is differentiated, the bell shape curve of Fig. 6.10 results. This is known as the probability density function, and for most random events has the form

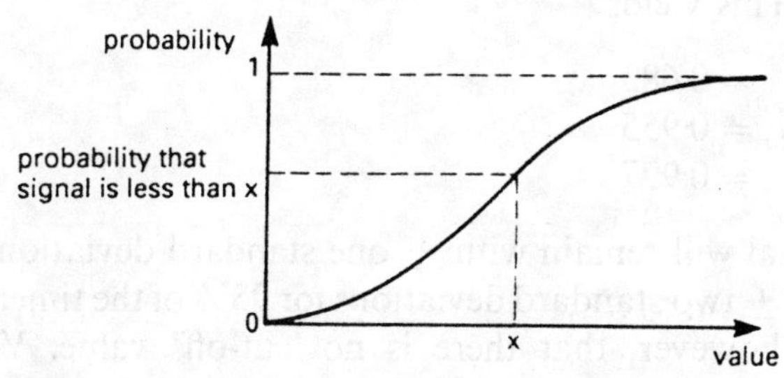

Fig. 6.9 Continuous cumulative probability function.

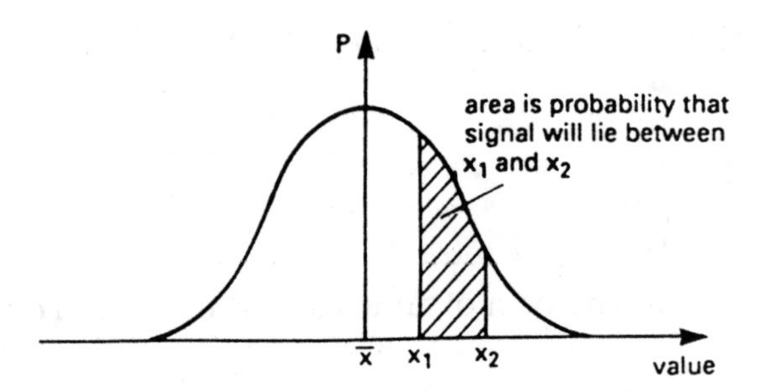

Fig. 6.10 The probability density function.

$$p(x) = \frac{1}{\sqrt{2\pi}\sigma} \exp\left[-\frac{(x - \bar{x})^2}{2\sigma^2} \right] \tag{6.11}$$

where σ is the standard deviation and $\bar{x}$ the mean value.

Equation 6.11 is remarkably pervasive, occurring in statistical studies as widely separated as traffic flows, life expectancies, experimental errors, Pascal's triangle and so on. It is known as the 'normal' or 'Gaussian' probability density function.

The area under the curve is unity, and it conveys similar information to the histogram of Fig. 6.8a. In particular, the probability of a signal lying between two values is simply the area under the curve between those two values. The shaded area of Fig. 6.10 is the probability that the signal will lie between x_1 and x_2. Mathematically this is equivalent to

$$\text{probability signal lies between } x_1 \text{ and } x_2 = P_{x_1 x_2} = \int_{x_1}^{x_2} p(x)\,dx \tag{6.12}$$

Equation 6.12 can be used to evaluate the probability that the signal lies in the range $\pm\sigma$, $\pm 2\sigma$, $\pm 3\sigma$, etc. where σ is the standard deviation. This yields

$$P_{-\sigma,\sigma} = 0{\cdot}683$$
$$P_{-2\sigma,2\sigma} = 0{\cdot}955$$
$$P_{-3\sigma,3\sigma} = 0{\cdot}997$$

i.e. the signal will remain with $\pm$ one standard deviation for 68% and within $\pm$ two standard deviations for 95% of the time. It should be noted, however, that there is no 'cut-off' value. Very large deviations are not impossible, just highly unlikely. This fact assumes great importance in the discussion of the effect of noise on digital data transmission.

6.2.2. Power spectral density

A periodic signal f(t) such as Fig. 6.11 can be represented as the sum of a series of sine and cosine waves with frequencies that are integer multiples (i.e. harmonics) of the fundamental frequency. If the period of the original signal is T sec (giving a fundamental frequency $\omega_1 = 2\pi/T$) we can write:

$$f(t) = a_0 + \sum_{n=1}^{n=\infty} a_n \cos n\omega_1 t + \sum_{n=1}^{n=\infty} b_n \sin n\omega_1 t \qquad (6.13)$$

$$\text{where} \qquad a_n = \frac{2}{T}\int_{-T/2}^{+T/2} f(t) \cos n\omega_1 t \; dt \qquad (6.14)$$

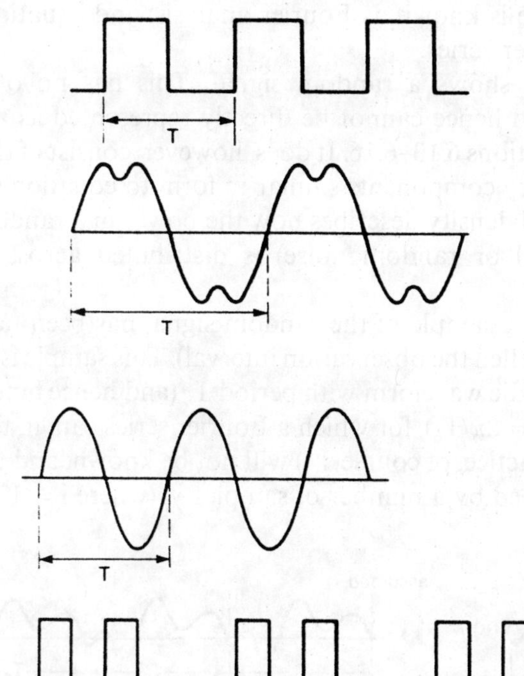

Fig. 6.11 Various periodic signals.

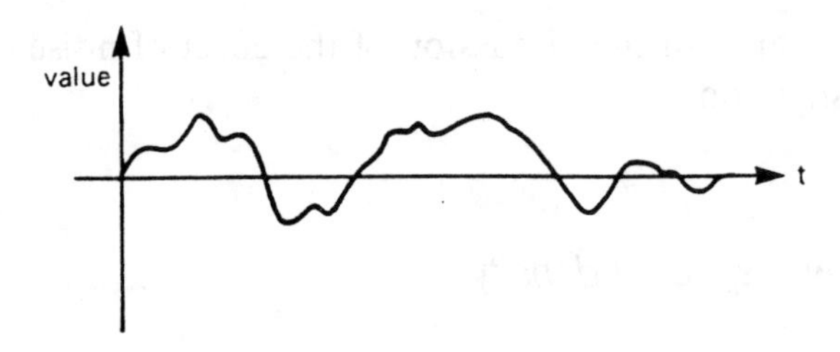

Fig. 6.12 A random signal with no observable periodicity.

$$b_n = \frac{2}{T}\int_{-T/2}^{+T/2} f(t) \sin n\omega_1 t \, dt \tag{6.15}$$

and
$$a_0 = \frac{1}{T}\int_{-T/2}^{+T/2} f(t) \, dt \tag{6.16}$$

$$= \text{mean value of f(t) over one period}$$

This technique is known as Fourier analysis, and equation 6.13 is called a Fourier series.

Figure 6.12 shows a random signal. This has no observable periodicity and hence cannot be directly represented as a Fourier series via equations 6.13–6.16. It does, however, consist of the sum of many frequency components similar in form to equation 6.13. The power spectral density describes how the power in a random signal (wanted signal or random noise) is distributed across different frequencies.

In Fig. 6.13 a sample of the random signal has been taken with duration T_o (called the observation interval). This sample is repeated to give a periodic waveform with period T_o (and hence fundamental frequency $\omega_1 = 2\pi/T_o$), for which a Fourier series can, in theory, be derived. In practice, of course, f(t) will not be known, and the signal will be described by a number of samples y_i (where i = 1 to N, the

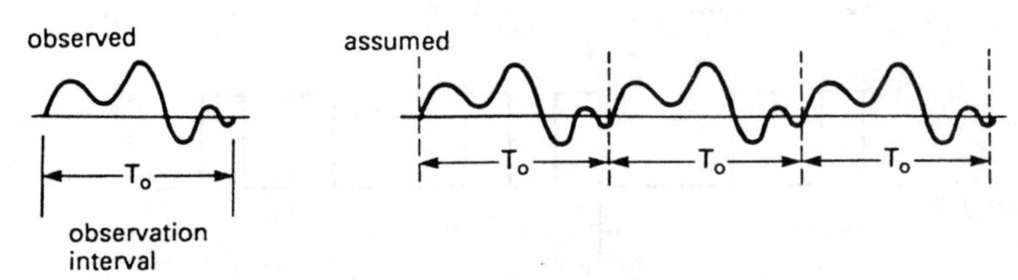

Fig. 6.13 A random signal with forced periodicity.

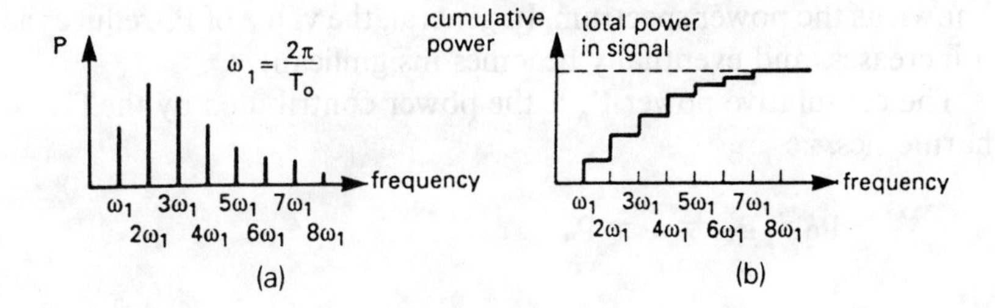

Fig. 6.14 Spectral representation of signal with observation interval T_o, (a) Spectral representation of power; (b) Cumulative power.

number of samples in the observation interval). Equations 6.14 and 6.15 can then be rewritten:

$$a_n = \frac{2}{N}\sum_{i=1}^{i=N} y_i \cos\left(\frac{2\pi ni}{N}\right) \tag{6.17}$$

$$b_n = \frac{2}{N}\sum_{i=1}^{i=N} y_i \sin\left(\frac{2\pi ni}{N}\right) \tag{6.18}$$

As before, a_o is the mean of the signal over the observation interval. Applying 6.17 and 6.18 allows the periodic signal of Fig. 6.14 to be represented as a Fourier series.

To use this series to describe the nature of the random signal we consider the power contributed by each harmonic.

If a voltage signal $a_n \cos(n\omega_1 t)$ is applied to a one ohm resistor, the mean power, in watts, dissipated over the observation interval is

$$P_n = \frac{1}{T_o}\int_0^{T_o} a_n^2 \cos^2 n\omega_1 t \, dt$$

$$= \frac{a_n^2}{2} \tag{6.19}$$

since $\omega_1 = 2\pi/T_o$ and n is an integer.

Similarly, the mean power for the sine wave component is $b_n^2/2$ watts. The power contributed by the nth harmonic (with frequency $n\omega_1$) is thus

$$P_n = \frac{1}{2}(a_n^2 + b_n^2) \tag{6.20}$$

The power in the signal can thus be represented by a series of spectral lines as Fig. 6.14a. Note that these occur at multiples of ω_1. This is known as the power spectrum. In general, the value of P_n reduces as n increases, and eventually becomes insignificant.

The cumulative power P_n is the power contributed by the first n harmonics, i.e.

$$P_n = p_0 + p_1 + \ldots + p_n$$

$$= \sum_{i=0}^{i=n} p_i \qquad (6.21)$$

This will have the form of Fig. 6.14b, tending towards a constant value as n becomes large. The limit represents the total power in the signal.

If the observation interval T_o is increased ω will decrease, and the cumulative power tends towards a continuous curve as Fig. 6.15a. This is called the cumulative power function (cpf) and is defined as

$$P(\omega) = \lim_{T_o \to \infty} P_n \quad \text{watts} \qquad (6.22)$$

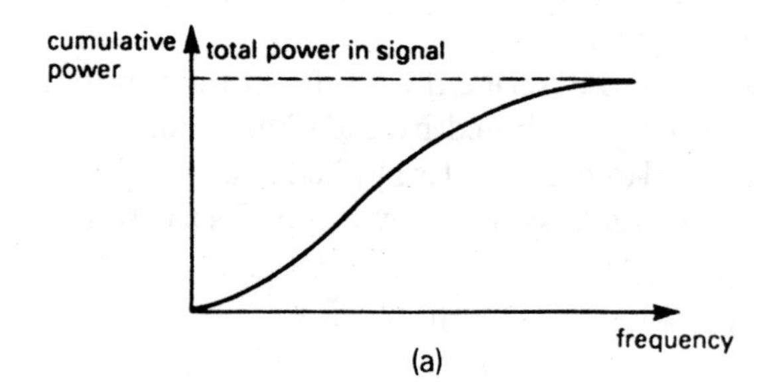

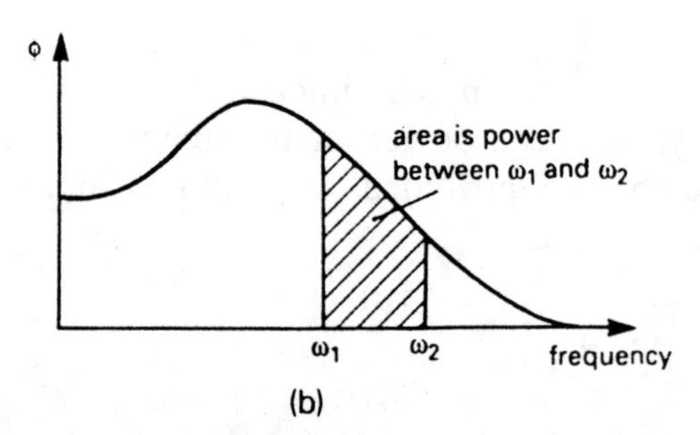

Fig. 6.15 Spectral representation as observation interval, T_o, becomes large, (a) Cumulative power function; (b) Power spectral density.

The continuous equivalent of Fig. 6.14a is the power spectral density, denoted by $\phi(\omega)$, which is defined as the derivative of the cumulative power function, i.e.

$$\phi(\omega) = \frac{dP}{d\omega} \qquad \text{watts/Hz} \tag{6.23}$$

This is shown in Fig. 6.15b.

The power spectral density can be used to calculate the power in a range of frequencies. The power in a frequency range is simply the area under the curve between the two frequencies, i.e.

$$P_{\omega_1 - \omega_2} = \int_{\omega_1}^{\omega_2} \phi(\omega) \quad d\omega \text{ watts} \tag{6.24}$$

Note that the power spectral density function is used to determine the relative power in frequency ranges. The power contributed by a single frequency is zero.

For specific signals of interest the mean power and the power spectral density are constant providing the observation interval T_o is sufficiently long. They can hence be used to describe the character of a random signal.

6.3. Noise

6.3.1. Signal to noise ratio (SNR)

Noise effectively adds to the wanted signal, so the effect and importance of noise is related to the relative amplitudes of the noise and signal over the bandwidth of interest. The signal to noise ratio is simply the ratio of signal and noise powers expressed in decibels, i.e.

$$\begin{aligned} \text{SNR} &= 10 \log_{10}\left(\frac{W_s}{W_n}\right) \text{dB} \\ &= 10 \log_{10}\left(\frac{V_s^2}{V_n^2}\right) \text{dB} \\ &= 20 \log_{10}\left(\frac{V_s}{V_n}\right) \text{dB} \end{aligned} \tag{6.25}$$

where W_s, W_n is the noise/signal power and V_s, V_n the noise and signal RMS voltage. For Gaussian noise with zero mean V_n^2 is the variance of the noise signal.

It is important to specify the bandwidth over which the SNR is defined if the signal bandwidth is limited. For a limited bandwidth signal the SNR will fall as the bandwidth increases because the noise power will increase but the signal power remain constant.

6.3.2. Types of noise

Noise is introduced from a variety of sources. In process control, the commonest type of noise is introduced as interference from external sources such as electric motors, relay coils, fluorescent lights, thyristor drives, etc. The process itself can introduce measurement noise (e.g. turbulence in flow measurement or surface noise in level measurement). Interference can be introduced via connection leads (i.e. the transmission media), via the power supply (mains borne noise) or via common impedances in the earthing system.

The spectrum and amplitude characteristics of the noise depend on the source. Mains hum (at 50 or 60 Hz dependent on the country) has a very sharp spectrum at a single frequency. Noise from thyristor drives and fluorescent lights has components at multiples of the supply frequency. Noise from switching of inductive loads tends to be impulsive, i.e. having a spiky waveform and a wide spectrum.

Noise can be introduced into a system as series mode noise (Fig. 6.16a) or common mode noise (Fig. 6.16b). Common mode noise can usually be eliminated by use of differential or isolation amplifiers. Series mode noise is more difficult to deal with as it effectively becomes part of the signal.

Noise generated internally within an electronic circuit becomes important with low level signals from thermocouples, strain gauges and similar sensors. Electronic components are inherently noisy. A resistor generates an open circuit random voltage across its terminals (called Johnson noise). The random noise is dependent on the temperature T (Kelvin) and the resistance R (ohms). It has a flat spectrum (i.e. equal power in each Hz of frequency). Noise with this flat characteristic is often called 'white noise'. The RMS value of Johnson noise is given by

$$V_n = \sqrt{4kTRB} \qquad (6.26)$$

where k is Boltzman's constant ($1{\cdot}4 \times 10^{-23}\,JK^{-1}$) and B the bandwidth (in Hz) over which the measurement is made. A 10 Kohm resistor, for example, at room temperature (293 K) has a noise

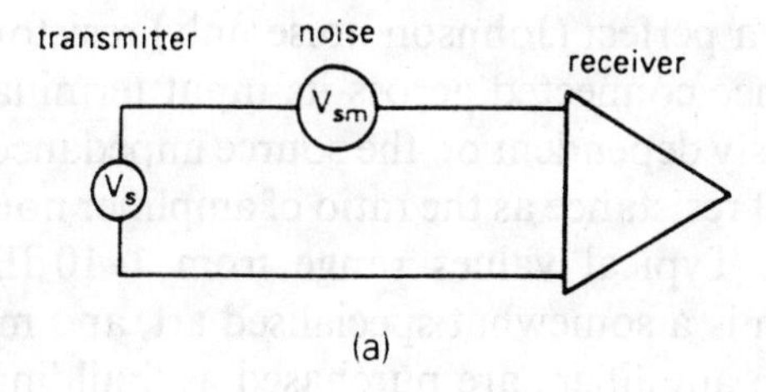

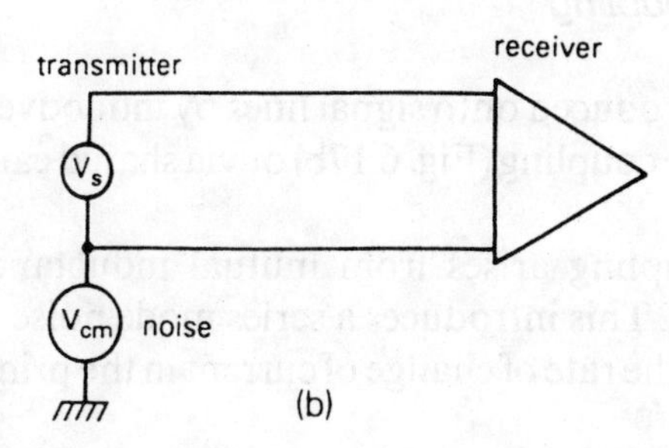

Fig. 6.16 Methods by which a signal can be degraded by noise, (a) Series mode noise; (b) Common mode noise.

voltage of just over 1 μV when measured with a bandwidth of 10 kHz.

Johnson noise is present in all resistors; in practice real resistors also exhibit small random fluctuations in resistance (called excess noise) which are proportional to the current flowing through them. These random changes in resistance produce random changes in voltage which add to Johnson noise. The noise depends on the construction of the resistor, and has a 1/f characteristic (i.e. equal power in equal decades of frequency, e.g. 1–10 Hz, 10–100 Hz). Noise with this characteristic is often called flicker noise or pink noise. Typical values range from 0·1 μV (wire wound resistors) to 3 μV (carbon resistors) per decade of frequency per applied volt.

An electric current is not a smooth flow similar to a liquid in a pipe, but is rather a passage of discrete charges. These are subject to random variations which becomes significant at low currents (typically below 1 μA). This effect is known as 'shot noise'.

All the above effects combine to degrade the signal to noise ratio in amplifiers used for low level signals. Low noise amplifiers are described by a noise figure (NF). Any real life signal source will have a source impedance, which will generate Johnson noise. The noise figure is defined as the ratio, in dB, of the noise output from the amplifier to the noise output from a perfect noiseless amplifier of the same gain with a perfect (Johnson noise only) resistor equal to the source impedance connected across its input terminals. The noise figure is obviously dependent on the source impedance, being worse for low values of resistance as the ratio of amplifier noise to Johnson noise is larger. Typical values range from 1–10 dB. Low noise amplifier design is a somewhat specialised art, and most high gain instrumentation amplifiers are purchased as 'building blocks'.

6.3.3. Noise coupling

Noise can be introduced onto signal lines by inductive coupling (Fig. 6.17a), capacitive coupling (Fig. 6.17b) or via shared earth impedances (Fig. 6.17c).

Inductive coupling arises from mutual inductance M between adjacent circuits. This introduces a series mode noise signal which is proportional to the rate of change of current in the primary circuit, i.e.

$$V_s = M\frac{di}{dt} \tag{6.27}$$

Thyristor drives, which cause very fast current changes are particularly prone to cause noise from inductive coupling. Because this is series mode noise it cannot be removed by a differential amplifier.

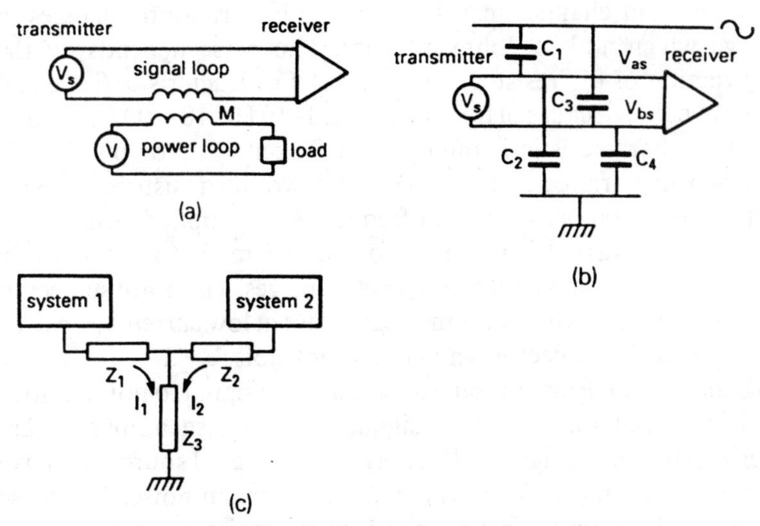

Fig. 6.17 Common methods of noise introduction, (a) Inductive coupling; (b) Capacitive coupling; (c) Shared earth impedances.

Capacitive coupling arises from the lumped capacitances between adjacent cables and earth. These form potential dividers to AC voltages in the primary circuit, producing noise voltages V_{as} and V_{bs}. This introduces a common mode noise signal (V_{bs}) and a series mode noise signal ($V_{as} - V_{bs}$). The series mode contribution is generally very small since $C_1 - C_3$ and $C_2 - C_4$ will be similar in value.

Earth borne noise arises from current changes through common impedances (Z_3 on Fig. 6.17c). If Z_3 has significant inductance, common mode voltages proportional to rate of change of current will be introduced. Earth routes should be arranged to avoid power and signal circuits having shared returns.

6.3.4. *Noise elimination*

Noise problems should be tackled in the following order:

(a) The cause of the noise.
(b) At the point of entry to the system (e.g. by increasing the distance between power and signal cables, avoiding parallel cable runs, the use of screening, etc.).
(c) Differential and isolation amplifiers at the ends of signal lines.
(d) As a last resort, filtering of the signal.

Inductive coupling can be reduced significantly by the use of twisted pair cables. The coupling between the primary circuit and the signal circuit alternates in phase between adjacent twists as shown in Fig. 6.18, effectively leading to total cancellation of the noise.

Capacitive coupling can be reduced by screening of devices and cables as shown in Fig. 6.19. The only capacitive divider effects that can occur arise from the very small screen to core capacitance. Since these will be equal for both cores, any remaining noise will be common mode which can be removed by differential amplifiers.

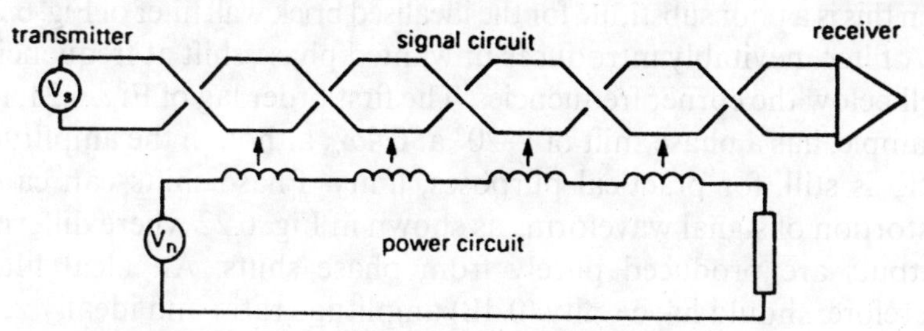

Fig. 6.18 Reduction of noise by twisted pair signal cable.

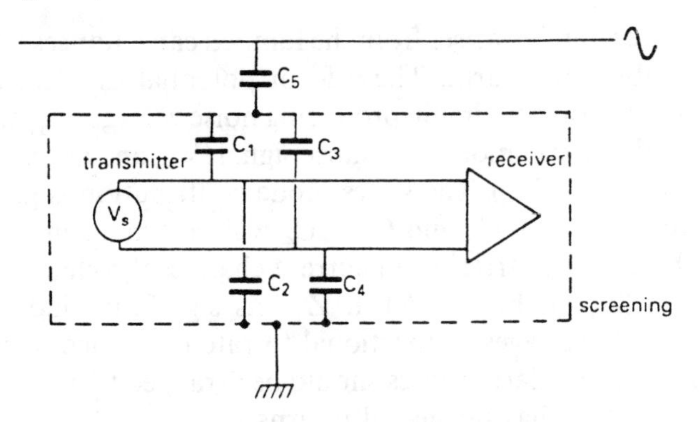

Fig. 6.19 The reduction of capacitive coupling by screening.

Note that the screening must connect to ground at one (and only one) point to avoid noise problems from ground loops.

In extremely electrically noisy environments (e.g. adjacent to electric arc furnaces or in welding shops) fibre optic cable offers a totally noise free transmission medium.

6.4. Filters

6.4.1. Introduction

If a signal and the interfering noise have spectra which do not significantly overlap as Fig. 6.20, the noise can be removed by a filter which passes the signal but blocks the noise. In Fig. 6.20 a low pass filter has been used.

A filter is generally designed to pass a range of frequencies, so the prime consideration is usually its amplitude ratio; the bandwidth (or passband) being defined as the range of frequencies over which the ratio does not fall by more than 3 dB. The −3 dB frequencies are often called the corner or break frequencies. Figure 6.21 shows the Bode diagram for a simple first order low pass filter, and as can be seen this is a poor substitute for the idealised brick wall filter of Fig. 6.20.

A filter inevitably introduces unwanted phase shift at frequencies well below the corner frequencies. The first order lag of Fig. 6.21, for example, has a phase shift of $-20°$ at $0{\cdot}4\omega_c$, although the amplitude ratio is still, for practical purposes, unity. Phase shifts can cause distortion of signal waveforms as shown in Fig. 6.22 where differing outputs are produced purely from phase shifts. An ideal filter, therefore, should have unity (0 dB) amplitude ratio and ideally zero phase shift in the pass band, with a sharp brick wall skirt beyond the corner frequencies.

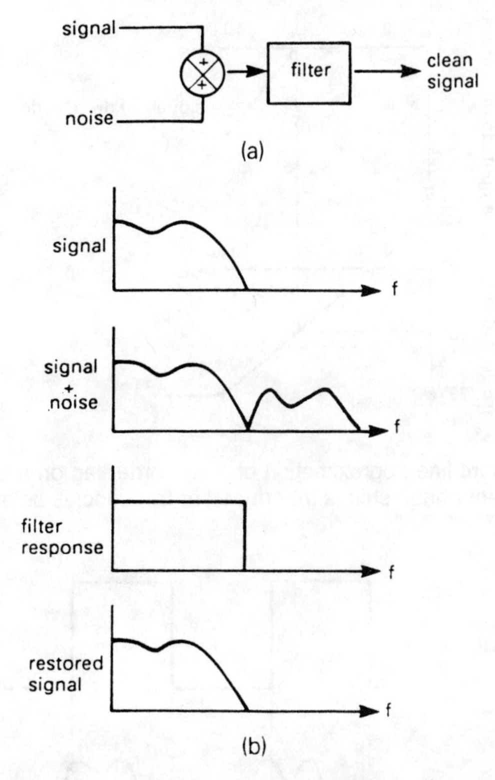

Fig. 6.20 A filter can be used to remove noise if the noise and signal spectra do not overlap significantly, (a) Filter action; (b) Noise removal by a low pass filter.

There is an alternative phase shift characteristic that can be acceptable in some circumstances. In Section 2.3.11 it was shown that a pure time delay has a phase shift which increases linearly with frequency. A filter which has the required amplitude ratio and a linearly increasing phase shift will remove unwanted noise whilst delaying (but not distorting) the wanted signal. Some of the filters of Section 6.4.3 have this characteristic, but it should be remembered that the introduction of a pure time delay into a control loop is generally destabilising.

6.4.2. Simple filter types

There are four basic filter types; low pass, high pass, bandpass and bandstop (also known as a notch filter). These are shown with various DC amplifier implementations in Figs. 6.23–6.26.

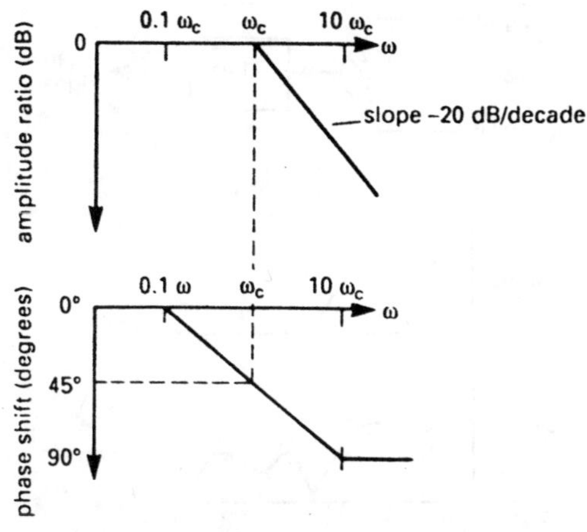

Fig. 6.21 A straight line approximation of a first order lag on a Bode diagram. Significant phase shift is introduced at frequencies below ω_c.

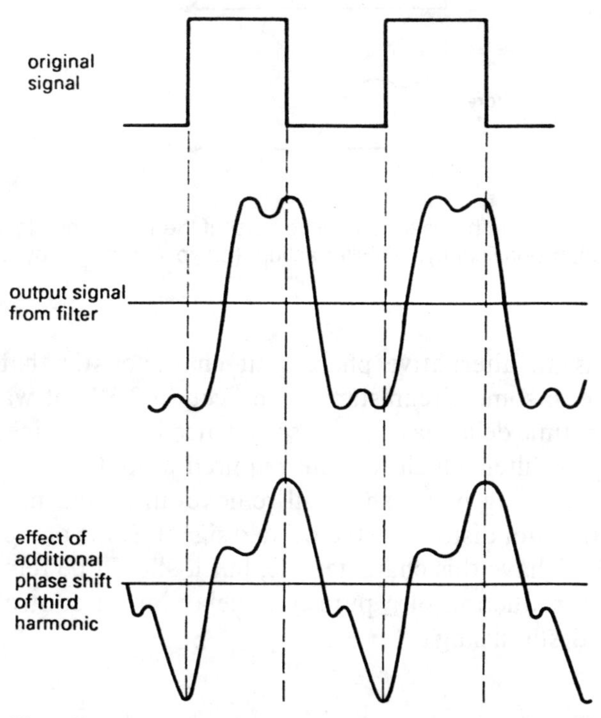

Fig. 6.22 The effect of phase shift. The bottom two signals have the same amplitude components, but different phase shifts.

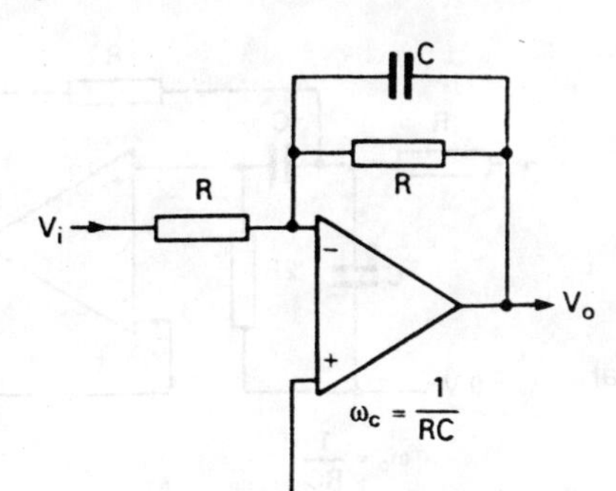

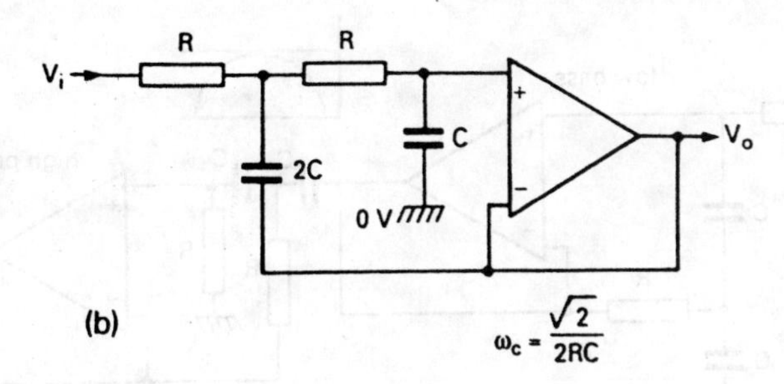

Fig. 6.23 Low pass filter circuits, (a) Simple first order filter; (b) Second order filter.

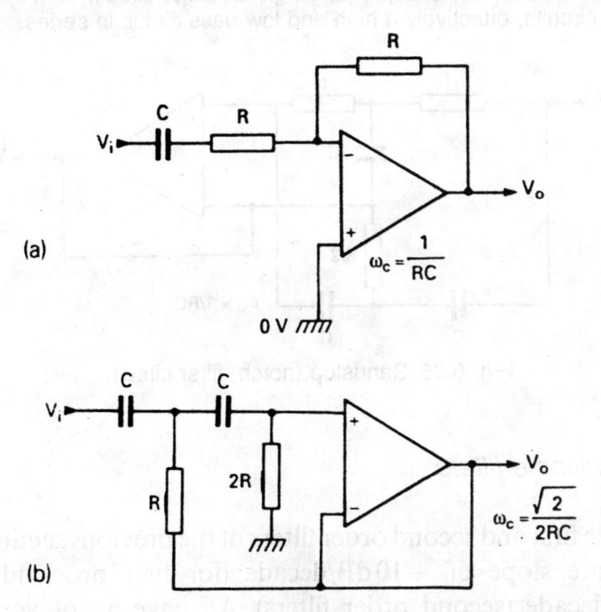

Fig. 6.24 High pass filter circuits, (a) Simple first order filter; (b) Second order filter.

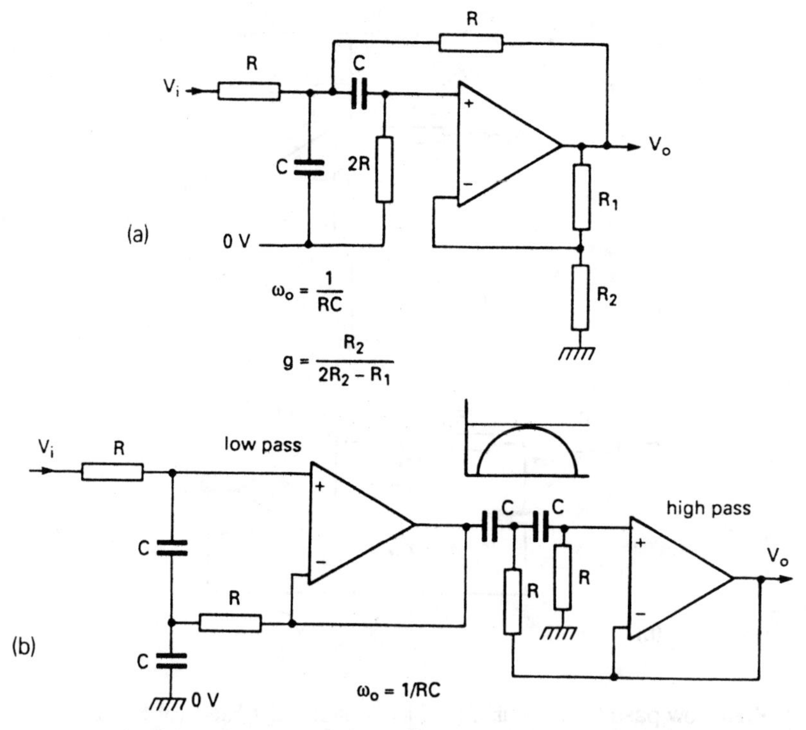

Fig. 6.25 Bandpass filter circuits, (a) Single amplifier circuit; (b) Two amplifier circuits, effectively a high and low pass circuit in series.

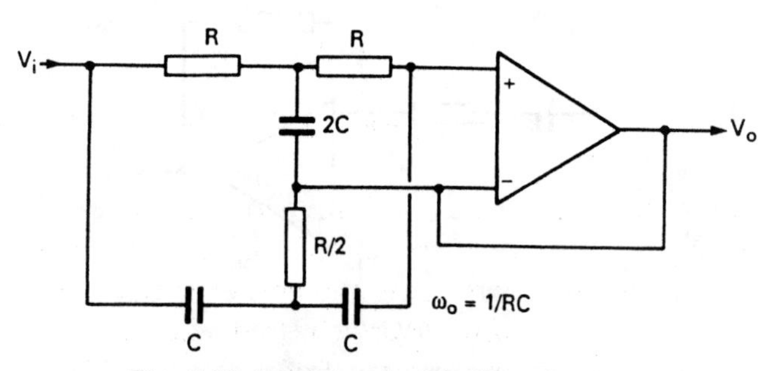

Fig. 6.26 Bandstop (notch) filter circuit.

6.4.3. Multipole filters

The simple first and second order filters of the previous sections have an ultimate slope of $-10\,\text{dB/decade}$ (for first order filters) or $-20\,\text{dB/decade}$ (second order filters). All have a not very sharp 'knee' at the corner frequency and do not really approach the ideal 'brickwall' filter characteristic.

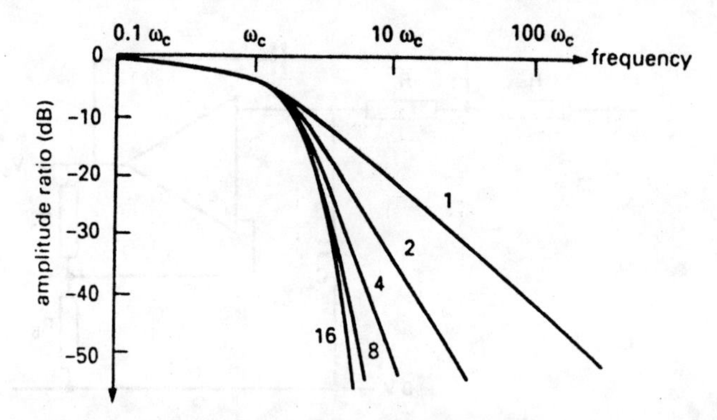

Fig. 6.27 Cascading filter circuits increase the ultimate slope, but not the sharpness of the corner.

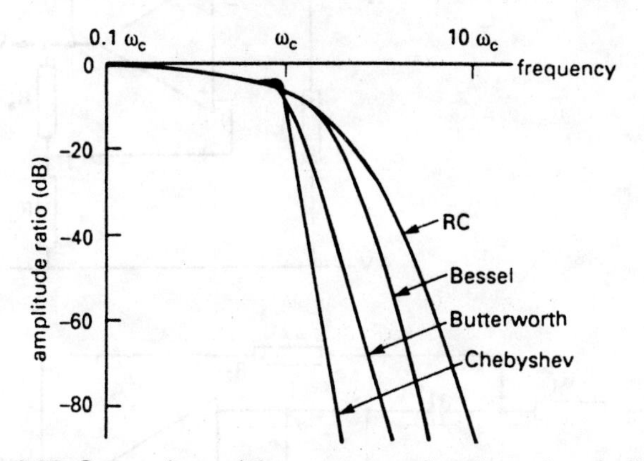

Fig. 6.28 Comparison of the response of various six pole filters.

Cascading filters increase the ultimate slope, as shown in Fig. 6.27, but does not improve the sharpness of the knee. It should also be noted that cascading filters shifts the corner frequency. Two first order low pass filters with a corner frequency of f will, when cascaded, have a new lower corner frequency of 0·6f (where each individual filter has an amplitude ratio of −1·5 dB). Figure 6.27 has been normalised for this effect.

Filters are generally classified by their number of 'poles' and their type (e.g. a two pole low pass filter). The number of poles determines the ultimate slope of 6N dB/octave (where N is the pole number). Generally there will be one capacitor or inductor per pole. A multipole filter can therefore give any desired ultimate slope, but simple cascading still gives a soft knee as shown above.

Three common filter types, shown on Fig. 6.28, approach the ideal

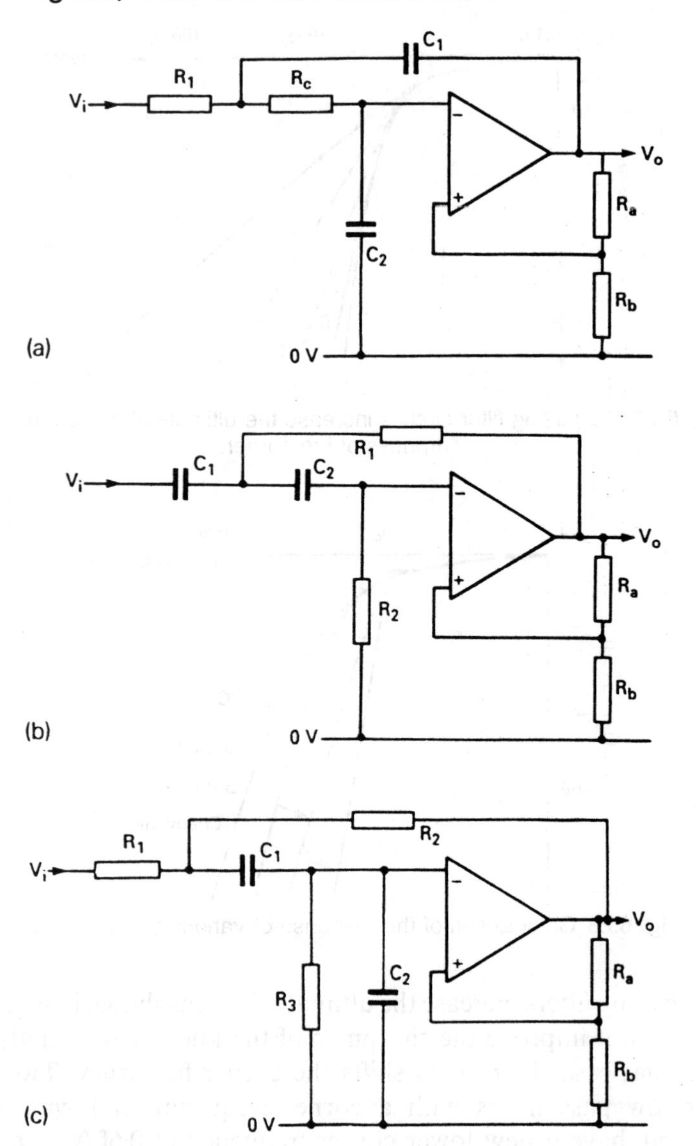

Fig. 6.29 Building block circuits for multipole filters, (a) Low pass; (b) High pass; (c) Bandpass.

brickwall characteristic. These have different design criteria; the Butterworth filter has a very flat amplitude ratio, the Chebyshev filter provides a very sharp knee but has 'ripples' in the amplitude ratio and the Bessel filter has a phase/frequency relationship which approaches a pure time delay.

These three filter types are constructed from the two pole filters of

Fig. 6.29. It follows that Bessel, Butterworth and Chebyshev filters can only have an even number of poles. The building block filters are similar to those in Figs. 6.23–6.25 but have gain K set by the values of R_a, R_b.

An N pole filter is constructed from N/2 building blocks of the required type in series. Each section (unlike simple cascading) is not, however, identical. A six pole Butterworth low pass filter, for example, is constructed from three low pass circuits of Fig. 6.29a with identical R and C values ($RC = 1/\omega_c$ where ω_c is the angular corner frequency), but gains of 1·1, 1·6 and 2·5 in the individual stages. Similarly a six pole Chebyshev low pass filter has three sections with corner frequencies of $0{\cdot}32\omega_c$, $0{\cdot}73\omega_c$ and $0{\cdot}98\omega_c$ and respective gains of 1·9, 2·6 and 2·9.

The design of Bessel, Butterworth and Chebyshev filters from first principles is somewhat complex, but tables (similar to the two preceding paragraphs) can be found in most standard electronics textbooks. Filter integrated circuits (such as the MFID manufactured by National Semiconductors) are also available with excellent application notes.

6.4.4. Signal averaging

Uncorrelated signals (e.g. noise) add as the square root of the sum of their RMS values, i.e.

$$V_{rms} = \sqrt{V_1^2 + V_2^2 + V_3^2 + \ldots} \qquad (6.28)$$

If a periodic signal is mixed with noise, addition of successive samples will cause the signal strength to increase linearly with the number of samples, but the noise will only increase as the square root of the number of samples, leading to an improvement in signal to noise ratio. The standard deviation of the signal plus noise reduces as the square root of the number of samples.

The technique can restore a signal which is totally buried in noise. This approach, called signal averaging, is not widely used in process control as it requires a signal which is periodic (or can be made periodic). It is a valuable tool, however, in scientific research where a regular stimulus can be applied to an experiment and the results averaged as Fig. 6.30. Signal averaging is also widely used in medical research.

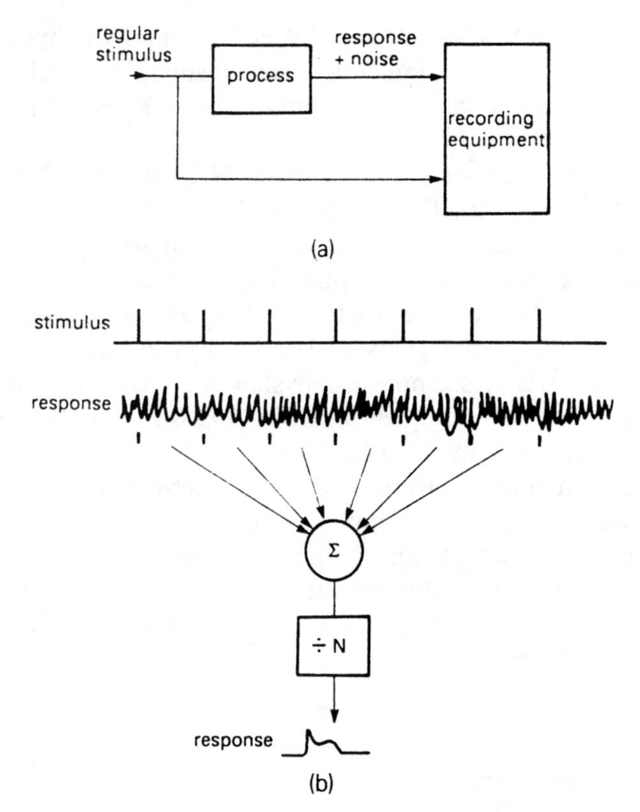

Fig. 6.30 Signal averaging, (a) Block diagram; (b) System operation.

6.5. Digital filters

Digital systems work with signals which are sampled and digitised at regular intervals. Figure 6.31 shows a representation of this procedure, where x(t) is a continuous signal, x_n is the most recent sample, x_{n-1} the previous sample and so on into the past history of x(t). The sample time is Δt, and the system can be considered as a form of shift register.

A first order filter with time constant T seconds can be represented in continuous form by

$$T\frac{dy}{dt} + y = x \tag{6.29}$$

In a sampled system, all that are available are sampled input values x_n, x_{n-1}, etc. and sampled output values y_n, y_{n-1}, etc. The rate of change of output signal approximates to

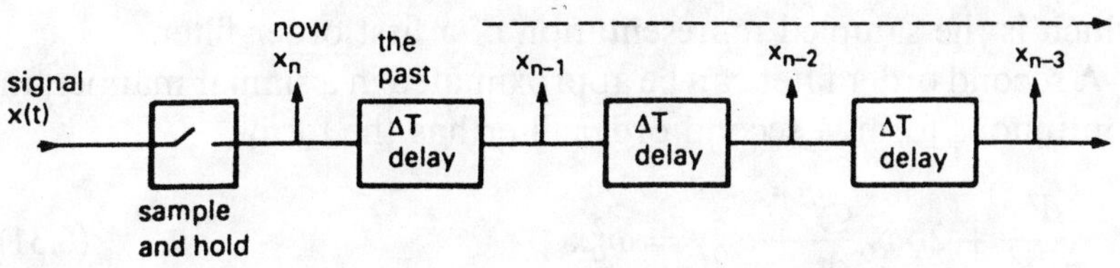

Fig. 6.31 Representation of a digital system.

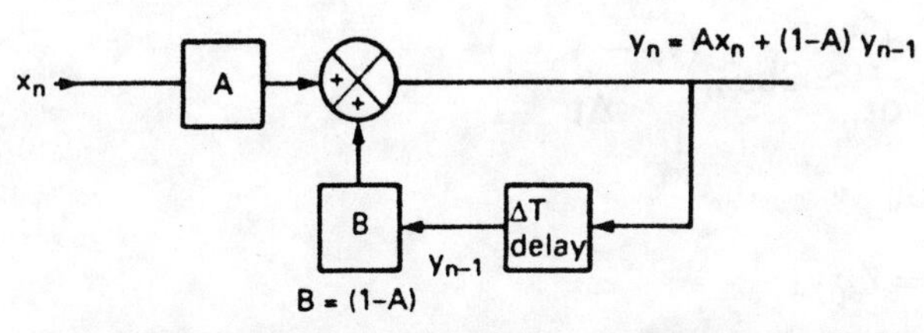

Fig. 6.32 Sampled representation of a first order filter.

$$\frac{dy}{dt} \simeq \frac{\Delta y}{\Delta t} = \frac{y_n - y_{n-1}}{\Delta t}$$

Substituting into equation 6.29 gives

$$Ty_n - Ty_{n-1} + y_n\Delta t = x_n\Delta t$$

Solving for y_n gives

$$y_n = \frac{y_{n-1}}{(1 + \Delta t/T)} + \frac{x_n\Delta t/T}{(1 + \Delta t/T)}$$

which has the form of Fig. 6.32 where A and B are constant multipliers

$$A = \frac{\Delta t/T}{(1 + \Delta t/T)} \qquad B = \frac{1}{(1 + \Delta t/T)}$$

If the sample time Δt is much less than the time constant T (as it should be in a well-designed filter), we can approximate

$$\left(1 + \frac{\Delta t}{T}\right)^{-1} = 1 - \frac{\Delta t}{T}$$

by ignoring terms in $(\Delta t/T)^2$. This leads to constant values:

$$A = \frac{\Delta t}{T} \qquad B = 1 - \frac{\Delta t}{T} = 1 - A$$

Hence $\quad y_n = Ax_n + (1 - A)y_{n-1} \qquad (6.30)$

which is the sampled representation of a first order filter.

A second order filter can be approximated in a similar manner. In continuous form, a second order filter has the form

$$\frac{d^2y}{dt^2} + 2b\omega_0\frac{dy}{dt} + \omega_0^2 y = \omega_0^2 x \tag{6.31}$$

where ω_0 is the natural frequency, and b the damping factor.

From the first order filter above, we can write

$$2b\omega_0\frac{dy}{dt} \simeq 2b\omega_0\frac{(y_n - y_{n-1})}{\Delta t} \tag{6.32}$$

and obviously

$$\omega_0^2 y = \omega_0^2 y_n \tag{6.33}$$

To express d^2y/dt^2 in sampled form, we note

$$\frac{d^2y}{dt^2} \simeq \frac{(\text{slope at } y_n) - (\text{slope at } y_{n-1})}{\Delta t}$$

$$= \frac{(y_n - y_{n-1}) - (y_{n-1} - y_{n-2})}{\Delta t^2}$$

$$= \frac{y_n - 2y_{n-1} + y_{n-2}}{\Delta t^2} \tag{6.34}$$

Combining equations 6.31–6.34 gives

$$\frac{y_n - 2y_{n-1} + y_{n-2}}{\Delta t^2} + \frac{2b\omega_0(y_n - y_{n-1})}{\Delta t} + \omega_0^2 y_n = \omega_0^2 x_n$$

Some laborious, but straightforward, reorganisation yields a value for y_n in terms of x_n and previous values of y_{n-1}, y_{n-2}:

$$y_n = \frac{2(1 + b\omega_0\Delta t)y_{n-1} - y_{n-2} + \omega_0^2\Delta t^2 x_n}{(1 + 2b\omega_0\Delta t + \omega_0^2\Delta t)}$$

which can be more simply represented as

$$y_n = b_1 y_{n-1} + b_2 y_{n-2} + a_0 x_n \tag{6.35}$$

where a_0, b_1 and b_2 are multiplication constants. Note that b_2 is negative. Equation 6.35 can be represented by Fig. 6.33.

In general, digital filters occur in two forms. Figure 6.32 and Fig. 6.33 are examples of recursive filters which have the generalised form

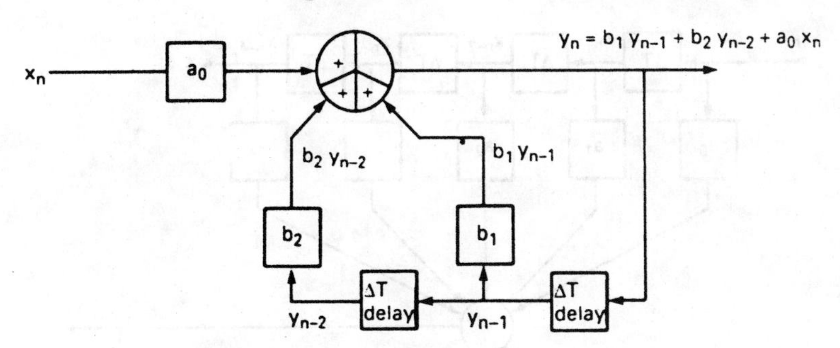

Fig. 6.33 Sampled representation of a second order filter.

of Fig. 6.34, where the output depends on previous samples of input and output. $a_0 - a_n$ and $b_1 - b_m$ are constants.

A simpler, non-recursive, form shown in Fig. 6.35 simply forms a weighted average of previous input values. Generally, but not necessarily, $a_0 > a_1 > a_2$ etc.

6.6. Modulation

6.6.1. Introduction

Modulation is a technique where the characteristics of a high frequency wave are modified by a lower frequency signal waveform. The signal information is conveyed by the higher frequency wave which is usually called the 'carrier'. The process can be summarised by Fig. 6.36. The terms 'high' and 'low' refer to the relationship between the signal and the carrier; the signal from a transducer with a bandwidth of a few tens of hertz can be modulated onto a carrier of 10 kHz. Speech has a bandwidth of around 3 kHz and requires a carrier of 100 kHz. Television signals, with a bandwidth of over 5 MHz, need at least a VHF carrier in the region of 160 MHz.

Modulation may be used for several reasons. In many applications it is necessary anyway because the transmission medium cannot support low frequency signals. Low frequency electromagnetic waves, for example, do not propagate far, so modulation of a high frequency carrier is essential for radio links. Similarly the public telephone network, designed for audio signals in the range 300 Hz–3 kHz, uses transformers and AC amplifiers. The telephone system, therefore, cannot pass digital signals directly since a long string of ones or zeros implies bandwidth down to DC (0 Hz).

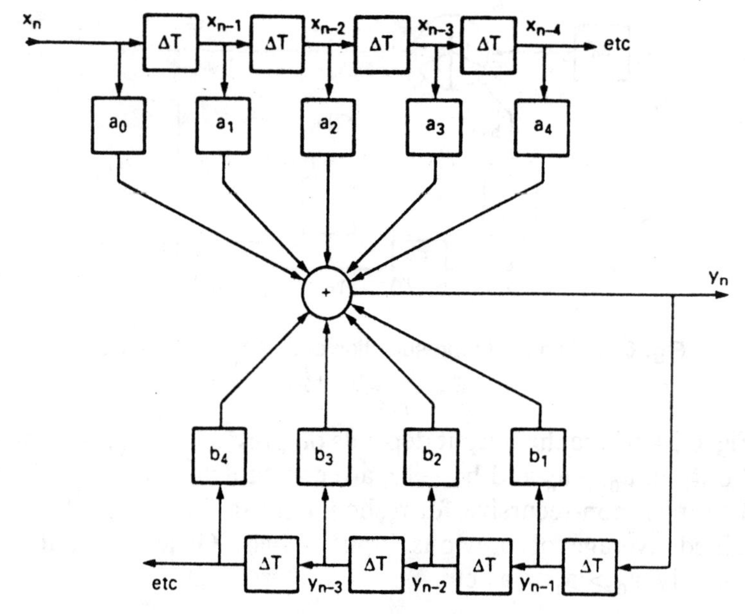

Fig. 6.34 General form of sampled recursive filter.

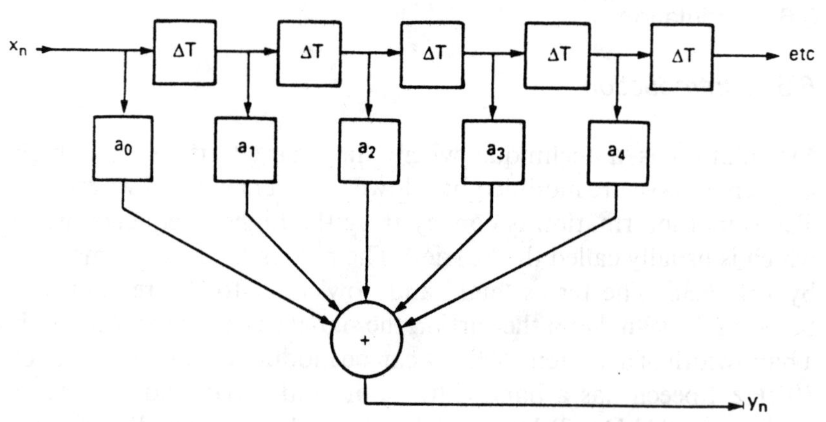

Fig. 6.35 General form of sampled non-recursive filter.

If the bandwidth of a signal is small compared with the available bandwidth on the transmission channel, modulation can allow the channel to be shared by several signals, with an obvious cost saving. As will be seen later, a modulated signal has a bandwidth similar to the original signal, but shifted to a higher frequency. Signals can thus be 'stacked' as Fig. 6.37. This technique is known as 'frequency domain multiplexing' or FDM.

Modulation can improve the signal to noise ratio by moving the

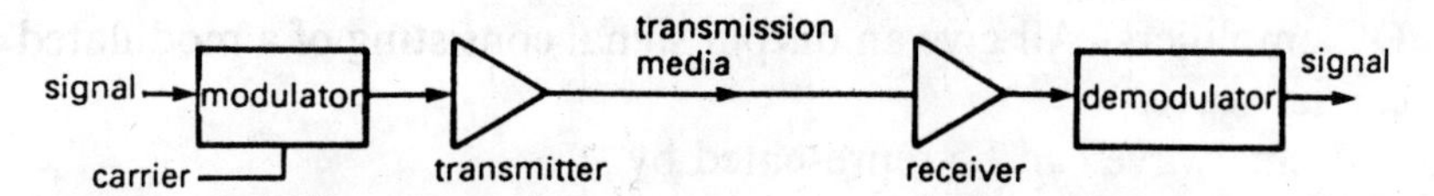

Fig. 6.36 A transmission system using modulation.

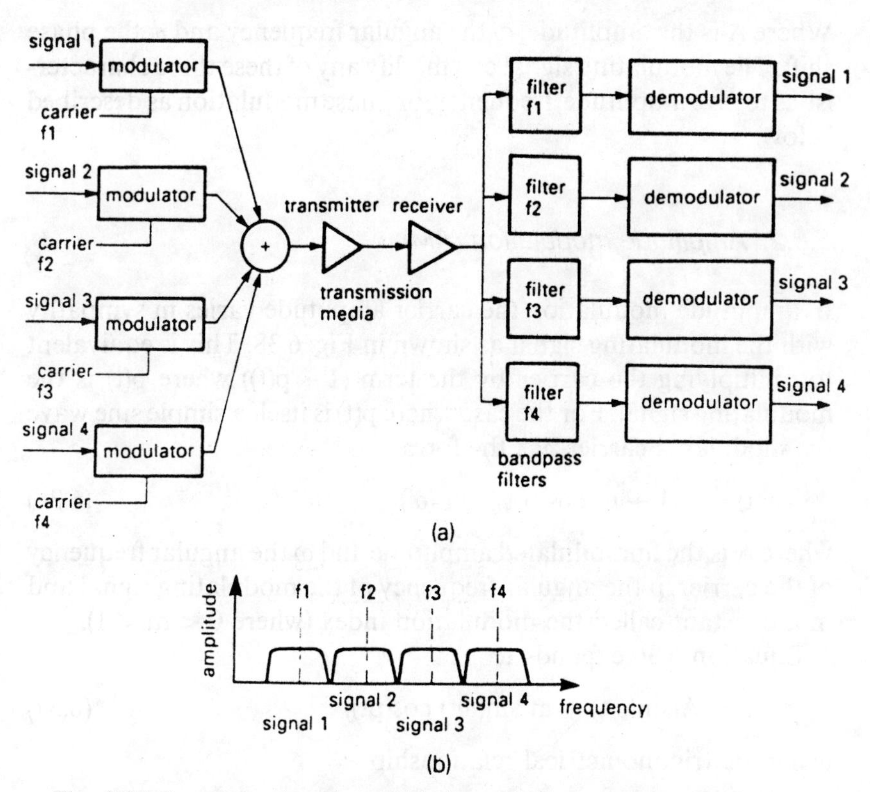

Fig. 6.37 Frequency domain multiplexing, (a) Block diagram; (b) Spectral distribution.

signal bandwidth away from a troublesome noise bandwidth. This is particularly effective if the noise has a 1/f characteristic. Some types of modulation (e.g. frequency modulation) have good inherent noise rejection.

A modulated signal has purely AC components (even if the original signal had bandwidth down to DC). Simple AC amplifiers can therefore be used to regenerate the signal without introducing problems of drift associated with high gain DC amplifiers.

Finally, some signals are inherently modulated in one form or another. Typical examples are LVDTs, resolvers or strain gauge bridges with an AC supply (sometimes used to preclude the need for

DC amplifiers). All give an output signal consisting of a modulated carrier wave.

A sine wave can be represented by

$$y(t) = A \sin(\omega t + \phi)$$

Where A is the amplitude, ω the angular frequency and ϕ the phase shift. The modulating signal can modify any of these three characteristics to give amplitude, frequency, or phase modulation as described below.

6.6.2. Amplitude modulation (AM)

In amplitude modulation the carrier amplitude varies in sympathy with the modulating signal as shown in Fig. 6.38. This is equivalent to multiplying the carrier by the term $(1 + p(t))$ where $p(t)$ is the modulating signal. For the case where $p(t)$ is itself a simple sine wave the modulated carrier has the form

$$y(t) = A(1 + m \cos(pt)) \sin(\omega) \tag{6.36}$$

where A is the unmodulated amplitude and ω the angular frequency of the carrier, p the angular frequency of the modulating signal and m a constant called the modulation index (where $0 < m < 1$).

Equation 6.36 expands to

$$y(t) = A(\sin(\omega t) + m \sin(\omega t) \cos(pt)) \tag{6.37}$$

using the trigonometrical relationship

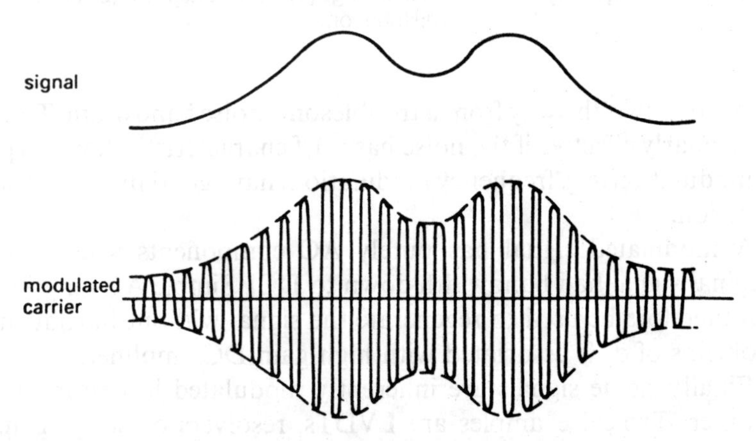

Fig. 6.38 Amplitude modulation (AM).

$$\sin A \cos B = \frac{1}{2}(\sin(A + B) + \sin(A - B))$$

Equation 6.37 can be rewritten

$$y(t) = A(\sin(\omega t) + \frac{m}{2}\sin((\omega + p)t) + \frac{m}{2}\sin((\omega - p)t)) \tag{6.38}$$

This represents three frequencies; the original carrier ω and two side frequencies spaced above and below the carrier frequency by the modulating frequency. This can be represented by Fig. 6.39.

A real life signal will not consist of a single spectral line, but will be represented by spectral power density as Fig. 6.39b. When a carrier is modulated by such a signal, bands of side frequencies occur spaced equally either side of the carrier (called, logically enough, sidebands). A simple AM carrier, therefore, has a bandwidth that is twice the highest frequency in the modulating signal.

Amplitude modulation is inefficient in its use of power. The power in a sine wave is proportional to the square of the amplitude. The information in equation 6.38 is contained in the sidebands. The total RMS power in the signal is

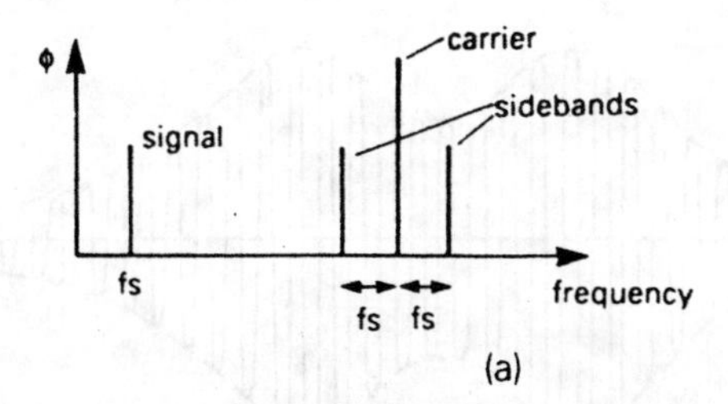

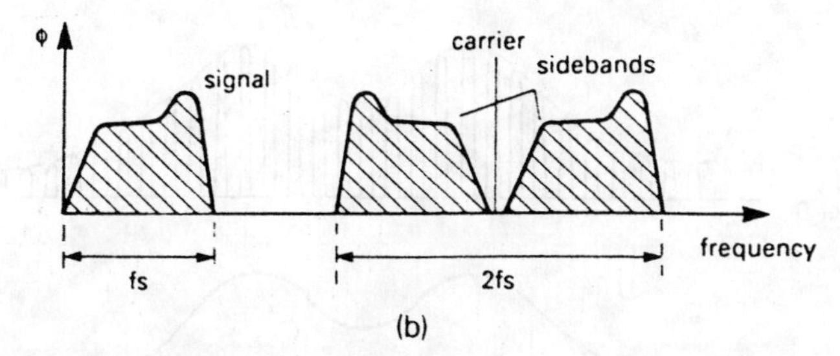

Fig. 6.39 Spectral power density of an amplitude modulated carrier, (a) Spectrum of carrier wave modulated by a sine wave signal; (b) Spectrum of carrier wave modulated by a multifrequency signal.

$$\text{power} \propto A^2 + \left(\frac{mA}{2}\right)^2 + \left(\frac{mA}{2}\right)^2$$

The power in the central carrier spectral line is A^2. The useful signal power is $(m^2A^2/2)$, i.e. only one third of the total transmitted power is used in conveying information for the maximum modulation index $m = 1$. At lower modulation indices the efficiency is even lower.

The proportion of useful power can be increased by suppressing one of the sidebands prior to transmission (called single sideband or SSB). This has the added bonus of reducing the total bandwidth of the modulated signal, thereby increasing the number of signals that can be carried on a given transmission channel. Further gains in the proportion of useful power can be made by suppressing the carrier signal. This, however, requires a more complex receiver as the carrier must be reinserted prior to demodulation at the receiver.

Demodulation of an AM signal is straightforward, as shown in Fig. 6.40. The modulated carrier is halfwave rectified, then filtered by a low pass filter to extract the original signal.

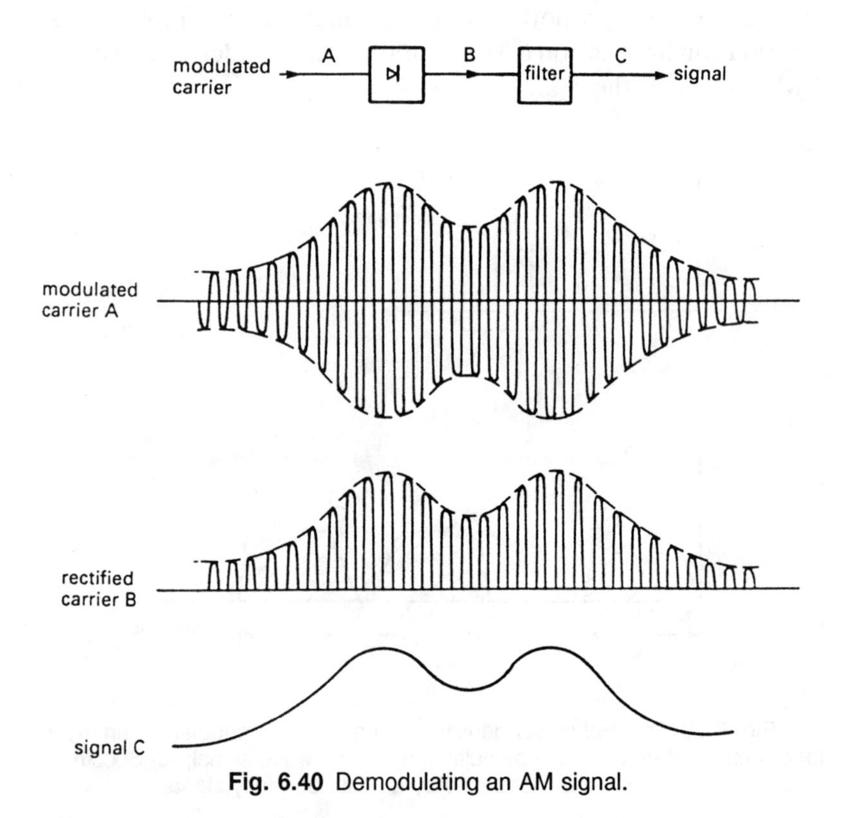

Fig. 6.40 Demodulating an AM signal.

6.6.3. Frequency modulation (FM) and phase modulation

Frequency and phase modulation both modify the angle $(\omega t + \phi)$ and are consequently closely related. Frequency modulation, in fact, implies instantaneous change in phase (and vice versa) and frequency modulation is the integral of phase modulation. Figure 6.41 shows a carrier waveform frequency modulated by a sine wave.

A carrier frequency modulated by a sine wave can be represented by

$$y(t) = A \sin(\int(\omega_0 + \Delta\omega \cos pt)dt) \quad (6.39)$$

where ω_0 is the centre frequency, p the modulating frequency and $\Delta\omega$ the frequency shift corresponding to the maximum amplitude of the modulating signal. Equation 6.39 becomes (assuming zero constant of integration):

$$y(t) = a \sin(\omega_0 t + m \sin pt) \quad (6.40)$$

where m is the modulation index $(\Delta\omega/p)$.

A phase modulated carrier can be represented by

$$y(t) = A \sin(\omega t + \phi_0 + m \cos pt) \quad (6.41)$$

where, as before, p is the modulating frequency, and ϕ_0 the phase shift for no modulating signal. If $\phi_0 = 0$

$$y(t) = A \sin(\omega t + m \cos pt) \quad (6.42)$$

Note the similarity between equation 6.40 (frequency modulation) and equation 6.42 (phase modulation). The difference is a 90° phase shift in modulation and a different interpretation of the meaning of the modulation index. In practice, frequency modulation is far more common than phase modulation in analog communication networks.

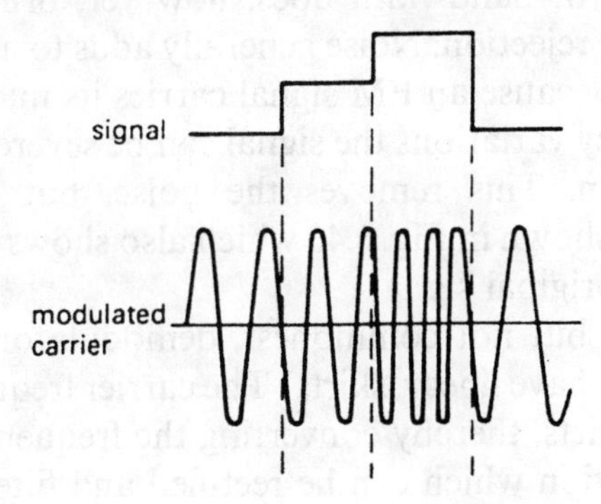

Fig. 6.41 Frequency modulation.

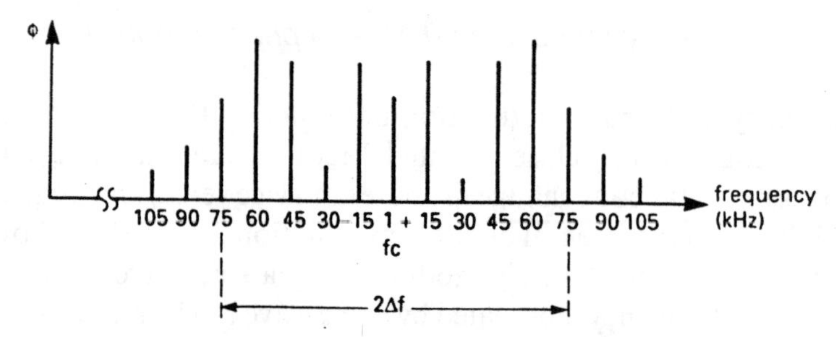

Fig. 6.42 Spectrum of a frequency modulated carrier based on 75 kHz ΔF and a modulating sine wave of 15 kHz.

The spectrum of a frequency (or phase) modulated signal is not easy to determine, requiring expansion of terms of the form sin(cos ωt). These can be solved by the use of Bessel functions, and expand to an infinite series. A typical spectrum for a single frequency modulating sine wave is shown in Fig. 6.42. If follows that the spectrum for a real life signal becomes very complex.

Although the spectrum of a frequency or phase modulated carrier extends theoretically to infinity, in practice it falls off rapidly outside the frequency band $\omega_0 \pm \Delta\omega$ (where $\Delta\omega$ is the frequency shift corresponding to maximum amplitude of the modulating signal, see equation 6.39). A useful rule of thumb, called 'Carson's rule', gives the bandwidth of a frequency modulated signal as

$$\text{BW} = 2 \times (\Delta\omega + \text{highest signal frequency}) \tag{6.43}$$

It follows that the bandwidth of an FM signal is wider than a corresponding AM signal (and over twice that of an SSB AM signal).

This increase in bandwidth does, however, bring a significant increase in noise rejection. Noise generally adds to a signal affecting the amplitude. Because an FM signal carries its information in the form of frequency variations the signal can be severely limited prior to demodulation. This removes the noise, but not the signal information, as shown in Fig. 6.43 which also shows a simple way of recovering the original signal.

The simplest, but not commonest, demodulator is a bandpass filter designed to have linear 'skirts'. The carrier frequency is centred on one of the skirts, thereby converting the frequency variation to amplitude variation which can be rectified and filtered to give the original signal as described previously.

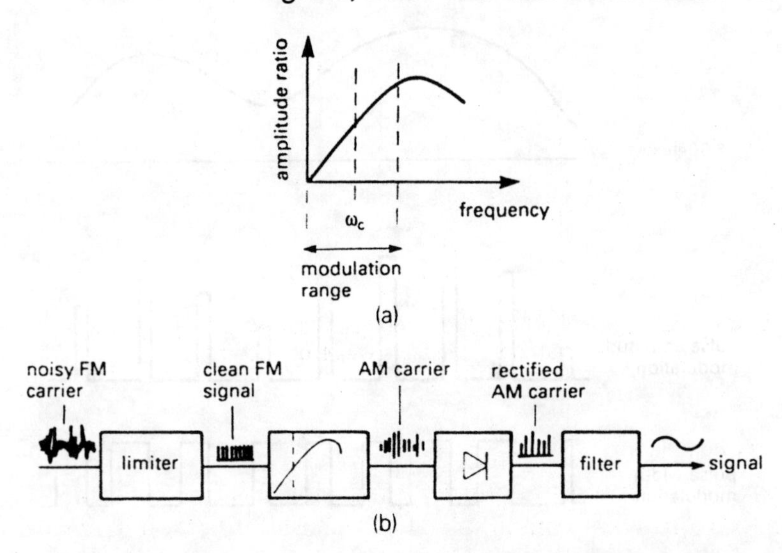

Fig. 6.43 The simplest (but not the commonest) FM demodulator, (a) Conversion of FM to AM; (b) Block diagram of demodulator.

6.6.4. Pulse modulation

In pulse modulation a series of narrow pulses (called a pulse train) is used as the carrier (rather than a sinusoid). There are three characteristics of the pulse that can be changed, as shown in Fig. 6.44.

Pulse amplitude modulation (PAM) effectively samples the signal at regular intervals to give a train of pulses of constant width and repetition rate but of varying height. The sampling must take place at a rate at least twice the maximum frequency of the signal to allow faithful reproduction at the receiver (Shannon's sampling theorem).

In pulse width modulation (PWM) the amplitude and repetition rate remain constant, and the width of the pulse is varied according to the modulation signal amplitude. In pulse position modulation (PPM) the position of the pulse is moved by the modulating signal.

In all forms of pulse modulation the pulse width is narrow compared with the repetition rate. This allows interleaving (or multiplexing) of several pulse modulated signals onto a single channel. Figure 6.45 shows a simplified representation of a four signal PAM system with the sampling and reconstitution being achieved by rotary switches.

Obviously some form of synchronisation between transmitter and receiver is required. This technique (which can be applied equally well to PWM and PPM) is called 'time division multiplexing' or TDM.

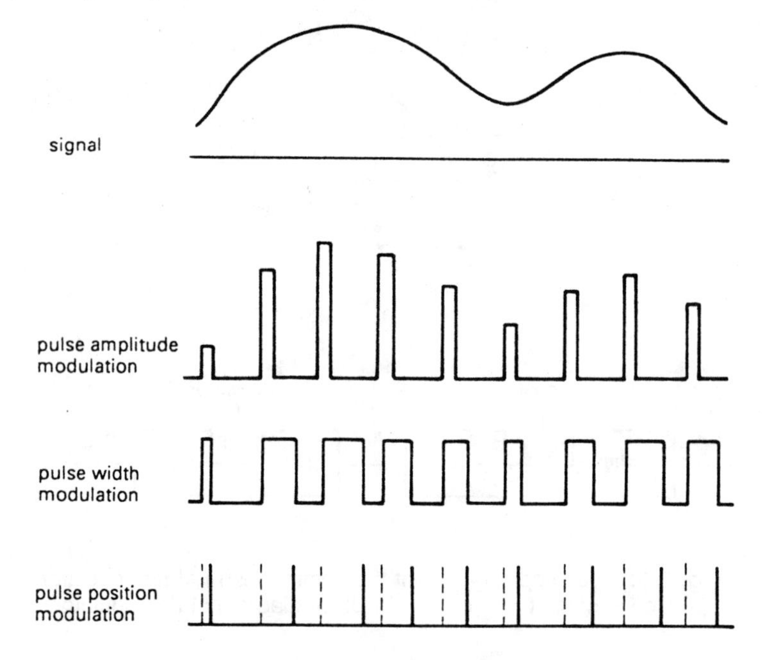

Fig. 6.44 Forms of pulse modulation.

The frequency spectrum of a pulse modulated signal can be determined from Fourier analysis. A typical pulse train is shown in Fig. 6.46b with pulse width δ sec and period T sec (repetition frequency 1/T Hz). The Fourier series of this train contains only cosine terms (because it is symmetrical about the zero time axis) and can be represented by

$$y(t) = b_0 + \sum_{n1}^{\infty} a_n \cos\left(\frac{n2\pi t}{T}\right) \tag{6.44}$$

where $b_0 = \delta/T$ (the mean value of the train) and

$$a_n = \frac{2}{\pi n} \sin\left(\frac{\pi n \delta}{T}\right) \tag{6.45}$$

Equation 6.44 represents a series of spectral lines at intervals of 1/T Hz (i.e. 1/T, 2/T, 3/T, etc.). These lines have amplitudes determined by equation 6.45, giving the spectrum of Fig. 6.46b. Note that the amplitude is zero at frequencies which are multiples of $1/\delta$ Hz ($1/\delta$, $2/\delta$, etc.).

Equation 6.44 and Fig. 6.46b imply the bandwidth of a pulse train is infinite. The power in each spectral line is proportional to the

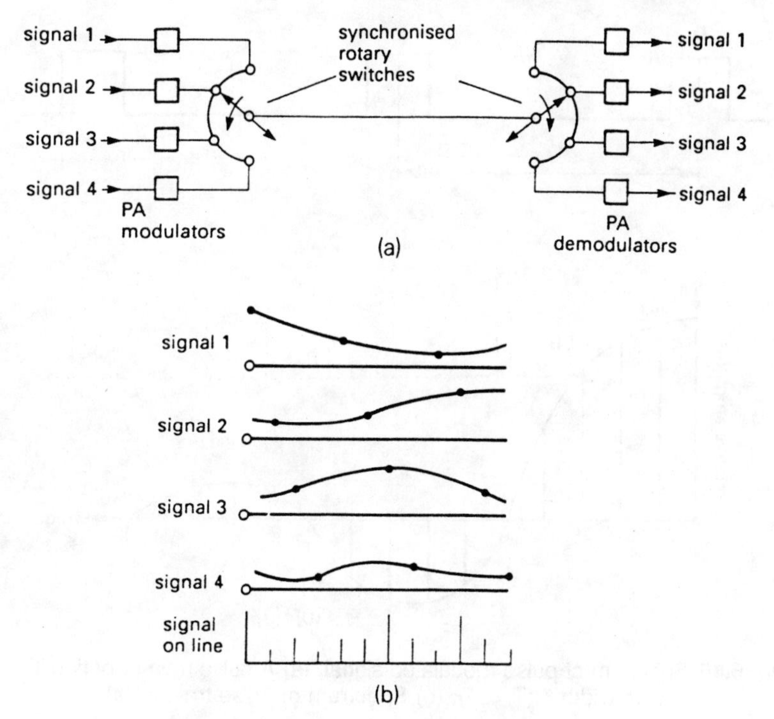

Fig. 6.45 Time division multiplexing.

square of the amplitude, and it is found that the majority of the power (about 95%) is contained in frequencies up to $1/\delta$ Hz (i.e. the first positive segment of Fig. 6.46b). In practice a bandwidth of $1/\delta$ Hz is sufficient to transmit a pulse train of pulse width δ seconds.

This result can also be deduced intuitively. Figure 6.47 represents a worse case for pulse transmission; a pulse δ seconds wide with a repetition period of 2δ seconds. A just acceptable representation would be a sine wave of period 2δ seconds, i.e. a frequency of $1/2\delta$ Hz. A bandwidth of $1/\delta$ Hz will therefore be just acceptable.

6.6.5. Pulse code modulation (PCM)

A PAM, PWM or PPM signal is still susceptible to noise and distortion on the transmission channel. Pulse code modulation converts a PAM signal to a binary coded signal by an analog to digital converter (ADC) as shown in Fig. 6.48. The transmitted signal now consists of a series of purely binary signals (i.e. zeros and ones).

The effect of noise on binary data transmission will be dealt with in

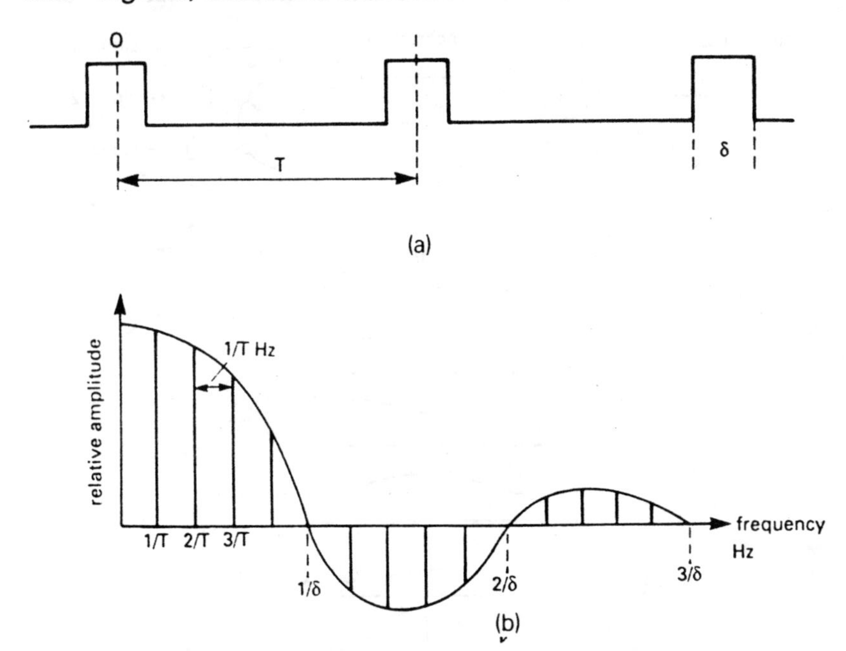

Fig. 6.46 Spectrum of pulse modulated signal, (a) A pulse train of period T and pulse width δ ($T = 5\delta$); (b) Spectrum of pulse train in (a).

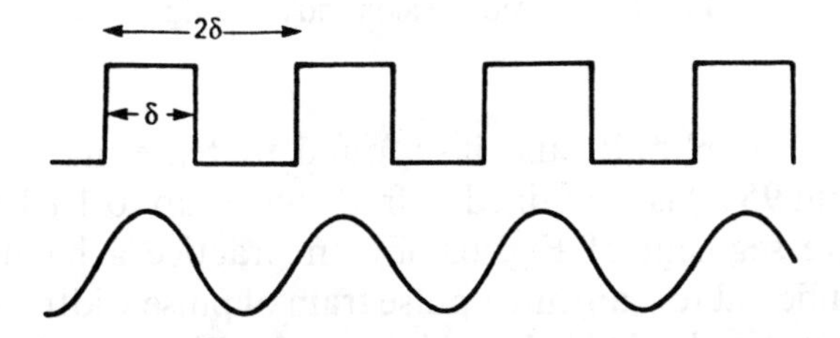

Fig. 6.47 Approximation of worse case pulse train by a sine wave.

more detail in the next section, but it is useful to note the main advantage of purely binary signals. Once noise is added to a purely analog signal it cannot be removed (except possibly by filtering). With a digital signal, providing the noise is not sufficient to turn a 'zero' into a 'one' (or vice versa), the original signal can be restored by a simple comparator. A digital signal can be regenerated as required for long distance transmission without amplifying the noise or degrading the signal as shown in Fig. 6.49.

Pulse code modulation does, however, introduce its own internal noise. Figure 6.48 used 3 data bits (and 8 signal levels represented in binary by 000–111). A more practical system would use 8 bits (256 levels) or 10 bits (1024 levels). The rounding of the signal to nearest

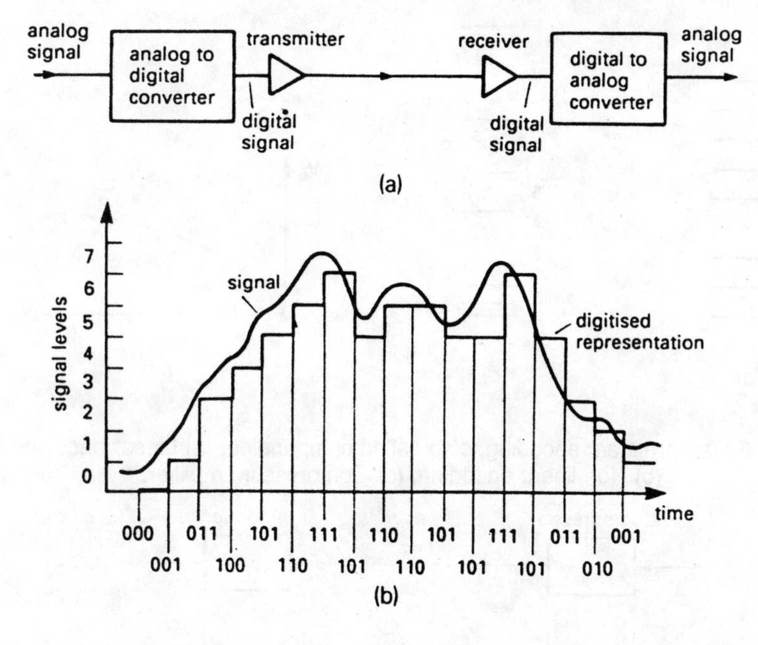

Fig. 6.48 Use of a digital link to transmit analog signals, (a) A digital transmission link using an ADC and DAC to convert between analog and digital signals; (b) Operation of a 3 bit (8 level) ADC.

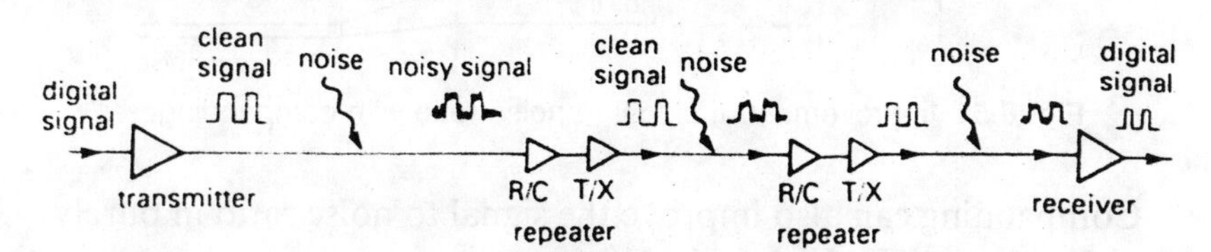

Fig. 6.49 The use of repeaters to regenerate digital signals on long lines.

level introduces an error called 'quantisation noise'. Obviously this is related to the number of data bits; decreasing with increasing number of bits (and greater resolution).

The magnitude of the quantisation noise is independent of the magnitude of the signal, and consequently affects low level signals more than high level signals. An improvement in SNR can be achieved by using non-linear encoding, with more quantisation levels for small signals as Fig. 6.50b. This technique is known as companding. Ideally the quantisation level should follow a logarithmic progression (e.g. each level twice the previous one). One common method of achieving logarithmic encoding is to precede a linear encoder by a logarithmic amplifier with the response of Fig. 6.50c.

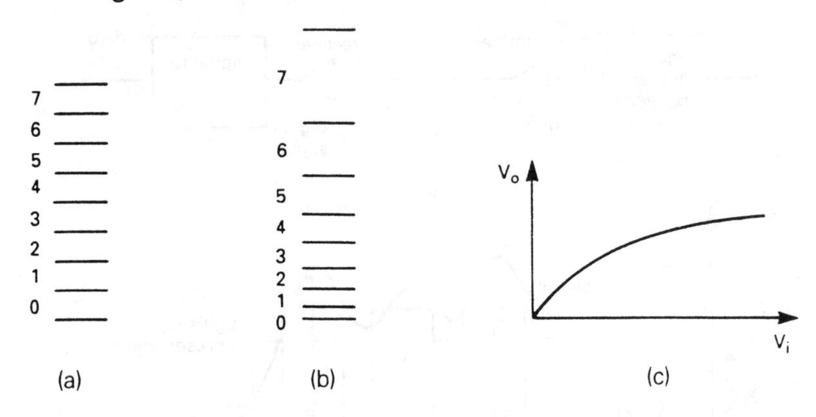

Fig. 6.50 Non-linear encoding, also called companding, (a) Linear encoding; (b) Non-linear encoding; (c) Compressor amplifier.

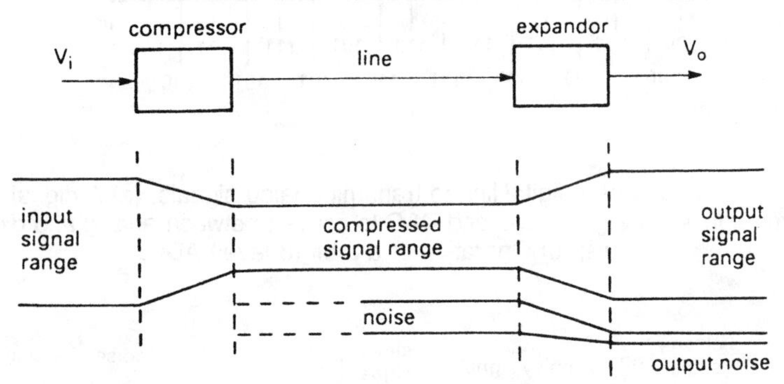

Fig. 6.51 Improvement of signal to noise ratio with companding.

Companding can also improve the signal to noise ratio in purely analog systems. In Fig. 6.51 the dynamic range of the signal is reduced prior to transmission, and the whole signal lifted in power above the noise. At the receiver the noise contribution is reduced as the signal is linearised by an amplifier with a response the inverse of Fig. 6.50c.

If successive signal samples do not vary greatly from one sample to the next, the number of data bits can be reduced by sending the change (positive or negative) from the previous sample. This is known as differential PCM, and can be achieved with the simple circuit of Fig. 6.52. The decoder and integrator at the transmitter match those at the receiver, so the signal should match the last signal at the receiver.

The simplest possible form of PCM is simply differential PCM with only two change signals; increase (1) and decrease (−1). The encoder of Fig. 6.52 thus becomes a simple comparator as shown in

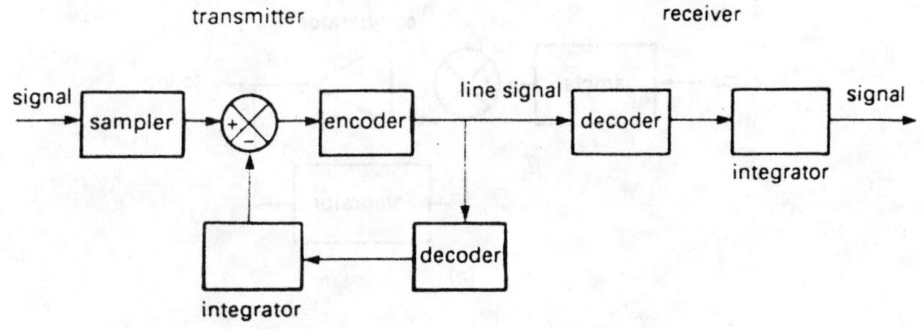

Fig. 6.52 Differential pulse code modulation.

Fig. 6.53. This system is known as delta modulation. Note that the steady state transmitted signal is alternate positive and negative pulses, and fast changes can lead to the received signal lagging behind the original signal.

6.7. Data transmission

6.7.1. Fundamentals

Digitally encoded information can be transmitted in two forms. In Fig. 6.54a one signal line is used for each binary digit (or bit). Limitations in the transmission media may result in 'skew' between the various bits as shown in Fig. 6.54b, so a separate strobe line is also required in the nominal centre of each bit group to gate the data into the receiver. This is known as parallel transmission.

In Fig. 6.54c, one signal line (or more pedantically a signal line and common) connects the transmitter and receiver, and the digital data is transmitted as a serial string of bits. Since computers, peripherals, instruments, etc. generally work internally in parallel, parallel to serial conversion is required at the transmitter, and serial to parallel conversion at the receiver. This is easily performed by a shift register into which data can be loaded in parallel and shifted out one bit at a time. Specialist ICs called UARTs (universal asynchronous receiver-transmitters) are available to perform the parallel/serial/parallel conversion along with numerous control functions not covered in the rather simplified representation of Fig. 6.54c. Not surprisingly this method of data communication is called serial transmission.

Serial transmission requires more complex transmitters and receivers (with dedicated LSI ICs this brings no great cost penalty) but gives an obvious cost saving in cabling. In many cases (e.g. radio

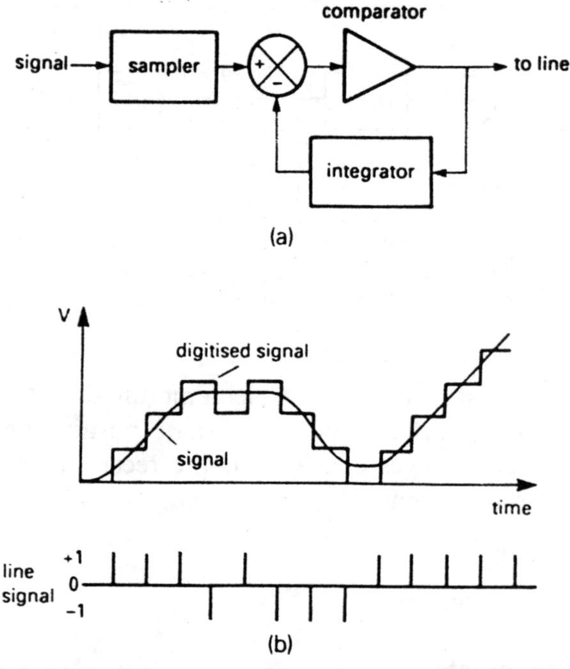

Fig. 6.53 Delta modulation, the simplest form of pulse code modulation, (a) Transmitter circuit; (b) Waveforms and signals.

telemetry, or transmission via media such as fibre optic or coaxial cables) serial transmission is the only practical method. Parallel transmission is faster (by a factor equal to the number of parallel lines). This section is primarily concerned with serial data transmission. Parallel systems are generally found in laboratory and research establishments where a number of scientific instruments are to be linked in a physically small area. A commercial parallel data transmission system is discussed further in Section 6.10.

Synchronisation between the transmitter and receiver is obviously essential in a serial link, and this can be achieved in two ways. A 'synchronous' system uses a separate clock line to synchronise the transmitter and receiver. Most systems, however, are 'asynchronous' and use separate clocks at the transmitter and receiver. Asynchronous signals have the form of Fig. 6.55.

The data to be transmitted is split into characters (typically 5 to 8 bits in length). The idle state of the line is a '1' (called a mark in telecommunications). The character starts with a '0' start bit (zeros are called a space) followed by the data bits, least significant bit first. An error correcting bit (called parity) is sometimes used after the data bits; this topic is discussed in Section 6.8 below. Finally the

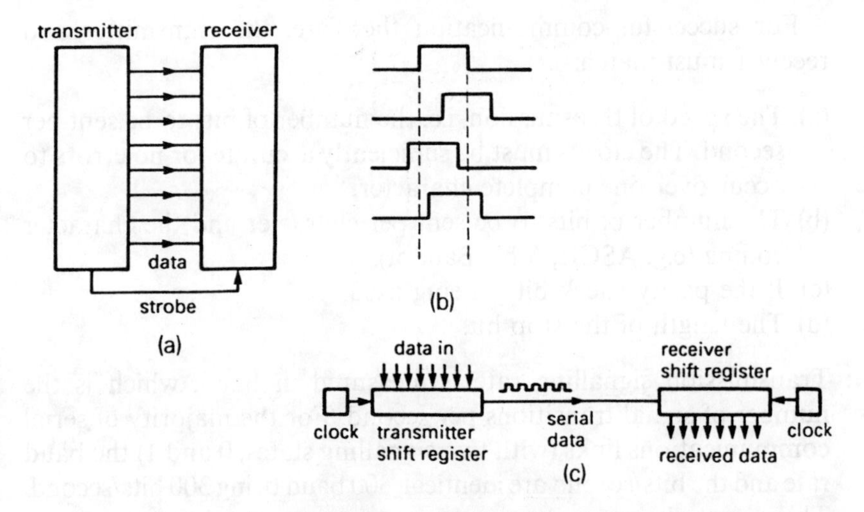

Fig. 6.54 Methods of data transmission, (a) Parallel data transmission; (b) Data skew; (c) Serial data transmission.

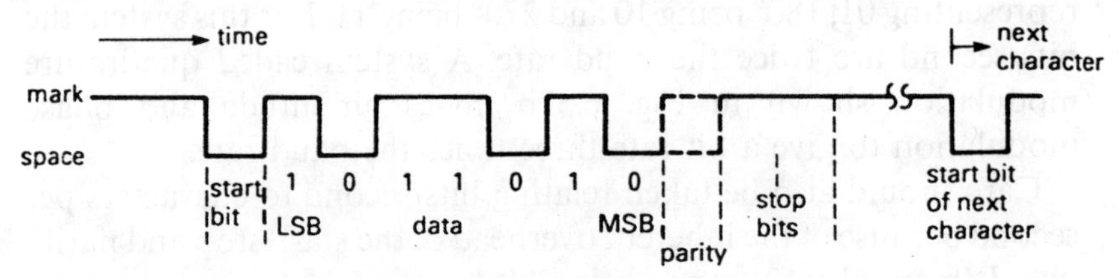

Fig. 6.55 An asynchronous data character.

signal must return to the idle mark state for a time before the next character can be sent. This is known as the stop bit, and, dependent on the system, can be 1, 1·5 or 2 bits in width. The next character can follow any random time after the end of the stop bit. By far the commonest coding is the ASCII code of Table 6.1.

The mark to space transition of the start bit synchronises the transmitter and receiver clocks which are nominally (but not exactly) the same frequency. Exact synchronisation is not needed, all that is required is for the error between clocks to be small over the number of bits in a character. With early mechanical teleprinters synchronous motors were used, and this limited the number of data bits to five (the Baudot code). With crystal oscillators, eight data bits can be used.

It may be thought that mark to space transitions in the data can be mistaken for start bits (a fault called a framing error). In practice if a framing error occurs the link will pull itself back into synchronisation in a few characters as shown in Fig. 6.56.

For successful communication therefore, the transmitter and receiver must match:

(a) The speed of transmission, i.e. the number of bits to be sent per second. The clocks must be sufficiently accurate for no errors to occur over one complete character.
(b) The number of bits to be sent per character and the character coding (e.g. ASCII, 5 bit Baudot).
(c) If the parity check bit is being used.
(d) The length of the stop bit.

Transmission signalling rate is measured in baud, which is the number of signal transitions per second. For the majority of serial communications links (with two signalling states, 0 and 1) the baud rate and the bits/second are identical, 300 baud being 300 bits/second. This is not always so, however.

Figure 6.57a represents a possible phase shift modulation system with four phase angles; 0° representing the binary pair 00, 90° representing 01, 180° being 10 and 270° being 11. For this system the bits/second are twice the baud rate. A system called quadrature modulation shown in Fig. 6.57b mixes amplitude and phase modulation to give a bit rate three times the baud rate.

Care should also be taken relating bits/second to characters per second because of the inherent overhead of the start, stop and parity bits. It may take 12 transmission bits to send an 8 bit character.

Synchronous serial communication requires a clock link between transmitter and receiver, either as a separate signal or extracted from the data. The latter method is commonly used with phase modulated

Table 6.1 *ASCII Codes*

Decimal	*Hex*	*Char*	*Decimal*	*Hex*	*Char*
0	00	NUL	14	0E	SO
1	01	SOH	15	0F	SI
2	02	STX	16	10	DLE
3	03	ETX	17	11	DC1
4	04	EOT	18	12	DC2
5	05	ENQ	19	13	DC3
6	06	ACK	20	14	DC4
7	07	BEL	21	15	NAK
8	08	BS	22	16	SYN
9	09	HT	23	17	ETB
10	0A	LF	24	18	CAN
11	0B	VT	25	19	EM
12	0C	FF	26	1A	SUB
13	0D	CR	27	1B	ESC

Table 6.1 *(Cont'd)*

Decimal	*Hex*	*Char*	*Decimal*	*Hex*	*Char*
28	1C	FS	78	4E	N
29	1D	GS	79	4F	O
30	1E	RS	80	50	P
31	1F	US	81	51	Q
32	20	space	82	52	R
33	21	!	83	53	S
34	22	”	84	54	T
35	23	#	85	55	U
36	24	$	86	56	V
37	25	%	87	57	W
38	26	&	88	58	X
39	27	’	89	59	Y
40	28	(	90	5A	Z
41	29	)	91	5B	[
42	2A	*	92	5C	\
43	2B	+	93	5D	]
44	2C	,	94	5E	^
45	2D	-	95	5F	_
46	2E	.	96	60	’
47	2F	/	97	61	a
48	30	0	98	62	b
49	31	1	99	63	c
50	32	2	100	64	d
51	33	3	101	65	e
52	34	4	102	66	f
53	35	5	103	67	g
54	36	6	104	68	h
55	37	7	105	69	i
56	38	8	106	6A	j
57	39	9	107	6B	k
58	3A	:	108	6C	l
59	3B	;	109	6D	m
60	3C	<	110	6E	n
61	3D	=	111	6F	o
62	3E	>	112	70	p
63	3F	?	113	71	q
64	40	@	114	72	r
65	41	A	115	73	s
66	42	B	116	74	t
67	43	C	117	75	u
68	44	D	118	76	v
69	45	E	119	77	w
70	46	F	120	78	x
71	47	G	121	79	y
72	48	H	122	7A	z
73	49	I	123	7B	{
74	4A	J	124	7C	|
75	4B	K	125	7D	}
76	4C	L	126	7E	~
77	4D	M	127	7F	DEL

Control characters can be obtained via the use of the CONTROL key and a character key. Control subtracts Hex 30. Backspace (BS) Hex 08, for example, is ctrl-H.

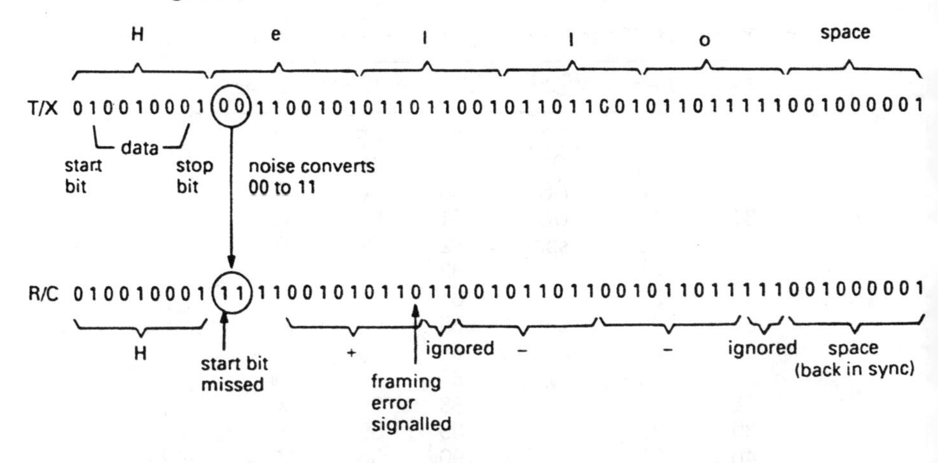

Fig. 6.56 Framing errors and the ability of an asynchronous transmission to recover from a fault. 7 data bits and a single stop bit are used with ASCII coding. Parity is not used.

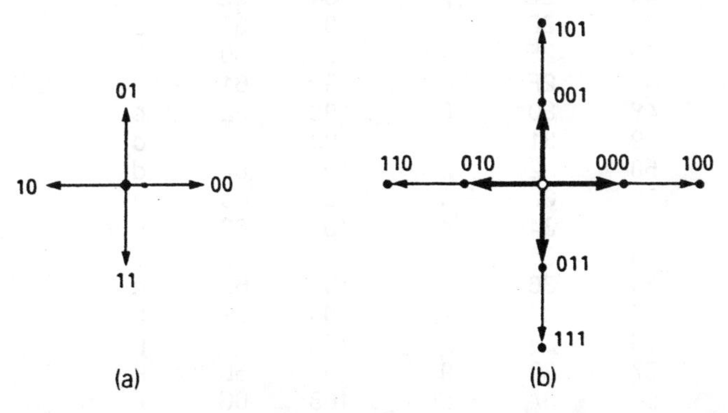

Fig. 6.57 Modulation systems in which baud rate and bit rate are different, (a) Four phase modulation; (b) Mixing phase and amplitude modulation (quadrature modulation) gives 8 signals and 3 bits 000–111.

signals as the clock can easily be extracted from the sidebands. Start and stop signals are not required, so synchronous transmission is more efficient. Synchronisation is still required at the start of a message, however. This is provided at the start of a message by a series of synchronisation characters (SYN; 028 octal in 7 bit ASCII, 228 octal in 8 bit ASCII) which have a bit pattern unlikely to occur in normal transmission. These are recognised by the receiver and used to 'frame' the succeeding characters.

Serial transmission can take place in one direction only, two directions but not simultaneously, or simultaneous in two directions. The first of these (of which a computer to printer link is an example) is called simplex. Non-simultaneous bidirectional transmission (of which a mobile radio with a press to talk button is typical) is called half duplex. Simultaneous bidirectional transmission (as found on a telephone) is called full duplex.

6.7.2. Noise and data transmission

It was shown earlier in Fig. 6.2 that the addition of noise to a digital signal does not necessarily corrupt the signal, and the original information can be regenerated providing the noise is not sufficiently severe to turn a '1' level into a '0' or vice versa.

Noise generally has a power density Gaussian distribution with zero mean as shown in Fig. 6.58. If the voltage difference between '1' and '0' levels is 'V', noise in region A will corrupt a '0' to '1', and noise in region B will corrupt a '1' to a '0'. The probability of an error is thus the area in regions A and B.

Their probability depends on the relative magnitude of the signal and the noise. The signal to noise ratio was defined earlier as

$$\text{SNR} = \frac{\text{mean square value of signal}}{\text{mean square value of noise}} \tag{6.46}$$

Because the tail of Fig. 6.58 tails off to infinity, there is no 'cut-off' level for noise, and it follows that there is a small probability of error whatever the value of SNR. The probability of error can be evaluated, and has the form of Fig. 6.59.

From Fig. 6.59 an SNR of 20 gives an error probability of 1 in 10^{-5}. A typical floppy disk has a capacity of 1·44 Mbytes, or 11·52 Mbits. If the entire contents of a floppy disk were transferred

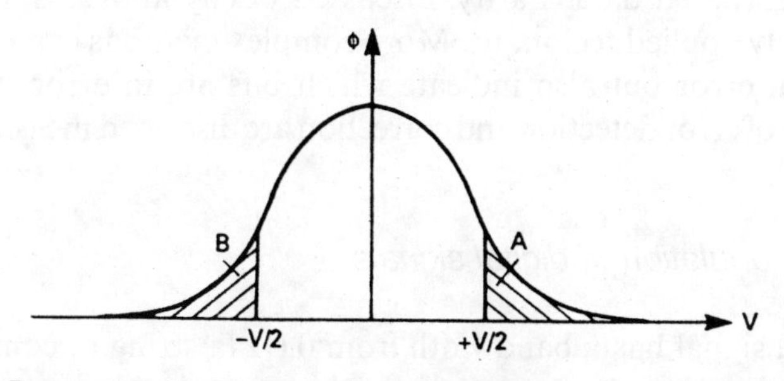

Fig. 6.58 Power density spectrum of a noise signal. Noise in regions A, B will corrupt a digital signal with V volts between a '1' and a '0'.

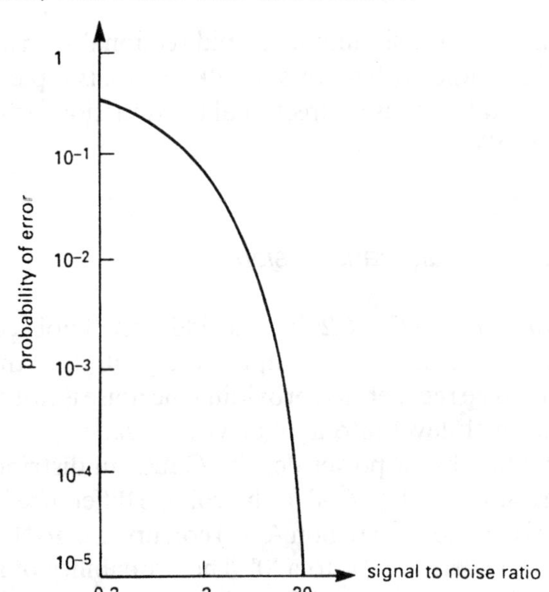

Fig. 6.59 Relationship between error probability and SNR.

over a serial link with an SNR of 20, some 115 bits would probably be corrupted.

It is interesting to note that as the system becomes swamped by noise (SNR < 1) the error probability tends towards 0·5 and not 1 (as might be intuitively thought). As noise dominates, a random string of '1's and '0's will be received, half of which will, on average, be correct purely by chance.

Even with very high SNRs, 100% reception cannot be guaranteed. As even one bit in error can cause severe problems (changing the sign of a number, for example), data transmission invariably includes methods for dealing with corrupted messages.

The simplest method is to detect errors and request a retransmission of the corrupted data. Parity, discussed briefly above, is a simple commonly applied technique. More complex methods can not only detect an error but also indicate which bits are in error. Various methods of error detection and correction are discussed in Section 6.8.

6.7.3. Modulation of digital signals

A digital signal has a bandwidth from 0 Hz (a string of continuous zeros or ones) to at least half the bit rate. A telemetry system sampling N signals, M times per second with a B bit word therefore

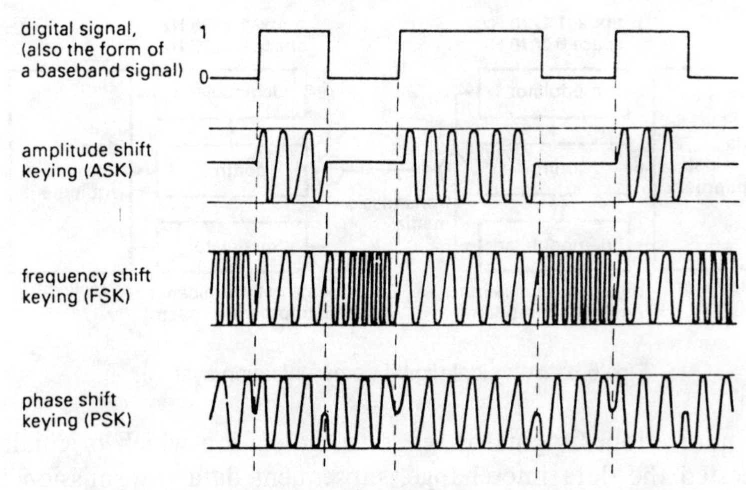

Fig. 6.60 Various forms of modulation for digital signals.

requires at least bandwidth from 0 Hz to 0·5N.M.B Hz (ignoring the need for any synchronisation or error checking signals).

With coaxial, or twisted pair, cable the digital signal can be sent in its raw form. This is termed 'baseband' transmission. Only one signal can be conveyed at a time unless TDM is used.

Many transmission media such as radio and the telephone network have inherent low frequency limitations. The standard telephone bandwidth, for example, is 300 Hz to 3·4 kHz so a digital signal needs modulating before it can be transmitted.

Modulation techniques were discussed earlier in Section 6.6. When applied to digital signals, amplitude modulation is known as amplitude shift keying (ASK), frequency modulation as frequency shift keying (FSK) and phase modulation as phase shift keying (PSK). These are summarised in Fig. 6.60. A signal which is modulated to overcome the limitation of the transmission media is sometimes said to be using 'broadband' transmission (sometimes the term is used to imply FSK). Several signals can, of course, be carried on the same channel if the modulation frequencies are separated. This is known as 'carrier band' operation.

Broadband and carrier band operation require the need for devices to interface the digital signal to the transmission media. These will modulate the signal at the transmitter and demodulate it again at the receiver. Such devices are known as modulators/demodulators or 'modems'.

Figure 6.61 shows a typical arrangement used with the public telephone network. Four tones are used to allow full duplex

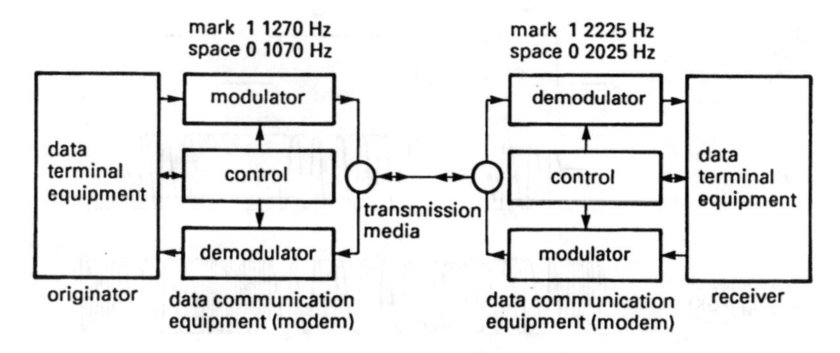

Fig. 6.61 A typical digital communication system.

communication. Originator refers to the station which originally requested the data interchange, subsequent data transmission is bidirectional.

Figure 6.61 consists of two pieces of digital equipment. These are referred to as data terminal equipment or DTE. They can be computers, data loggers, printers, VDUs, programmable controllers, etc. They are connected by the modems, known as data communication equipment, or DCE (one notable aspect of data communications is a plethora of abbreviations). A point to point communication system thus consists of two DTEs, two DCEs and a transmission media.

Obviously there must be some form of standardisation to ensure that the link is established in an orderly and controlled manner. Various data communications standards are discussed in the next section.

Modulation and demodulation of digital signals is far simpler than for purely analog circuits, generally involving little more than electronic switches. Frequency modulator circuits generally use voltage controlled oscillators or switch between two oscillators. The rather misnamed product multiplier circuit is used to demodulate PSK circuits, and uses the trig relationship

$$\cos x \cos y = \frac{1}{2}(\cos(x + y) + \cos(x - y)) \tag{6.47}$$

$$\text{Hence } \cos(\omega t + \phi)\cos(\omega t) = \frac{1}{2}\cos(2\omega t + \phi) + \frac{1}{2}\cos \phi \tag{6.48}$$

The $\cos(2\omega t + \phi)$ signal is filtered out leaving the $\cos \phi$ term indicating the phase and hence the digital data.

6.7.4. Standards and protocols

For successful communication to take place, a set of rules must exist which govern the transmission of data. These rules can be split into standards governing the connection and operation of the DTE and DCE interface, and protocols which determine the content and control of the message.

Standards (of which the ubiquitous RS232 is the best known) contain details of the electrical signal levels and characteristics (e.g. edge speeds), the mechanical characteristics of the connectors and the function of the various handshake and control signals.

Much of the early work on data transmission was undertaken by the Bell laboratories in the USA, and the result of their research was formalised by the Electrical Industries Association (EIA) into a standard for 'interface between data terminal equipment and data communication equipment employing serial binary interchange'. This standard is RS232 usually used at revision C.

Worldwide standards are set by the Comité Consultatif International Téléphonique et Télégraphique (CCITT) which is a part of the United Nations International Telegraph Union. The CCITT publishes standards and recommendations; those prefixed by letters V and X being related to data transmission. Standard V24 is, for all practical purposes, identical to RS232C.

Figure 6.62 shows a typical connection using DTEs and DCEs with the minimum control signals and their designation in the RS232 standard. Figure 6.63 shows the steps involved in establishing communication. These are only a subset of the full standard (which is a rather lengthy document).

Signal levels defined for RS232 and V24 are +6 V to +12 V at the source for a space (zero) and −6 V to −12 V for a mark (one). These are allowed to degenerate to +3 V and −3 V at the receiver. Other characteristics such as line capacitance and edge speeds are also defined.

Confusion often arises when interconnecting 'RS232 compatible' equipment. Most of the problems arise because the standards define the connection and operation of the interface between a DTE and a DCE. If, say, an intelligent instrument with an RS232 output is to transmit data direct to a computer, the standard is being used for a purpose for which it was not designed, i.e. linking two DTEs.

The first problem is generally one of distance. The standard was designed for a DTE and DCE in close proximity, usually within the same room, although much greater distances are feasible at low baud rates.

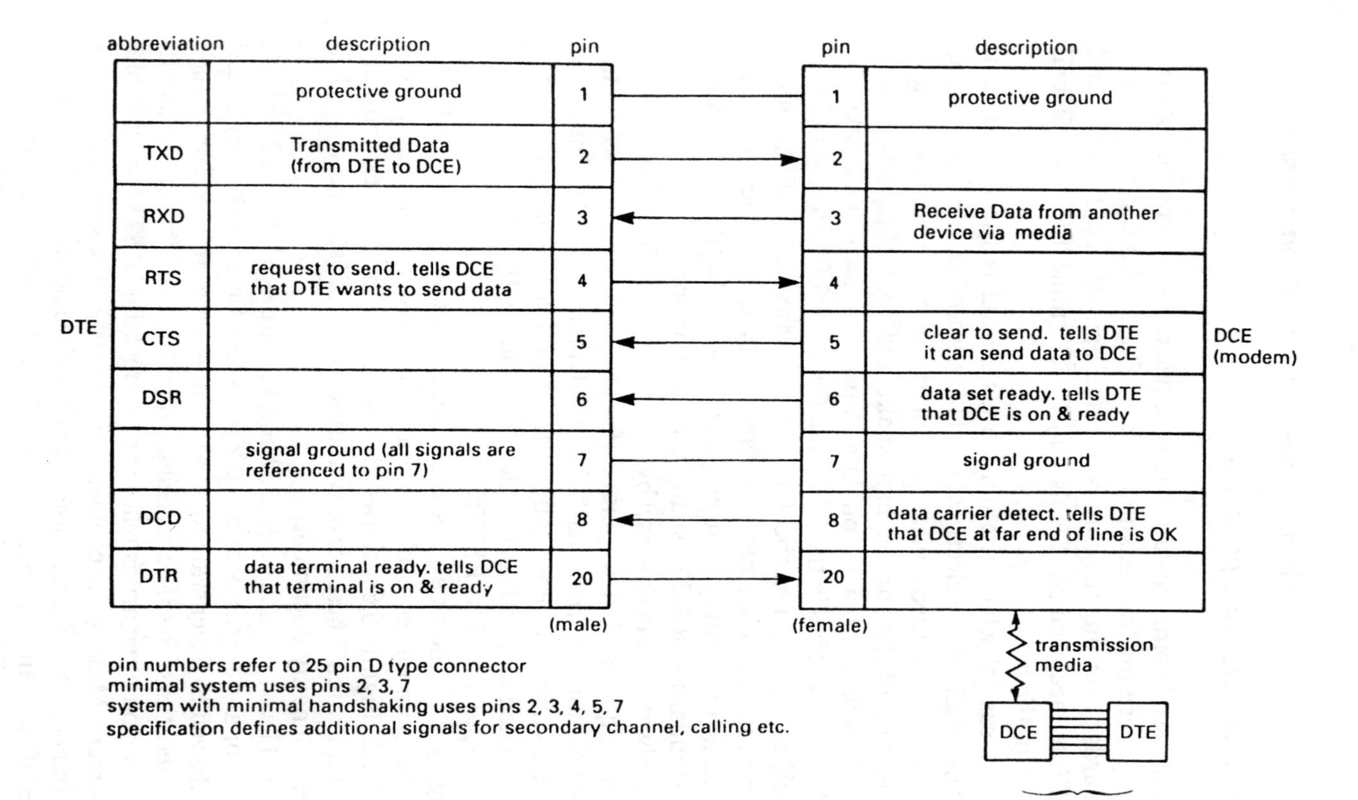

Fig. 6.62 Connections between DTE and DCE as specified by RS232C.

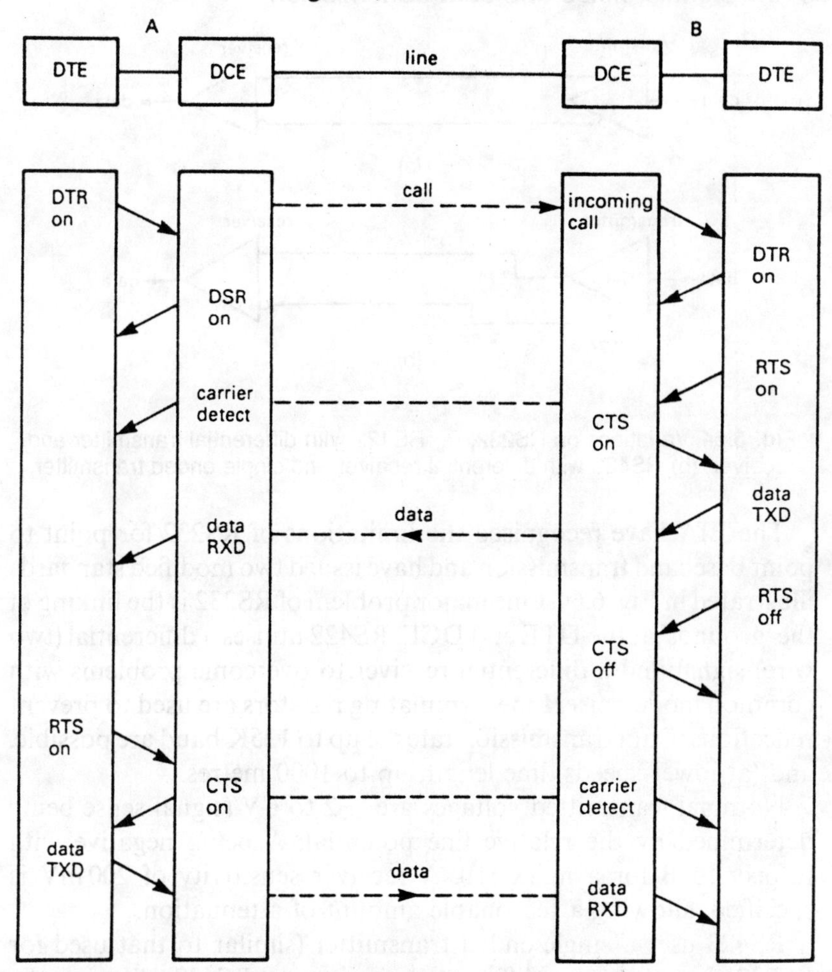

Fig. 6.63 Steps and signals involved in establishing a data link. A makes the initial link, then B sends data, followed by A sending data.

The second problem is the meaning attached to the various signals. To some extent this is similar to the problem of copying between domestic tape recorders; do the connectors and leads need a mirror image cross over, or a straight pin to pin link? Some manufacturers interchange pins 2 and 3 (transmit and receive data) and assign all sorts of peculiar interpretations to the handshake and control signals (including totally ignoring them!). A 'breakout box' which monitors signal lines on LEDs should be an essential part of the tool kit of any engineer involved in data transmission. Often all that is meant by 'RS232 compatible' is the use of +12 V for '0' and −12 V for '1' on the signal lines.

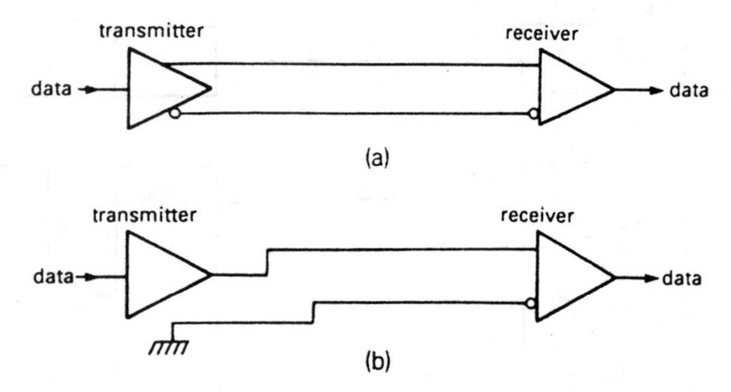

Fig. 6.64 Variations on RS232, (a) RS422 with differential transmitter and receiver; (b) RS423 with differential receiver and single ended transmitter.

The EIA have recognised the limitations of RS232 for point to point baseband transmission and have issued two modified standards illustrated in Fig. 6.64. One major problem of RS232 is the linking of the 'grounds' at the DTE and DCE. RS422 utilises a differential (two wire) signal and a differential receiver to overcome problems with common mode noise. Line terminating resistors are used to prevent reflections. Data transmission rates of up to 125K baud are possible, and (at lower speeds) line length up to 1000 metres.

Nominal transmitted voltages are ± 2 to 6 V (signal sense being determined by the relative line potential, A being negative with respect to B for a mark (1)). A receiver sensitivity of 200 mV is specified, allowing a reasonable amount of attenuation.

RS423 uses a single ended transmitter (similar to that used for RS232) but the same differential receiver as RS422 allowing the permitted distance of a standard RS232 transmitter to be increased provided the difference in ground potential does not exceed 4 V. A terminating resistor in excess of 450 ohms is permitted. The mechanical (37 pin connector) details of RS422 and RS423 are defined in RS449.

An unofficial 'standard' is the 20 mA loop. This originated with the early mechanical teleprinters, and uses a current source, a switch (electronic or mechanical) driven by the data and current sensor at the receiver. Presence of current is a mark (1) and absence a space (0). The current source, isolated from earth, gives good common mode noise immunity and overcomes difficulties with different ground potentials between the transmitter and receiver. Initially designed for use with 110 baud and mechanical switches and solenoids it can be used at far greater speeds and distances up to 1000 metres.

Unfortunately there is no standard as to at which end of the loop

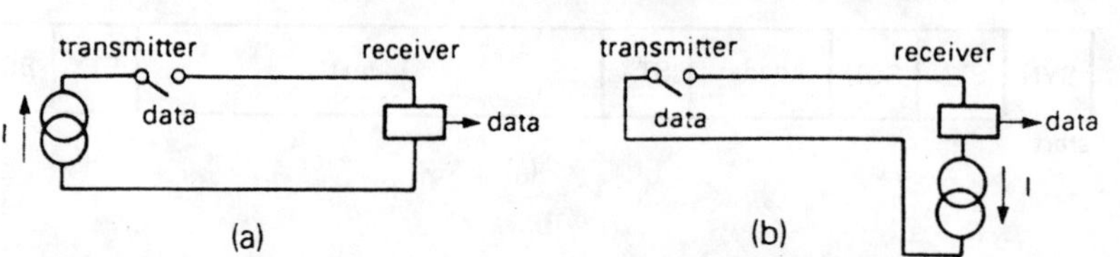

Fig. 6.65 The two forms of the 20 mA loop, (a) Active transmitter, passive receiver; (b) Passive transmitter, active receiver.

the current source is located. Figure 6.65a is called active transmitter/passive receiver and Fig. 6.65b passive transmitter/active receiver. Obviously transmitter and receiver must match. Mechanical teleprinters are generally passive for both transmit and receive.

Standards, described above, govern the 'mechanics' of data transmission. The message content is defined by the protocol used. In addition to defining the form of a message (i.e. what group of bits form characters, and which groups of characters form a message), the protocol must also define how communication is initiated and terminated and what happens if the link is broken during a message (it is obviously undesirable to have the receiver 'hang' awaiting the completion of the message). Error control is also covered by the chosen protocol, i.e. how errors are detected, and what action, if any, is taken.

There are, essentially, three types of protocol illustrated in Fig. 6.66. Character based protocols use control characters such as STX (start of text), SOH (start of header), ETB (end of transmission block) to format the message. This type of protocol is widely used for serial linking of instruments to a computer and a typical example is described later. Most character based protocols are derived, to some extent, from IBM's BISYNC protocol.

Bit pattern based protocols have the form of Fig. 6.66b. Special flag characters define the start and end of the message, with the end flag being preceded by error checking characters. On receiving the start flag the receiver accepts characters until the stop flag, at which point the error checking characters (preceding the stop flag) are used to validate the message. The protocol must ensure that the flag characters (usually the bit pattern 01111110) cannot occur in the message or error checking characters. Common protocols of this type are IBM's SDLC, ISO's HDLC and CCITT X25 Protocol.

The final type of protocol uses a byte count. The start of the message is signalled by a start flag character followed by a count showing the number of characters in the message. The message itself is followed by the error checking characters. The receiver 'counts in'

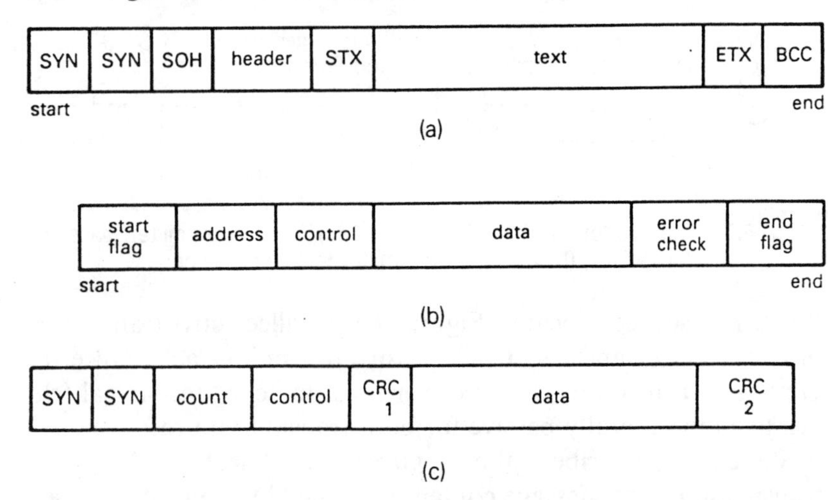

Fig. 6.66 The three types of protocol commonly used in data communication, (a) The basic form of a character based protocol; (b) The basic form of a bit pattern based protocol; (c) The basic form of a byte count protocol. Note that two error checks are used; one for the header (count and control) and one for the data.

the message characters and validates the message with the error checking data. A common example of this type of protocol is DEC's DDCMP.

Of the above protocols, variations on BISYNC (sometimes called BSC for binary synchronous communication) are most commonly used in process control to link intelligent instruments to a supervisory computer. Although not the most efficient or flexible protocol, BSC is easy to implement and has the great attraction that its operation is easy to monitor with a simple terminal which can be hung across the signal lines.

Character based protocols such as BISYNC are built around control characters from the ASCII set. The most important are:

Hex 16 SYN	Synchronising character, establishes synchronisation between transmitters and receiver and sometimes used as a fill character.
Hex 05 ENQ	Enquiry; used to bid for the line in a multidrop system.
Hex 02 STX	Start of text; characters following are the message.
Hex 01 SOH	Start of header; characters following are header information, e.g. message type, routine or priority.
Hex 17 ETB	End of transmission block. Text commenced with SOH or STX is complete.

Hex 03 ETX End of text. Text commenced with SOH or STX is complete and also the end of a sequence of blocks. It is normally followed by error checking information, and the receiver should reply with one of the following two codes:

Hex 06 ACK Positive acknowledgement; message received error free and I am ready for more data. Also used to acknowledge selection on multidrop;

Hex 15 NAK Negative acknowledgement; message faulty, please retransmit. Also used to give a negative reply to multidrop selection.

Hex 04 EOT End of transmission (often used as a reset signal).

Figure 6.67 shows a typical system with intelligent controllers connected by a full duplex (two twisted pair) link to a host computer. The link is used to retrieve information from the controllers for mimic displays and data logging, and to send information such as set points to the controllers. The method of communication described below is loosely based on that used by TCS to link their 6300 series of controllers.

Each controller has a unique address on the line consisting of a group and instrument unit number. This is set up on switches inside the instrument.

The read sequence (where data is retrieved from inside the instrument) is illustrated in Fig. 6.68. Data within the instrument is identified by a two letter mnemonic (e.g. PV for process variables, SP for set point, etc.). The host computer sends the message

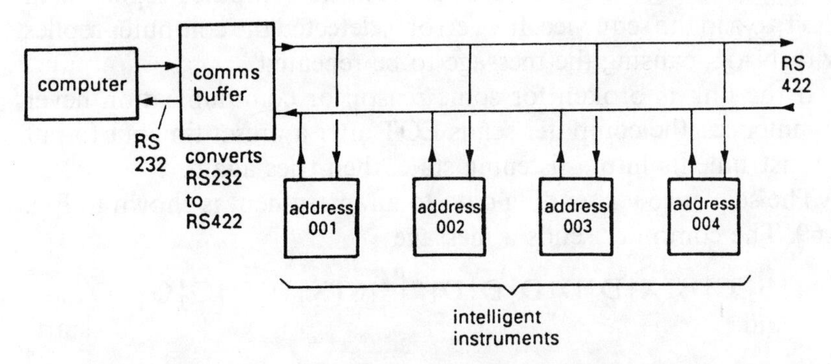

Fig. 6.67 A typical industrial data communications system, with several intelligent instruments connected via a serial link to a host computer.

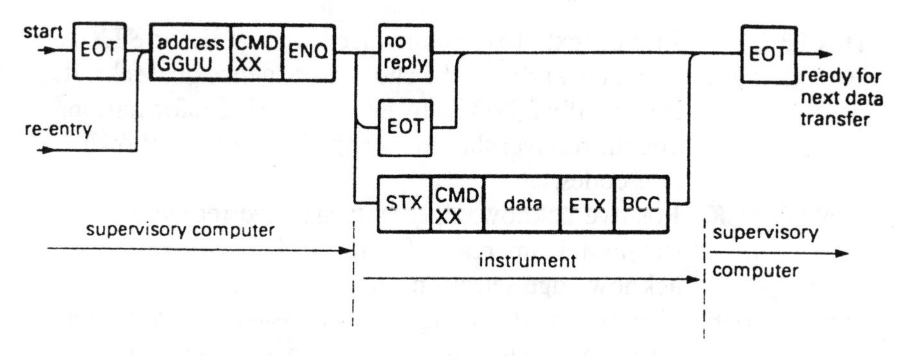

Fig. 6.68 The read sequence for a single piece of data.

```
   8   7 6 5 4 3 2   1   Byte
 |ENQ|*|*|U|U|G|G|EOT|→
end                 start
```

where G represents the instrument group address, U the unit address (both repeated for security) and ** the two digit mnemonic requested. EOT (sent as the first character) clears the line and sets all instruments to a 'listen' mode.

If the instrument recognises the address and the mnemonic it replies with

```
 |BCC|EXT|D|D|D|D|D|*|*|STX|→
end                       start
```

where ** is the requested mnemonic returned (again for security) and D the digits of the data requested. BCC is the error checking code, described in the next section.

If the message is received error free, the computer replies with EOT to end the sequence. If an error is detected the computer replies with NAK, causing the message to be repeated.

If the link is broken for some reason or communication never commences, the computer sends EOT after a preset timeout to put all instruments into a listening state, then tries again.

The sequence for sending data to an instrument is shown in Fig. 6.69. The computer sends a message

```
|BCC|ETX|D|D|D|D|D|*|*|STX|U|U|G|G|EOT|→
end                                   start
```

where G, U, *, BCC have the meanings above and D is the data to be updated for the parameter *, *.

The instrument can reply with ACK (data received OK) or NAK

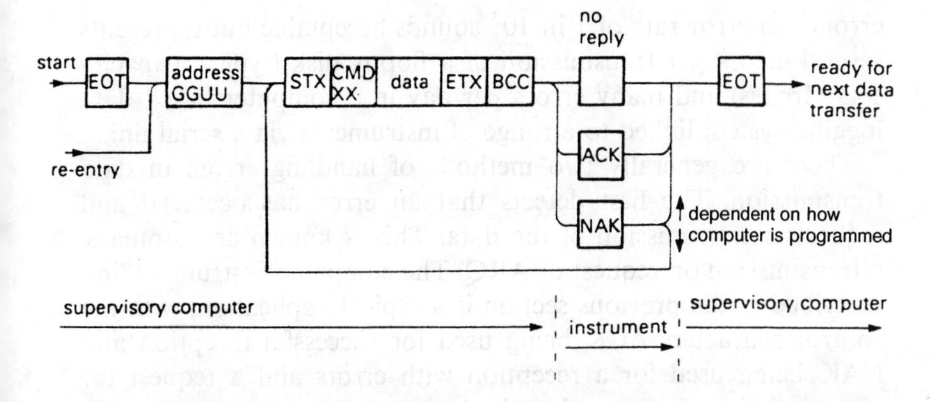

Fig. 6.69 Sequence for sending data from a computer to an instrument.

(data in error, or mnemonic unrecognised or an attempt made to update read only value, e.g. process variable value). The computer can choose to retry after a NAK dependent on the application. The sequence is terminated by an EOT from the computer.

6.8. Error control

6.8.1. Introduction

Data communication is particularly vulnerable to errors introduced by noise. Figure 6.59 shows that even at high signal to noise ratios there is still a small, but finite, probability of an error occurring. Error rates are generally quoted as the number of good bits per error, a typical modem link, for example, could have an error rate of about 1 in 10^5.

Error rates do not, however, give a full picture of what is going on, as an error rate of 1 in 10^5 implies a single error bit followed by 99 999 correct bits. Anyone using a telephone will be aware that interference generally has the form of 'clicks' and 'pops' introduced by the switching of inductive loads close to the transmission media. This form of noise is similar to that found on most data transmission lines. A click lasts about 0·05 seconds and is ignored in speech, but it represents the demise of 60 bits of data at 1200 baud. Noise, therefore, tends to introduce bursts of errors separated by long periods of error free transmissions, and the error rate represents the average over a long period of time.

A data transmission system must have some way of dealing with errors. An error rate of 1 in 10^5 sounds acceptable but represents several errors per transmission of a floppy disk (typical capacity 1·44 Mbytes) and many errors per day in a computer based data logging system linked to a range of instruments via a serial link.

There are generally two methods of handling errors in data transmission. The first detects that an error has occurred and requests retransmission of the data. This is known as automatic retransmission on request or ARQ. The computer–instrument link described in the previous section is a typical application, with the control characters ACK being used for successful reception and NAK being used for a reception with errors and a request for retransmission. The second method attempts to correct errors at the receiver and is known as forward error control (FEC). Both methods require the data to be transmitted in blocks which have additional error control bits added. Typically these 'redundant' error control bits will increase the message length (and hence decrease the useful transmission rate) by 5–20%.

6.8.2. Error detection methods

The simplest (and probably the commonest) error detection method is the addition of a parity bit to each character (called the vertical parity check). In ASCII coding, characters are represented by 7 bits, and the parity bit increases the length to 8 bits. The parity bit is added to ensure that the number of bits in the character is always odd as shown in Fig. 6.70a. Even parity (bits in the character made even by the parity bit) is equally feasible, but odd parity is more commonly used. The parity bit is easily calculated with simple logic circuits using exclusive OR gates as shown in Fig. 6.70b. Parity calculating ICs (such as the 74180 and 4531) are readily available.

Parity will detect single (or 3, 5, 7) errors, but not an even number of errors. In an error burst there is a distinct possibility that all the introduced errors will not be caught by the parity check.

Additional protection can be given by breaking the message down into blocks, each character of which is protected by a parity bit, and following the block with a block check character (BCC) which contains a single parity bit for each column bit position as shown in Fig. 6.71. Normally even parity is used for these column parity bits. This is known as longitudinal parity checking. The block check character has its own odd parity bit which is calculated from the number of bits in the BCC and not the parity of the parity bits in the text.

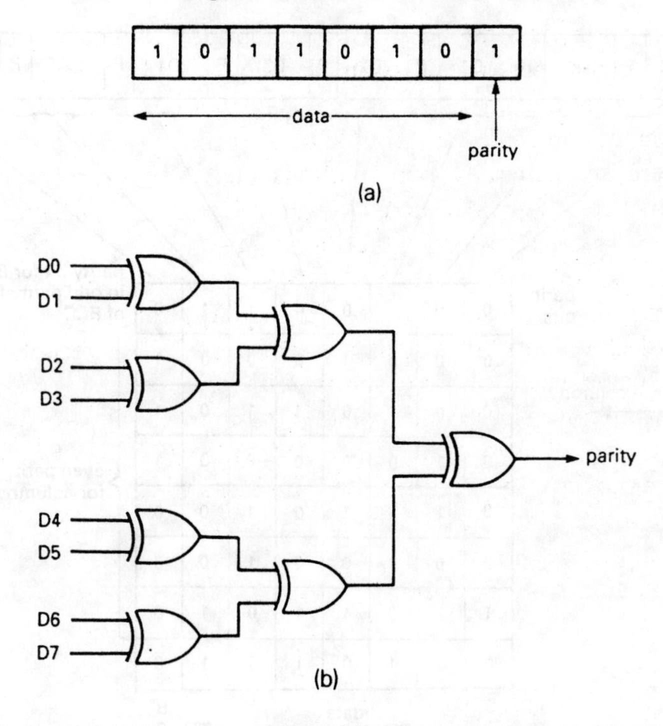

Fig. 6.70 Error checking with a parity bit, (a) The parity bit makes the number of bits in the word an odd number; (b) Parity circuit for an 8 bit word using exclusive OR gates.

The initial STX or SOH is excluded from the BCC calculation, but all succeeding characters including the terminating ETB or ETX are used as shown in Fig. 6.72.

Block check characters and longitudinal parity checking can detect odd numbers of errors, two bit errors and many even numbers of error. It is defeated by even numbers of errors which are rectangularly spaced around the block.

The most powerful error detection method is known as the cyclic redundancy code (CRC). Like the BCC method, the message is split into blocks, and the block of data is treated as a binary number which is divided by a predefined number. The remainder is sent after the block as two 8 bit numbers. The same calculation is performed at the receiver, errors being detected by differences in the CRC.

Typical circuits are based on exclusive OR logic gates and shift registers as shown in Fig. 6.72 which contains the common CRC implementations. In Europe the CRC–CCITT standard is commonly used.

CRC checks are very powerful and can detect long error bursts.

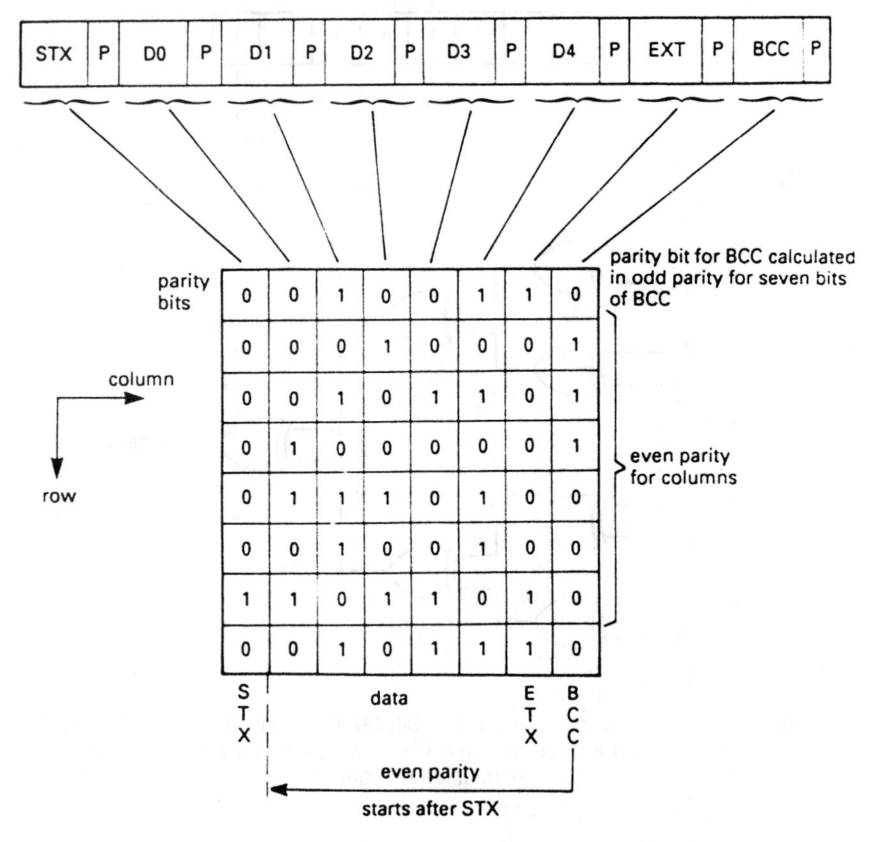

Fig. 6.71 Block check character used for error checking.

CRC–CCITT can detect error burst up to 12 bits in length and 99% of longer bursts. CRC–16 can detect error bursts up to 16 bits in length and 99% of longer bursts.

The use of CRC blocks can greatly improve the error rate on a link. Typically improvements of the order of 10^5 are achieved, giving an undetected error rate of 1 in 10^9 for a circuit with a basic error rate of 1 in 10^4.

6.8.3. Error correcting codes

Error detecting methods give an acceptable performance in most applications when coupled with an ARQ protocol. Occasionally it is not feasible to have ARQ, and in these circumstances it is desirable to use error correcting codes which indicate where the error has occurred. One obvious example is satellite and rocket probe data

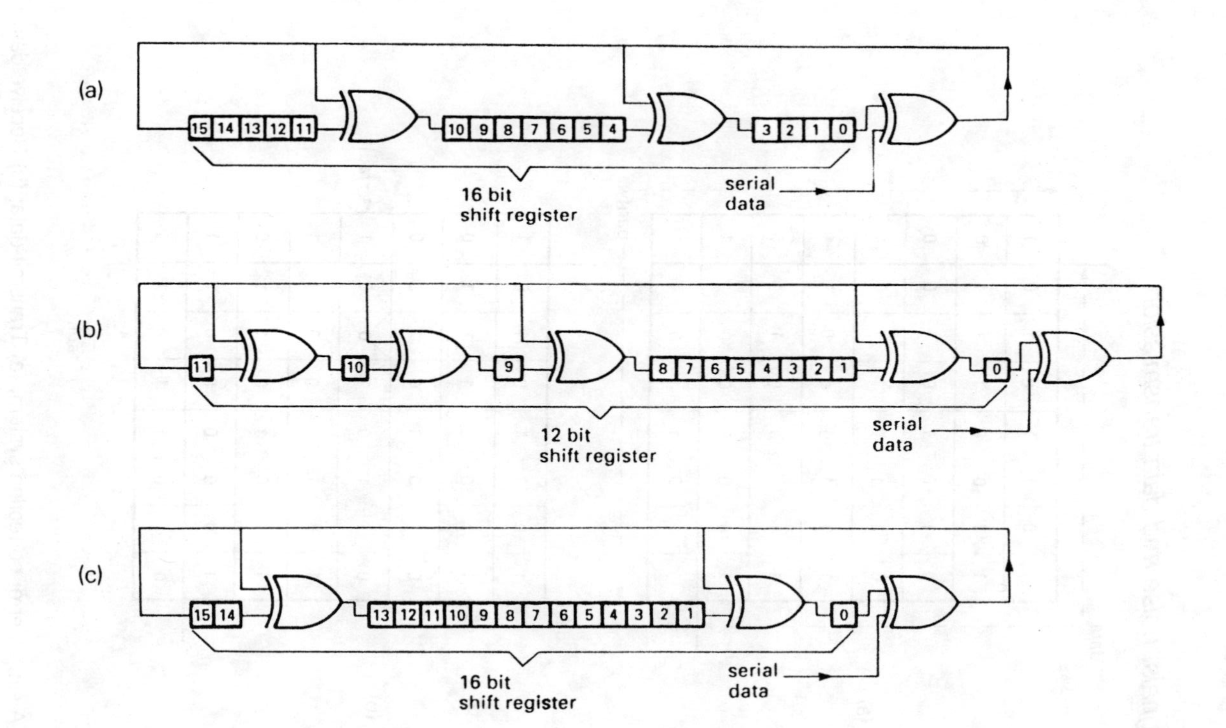

Fig. 6.72 Common cyclic redundancy code circuits, (a) Circuit for CRC-CCITT; (b) Circuit for CRC-1; (c) Circuit for CRC-16.

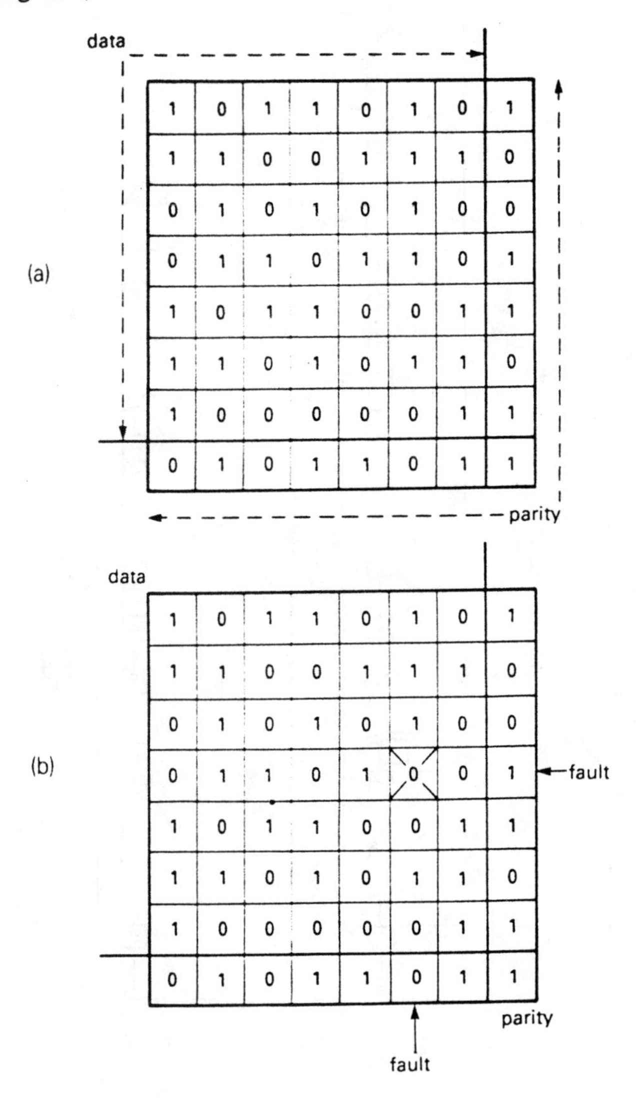

Fig. 6.73 A simple error correction system, (a) Transmit data; (b) Receive data with error.

links where the time taken for a signal to reach the receiver prevents the use of ARQ. Error correcting codes are not widely used in instrumentation and process control as error detection is generally more efficient.

A simple error correcting code is shown in Fig. 6.73. A square number of data bits (49 in this code) is arranged as a square with parity bits for each row and column. If an error occurs the location is

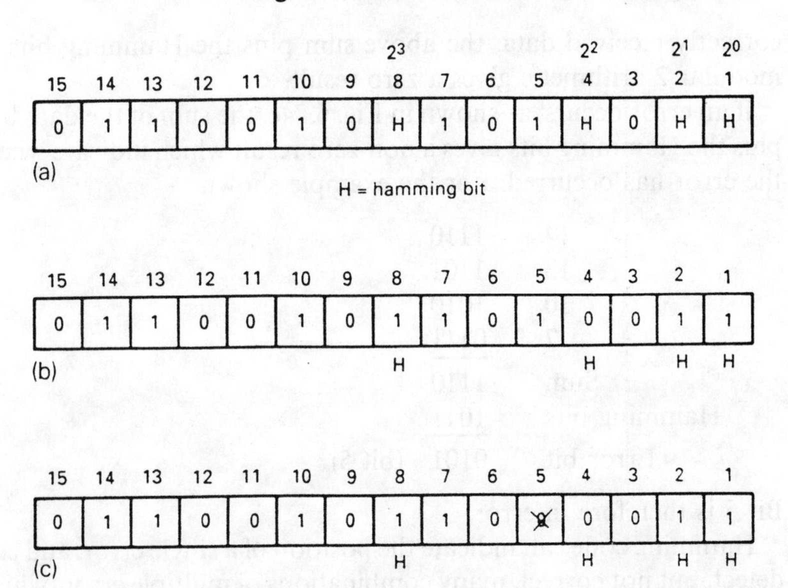

Fig. 6.74 The Hamming error correcting code, (a) A 15 bit word with 11 data bits and 4 Hamming bits; (b) Hamming bits added for the above data; (c) Word with a single error.

indicated by the interception of the incorrect row and column faulty parity bits as shown in Fig. 6.73b.

The scheme can be fooled, however, if two or more errors occur in a single row or column. It can give error detection but not correction for many other error conditions.

A more efficient system is the Hamming code illustrated in Fig. 6.74. The example shown has 11 data bits which require 4 check bits arranged in positions corresponding to a power of two. In practice data is arranged in larger blocks which require more check bits.

The check bit states are determined by modulo 2 (no carry) addition of the binary representation of the bits containing a binary one. In Fig. 6.74 bits 14, 13, 10, 7 and 5 are one, so the Hamming bit pattern is given by

14	1110
13	1101
10	1010
7	0111
5	0101
Hamming bits	1011

i.e. bits 8, 2 and 1 are required to be one as shown in Fig. 6.74b. For correctly received data, the above sum plus the Hamming bits in modular 2 arithmetic gives a zero result.

If an error occurs, as shown in Fig. 6.74c, the sum of the data bits plus the Hamming bits gives a non-zero result which indicate where the error has occurred. For the example shown:

14	1110	
13	1101	
10	1010	
7	0111	
Sum	1110	
Hamming bits	1011	
Error bit	0101	(bit 5)

Bit 5 is therefore in error.

Hamming code can indicate the position of a single error, and can detect, but not correct, many combinations of multiple error (which give an error indication of one of the Hamming bits or an unused position).

6.9. Area networks

6.9.1. Introduction

So far we have mainly considered point to point links. For true distributed control we need a method where several instruments, PLCs or computers can be linked together to allow communication to freely take place between any members of the system.

To achieve this we need to establish a connection topology, i.e. some way of sharing the common network that prevents time wasting contention plus an address system that allows messages to be sent from one member to another. Such systems are known as local area networks (LANS) or wide area networks (WANS) dependent on the size of the area and the number of stations. Some organisations use the term WAN to denote off-site links via public or private lines.

6.9.2. Transmission lines

Any network will be based, to some extent, on cable, and at the high speeds used there are aspects of transmission line theory that need to be considered.

Consider the simple circuit of Fig. 6.75a. At the instance that the switch closes, the source voltage does not know the value of the load at the far end of the line. The initial current step, i, is therefore determined not by the load, but by the characteristics of the cable (dependent on the inductance and capacitance per unit length). A line has a characteristic impedance, typically 75 ohms or 50 ohms for coax, and 100 to 300 ohms for biaxial or screened twisted pair. The initial current step will therefore be V/Z where Z is the characteristic impedance of the cable.

After a finite time, this current step reaches the load R, and produces a voltage step iR. If R is not the same as Z, this voltage step will not be the same as V, and a reflection will result. We can define a reflection coefficient ρ where

$$\rho = \frac{\text{reflected voltage amplitude}}{\text{incident voltage amplitude}}$$

It can be shown that

$$\rho = \frac{R - Z}{R + Z}$$

Typical results are shown on Fig. 6.75b.

This effect occurs on all cables and is normally of no concern as the reflections only persist for a short time. If, however, the propagation delay down the line is similar to the maximum frequency rate of the signal, the reflections can cause problems. It follows that a transmission line should be terminated by a resistance equal to the characteristic impedance of the line. Normally, devices for connecting onto a transmission line have a high input impedance to allow them to tap in anywhere, with terminating resistors being used at the ends of the line.

A side effect is that T connections, or spurs, are not allowed (unless the length of the spur is short). In Fig. 6.75c a T has been formed. To the signal, coming from the left, the two legs appear in parallel giving an apparent impedance of Z/2 and a reflection.

6.9.3. Network topologies

From the previous section it should be apparent that any network can only be based on a ring (which needs no terminating resistors) or a line (with a terminating resistor at each end). More complex

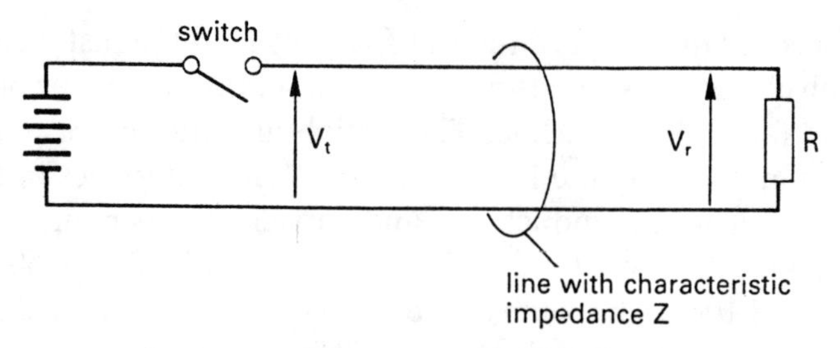

(a) A transmission line

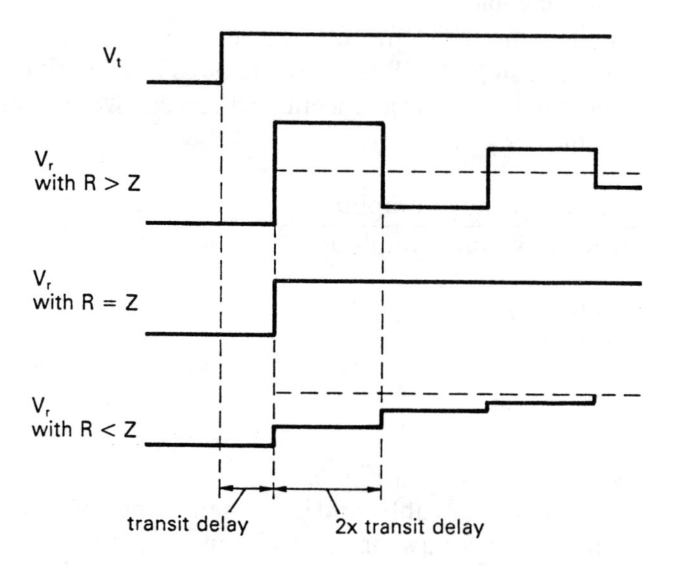

(b) Effect of terminating resistor

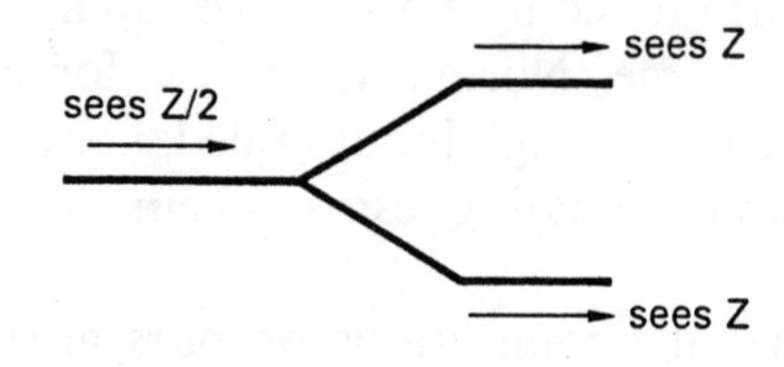

(c) The effect of a branch

Fig. 6.75 Transmission lines and characteristic impedance.

topologies can, of course, be constructed with bidirectional amplifiers acting as repeaters, splitters or buffers.

Figure 6.76 is a master/slave system where a common master wishes to receive or send data from/to slave devices, but the slaves never wish to talk to each other. All the slaves have addresses, which allows the master to issue commands each as 'Station 3; give me the

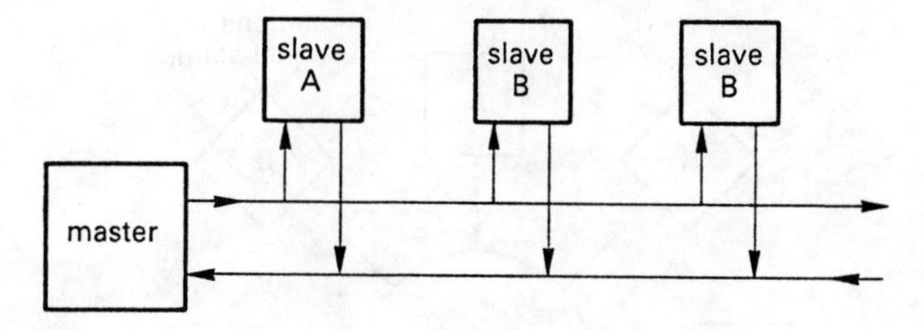

Fig. 6.76 A master/slave network.

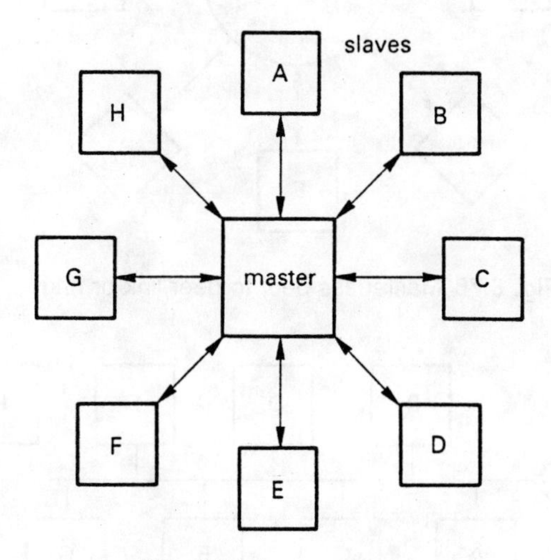

Fig. 6.77 A star network.

value of analogue input 4' or 'Station 14; your set point is 751.2'. Such systems are often based on RS422. The system described in section 6.7.4 and Fig. 6.67 is of this type.

The star network of Fig. 6.77 is again based on a master with a point to point link to individual stations. This arrangement is commonly used for high level mainframe computer systems. Communication control is performed by the master station. Station to station communication is possible via, and with the cooperation of, the master.

In Fig. 6.78 all the stations have been connected in a ring. There is no master, and all stations can talk to any other station and all have equal right of access. The term peer to peer link is often used for this arrangement. With Figs. 6.76 and 6.77, control was firmly in the hands of the master. With the ring, some technique is needed to avoid clashes when two stations wish to use the line at the same time. We will discuss this in the following section.

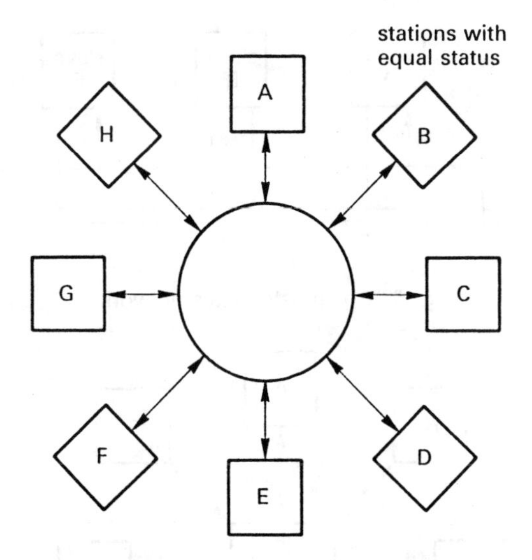

Fig. 6.78 Masterless peer to peer link or ring.

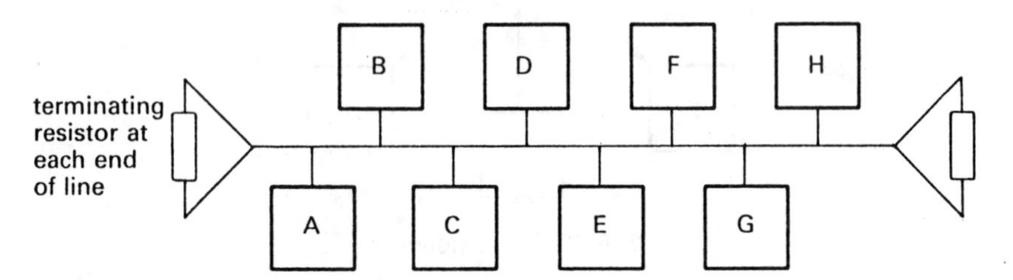

Fig. 6.79 Peer to peer link arranged as a bus or highway with terminating resistors.

Figure 6.79 is probably the commonest type of network in industrial networks. It is a single line with terminating resistors and, like the ring, is a peer to peer link where all stations have equal standing. It is usually called a bus (for busbar) network.

6.9.4. Network sharing

A peer to peer link allows many stations to use the same network. Inevitably two stations will want to communicate at the same time. If no precautions are taken, the result will be chaos. Various methods are used to govern access to the network.

One idea is to allocate time slots into which each station can put its messages. This is known as time division multiplexing, or TDM. Whilst it prevents clashes, it can be inefficient as a station will have to

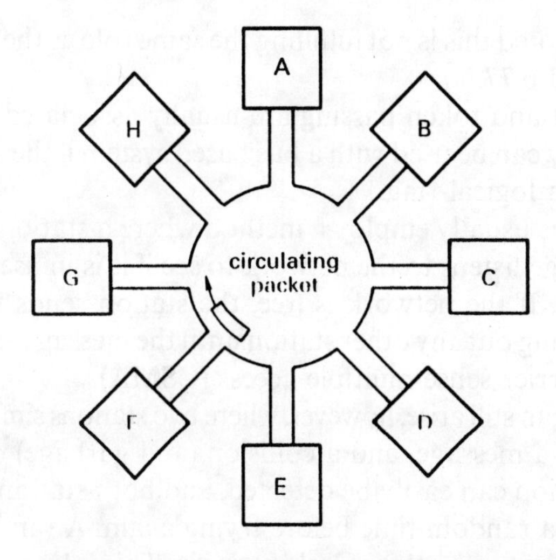

Fig. 6.80 Empty slot and token passing.

wait for its time slot even if no other station has a message to send. To some extent a mismatch between the frequency of messages from different stations can be overcome by giving more slots to hardworking stations. With a five station network and stations labelled A to E, if A has a high workload an order ABACADAEAB etc. might be adopted. This is sometimes known as statistical TDM.

The empty time slot of Fig. 6.80 uses a packet which continuously circulates around the ring. When a station wishes to send a message it waits for the empty slot to come round, when it adds its message. In Fig. 6.80, station A wishes to send a message to station D. It waits until the empty packet comes round. Then it puts its message onto the network along with the destination address D. Stations B and C pass the message but ignore it because it is not for their address. Station D matches the address, reads the contents (and appends that it has received the message). Stations E–H ignore it, but pass it on. Station A receives the message back again, sees the acknowledgement and removes its message leaving the empty packet circulating round the ring again.

A similar idea is token passing, where a 'permit to send' token circulates round the network. A station can only transmit when it is in possession of the token, which is released when the acknowledgement that the message arrived is received.

Both empty slot and token passing require some way of reinstating the packet or token if the network is corrupted by noise or broken. This is usually provided by a master station, or monitor station, but it should be noted this is not fulfilling the same role as the masters on Figs 6.76 and 6.77.

Empty slot and token passing are usually associated with rings, although they can be used with a bus based system if the stations are arranged as a logical ring.

Bus systems usually employ a method where a station wishing to send a message listens to the network to see if it is in use. If it is, the station waits. If the network is free, the station sends its message (thereby locking out any other station until the message ends). This is known as carrier sense multiple access (CSMA).

Situations can still arise, however, where two stations simultaneously start to send a message, and a collision (and garbage) results.

This situation can easily be detected, and both stations then stop and wait for a random time before trying again. A random time is used to stop the two stations clashing again. This is known as carrier sense multiple access with collision detection (CSMA/CD).

There is a fundamental difference between CSMA and TDM, empty slot or token passing. With the latter methods there is a certain amount of time wasting, but every station is guaranteed access within a specified time. With CSMA there is little time wasting, but a station can, in theory, suffer repeated collisions and never get access at all.

A useful analogy is to consider motor car traffic control. TDM/token passing approximates to traffic lights, CSMA to roundabouts. In heavy traffic the best solution is traffic lights; everyone gets through and the waiting is shared evenly. Roundabouts can 'lock out' one road when the traffic flow is heavy and uneven from one direction. In light traffic, however, roundabouts keep the traffic flowing smoothly; and there are few things more annoying than being brought to a halt by a red light, then have nothing go past in the other direction.

6.9.5. A communication hierarchy

Early process control systems tended to be based on a single large computer or PLC. The advent of cheap devices (such as PLCs or intelligent instruments) with good communications has led to the development of a hierarchy of machines which split the tasks between them.

This is generally arranged as Fig. 6.81 with a hierarchy split into four levels.

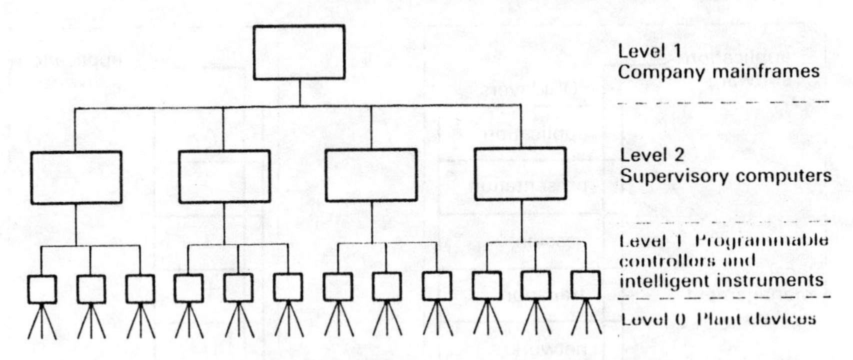

Fig. 6.81 Communication hierarchy.

Level 0 is the actual plant, with intelligent devices linking to the next level by direct wiring or simple RS232/422 serial links.
Level 1 consists of PLCs and small computers directly controlling the plant.
Level 2 is microcomputers, such as the DEC VAX, acting as supervisors for large areas of plants.
Level 3 is the large company mainframes, such as IBM's AS400.

Usually the layout is not as clear cut as Fig. 6.81 implies. There are also differences between different companies, some number the layers from top to bottom and some ignore level 0.

There are many advantages to distributed systems. The resulting tree is conceptually simple, and as such is easy to design, commission, maintain and modify.

A correctly designed system will also, for short periods, be fault tolerant and can cope in a limited mode with the failure of individual stations. A distributed system can also bring about an increase in performance as lower level machines take the work off higher level machines.

6.9.6. The ISO/OSI model

Neat as Fig. 6.81 is, the interconnection between different machines can bring even more problems than linking two 'RS232 compatible devices'. Common problems are different baud rates, flow control, routing and protocols.

In 1977 the International Standards Organisation (ISO) started work on standards to try to ensure compatibility between different

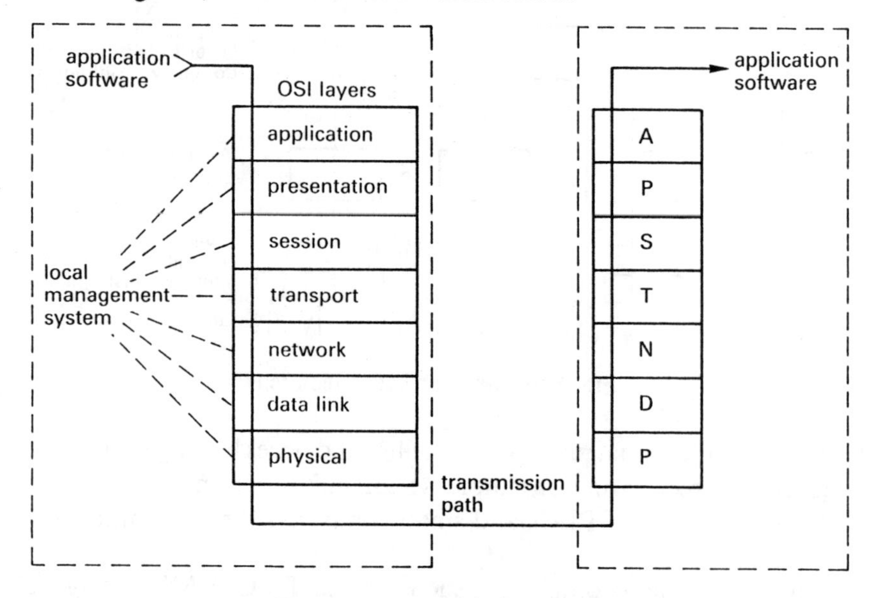

Fig. 6.82 The ISO/OSI model.

manufacturers' equipment. This is known as the Open Systems Interconnection (OSI) model, and is primarily concerned with communication between level 2/3 systems on Fig. 6.81.

It consists of definitions for the seven layers of Fig. 6.82. Each layer at the transmission end has a direct relationship with the same layer at the receiving end. The function of each layer is, from the bottom:

(a) The physical link layer – which is concerned with the coding and physical transmission of the message. Requirements such as transmission speed are covered.
(b) Data link layer – controls error detection and correction. It ensures integrity within the network and controls access to it by CSMA/CD or token passing.
(c) Network layer – performs switching and makes connection between nodes.
(d) Transport layer – provides error detection and correction for the whole message by ARQ, and controls message flow to prevent overrun at the receiver.
(e) Session layer – provides the function to set up, maintain and disconnect a link, and the methods used to re-establish communication if there are problems with the link.
(f) Presentation layer – provides the data in a standard format

(which may require the data to be converted from its original form in the initiating application).

(g) Application layer – links the user program into the communication process and determines what functions it requires.

As a very rough analogy, consider the placing of a verbal order by telephone. This analogy is based on Siemens material published in their brochure 'Communications Setting the Pace in Automation'.

(a) Physical link layer – the phone is lifted and connected to the telephone network. A dialling tone is heard.
(b) Error detection and control – it's a good line with no noise.
(c) Network layer – The number is dialled, 9 for an outside line then the number. The phone rings at the other end.
(d) Transport layer – The telephone is lifted at the receiving end. 'This is ACME products, could you hold on please I'm handling another call. OK, go ahead now. Sorry, I didn't get that could you repeat please'.
(e) Session layer – 'This is Aphrodite Glue Works, I have a verbal order for your number CAP4057, my account number is 3220'. The receiver makes a note of these details in case the call is broken prematurely.
(f) Presentation layer – 'I am using order numbers from the June 1995 catalogue'.
(g) Application layer – 'I require 100 off 302-706 and 50 off 209-417, delivery by datapost.' 'OK, 100 off 302-706 and 50 off 209-417 will be despatched by datapost this afternoon. Total cost £147.20, invoice to follow'.

At any stage, the lower layers can interact. A burst of noise on the line, for example, will cause the transport layer to ask for a repeat of the last message.

It can be seen that layers (a) to (d) are concerned with the communication and layers (e) to (g) are concerned with processing functions for the particular applications.

6.9.7. Ethernet

Ethernet is a very popular bus based LAN originated by DEC, Xerox and Intel and commonly used to link the computers at level 2 in Fig. 6.81. It uses 50 ohm coaxial cable, with a maximum cable length of 500 m (although this can be extended with repeaters). Up to

1024 stations can be accommodated, although in practical systems the number is far lower. Baseband signalling is used with CSMA/CD access control. The raw data rate is 10 Mbaud, giving very fast response at loading levels up to about 20–30% of the theoretical maximum. Beyond this, collisions start to occur.

Stations are connected onto the cables by transducers known as nodes on the network. Commonly 'Vampire Technology' is used where the transceiver clamps onto the cable, with a sharp pin piercing the cable and contacting the centre conductor. An insulated ring on the pin contacts the screen. The arrangement of the pin/ring prevents short circuits and allows transceivers to be added, or removed, without disturbing the rest of the network.

To avoid reflections (as discussed earlier in Section 6.9.2) a minimum spacing of 2·5 m must be maintained between nodes. To assist the user, Ethernet cable has 'tap in' points marked on its sleeving.

An alternative to the Vampire transceivers is plug in transceivers. These use in line coaxial connectors and are obviously more secure, but have the disadvantage that the network must be broken and hence disrupted if a station is added or removed.

The transceivers are connected to a local controller which performs the access control. Ethernet has three layers shown on Fig. 6.83 which approximate to the functions performed by the same layers in the OSI model discussed in Section 6.9.6.

6.9.8. Towards standardisation

We have already discussed the difficulties of linking different equipment. There is normally little problem linking devices to higher level computers. Manufacturers publish their message format and protocols, and interfacing software (called 'drivers') has been written for all common computers. The difficulty comes when two devices at level 0 or 1 in Fig. 6.81 have to be linked. In many cases, the only economical solution is to do it through the computers and the higher level link.

General Motors (GM) in the USA were faced with this problem and attempted to specify a LAN for industrial control. This was called MAP (Manufacturing Automation Protocol). A similar office based LAN called TOP (Technical Office Protocol) was conceived at the same time. With GM's purchasing muscle, it involved several automation equipment manufacturers. A firm commitment to the OSI model was made, and the network based on broadband token

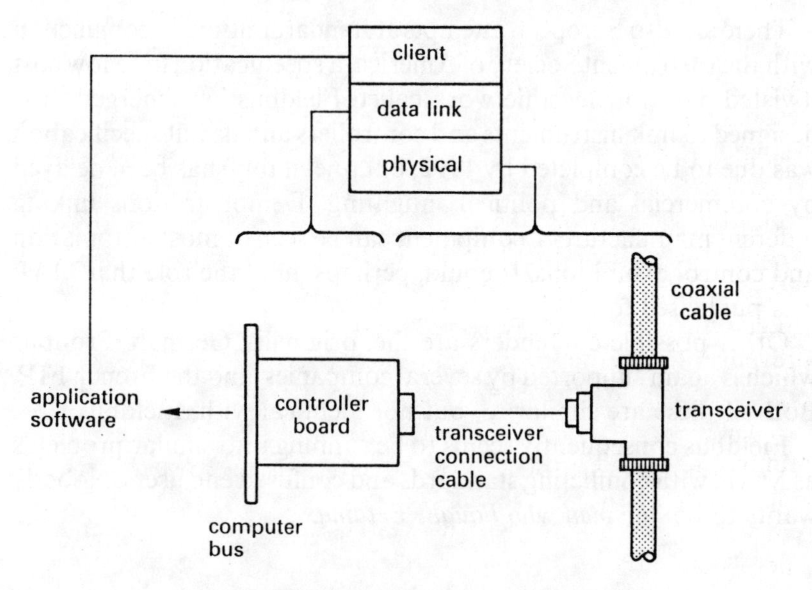

Fig. 6.83 The Ethernet architecture.

bus as specified in IEEE 802.4 (compare Ethernet, baseband CSMA/CD to IEEE 802.3).

The token bus was chosen as it is deterministic, i.e. the response time can be predicted (see earlier discussion on roundabouts and traffic lights in section 6.9.4).

In the mid 1980s MAP was going to be the common standard of industrial control. MAP systems have been installed, both in Europe and the USA, but it has not yet achieved anything like acceptance. There appear to be several reasons for this distinct lack of enthusiasm. The first is a bureaucratic organisation and a changing specification. The term 'moving target' has been used independently on several occasions by different manufacturers. The second reason is cost; MAP links often cost far more than the device to which they are connected. The expression 'Designed by big organisations, for big organisations' used by many suppliers seems apt. The third reason is speed; by using token passing MAP is slow by comparison with Ethernet and is not really designed for time critical applications. The non-deterministic nature of CSMA/CD does not seem to cause any problems up to about 30% of the theoretical maximum loading, and real systems normally operate below 10% loading. The final, and perhaps most crucial, fact is that MAP seems to have settled at a level where it is in direct competition with established LANs such as Ethernet rather than the proprietary systems at level 1 of Fig. 6.81.

There are also European attempts at standardisation. In conjunction with the Instrument Society of America, a specification for a low cost (twisted pair) low level network called Fieldbus has emerged. It is designed to link instruments and controllers and its full specification was due to be completed by 1992, but (inevitably) has been delayed by commercial and political infighting. Demonstrations linking different manufacturers' equipment can be seen at most automation and control exhibitions. It could, perhaps, fulfil the role that MAP was publicised for.

Other possible contenders are the, originally German, Profibus which is again supported by several companies, and the French FIP. Both of these are similar to, but not identical with, Fieldbus.

Fieldbus consequently seems to be running into similar problems as MAP, with conflicting standards and confused end users. Nobody wants to be *'the man who bought Betamax'*.

6.9.9. Safety and practical considerations

Figure 6.84 shows a fairly common situation where a switch connected to PLC_A is, via a serial link, causing a motor to run in another PLC_B. Supposing the motor is started and the link is severed. The bit corresponding to 'Motor Run' which is set inside PLC_B will not be cleared by the link failure, and PLC_A will be unable to stop the motor.

When the switch is turned off, the serial link control in PLC_A will signal an error, but this is of no use to PLC_B which does not know that PLC_A is trying to communicate with it. This may, or may not, be a problem depending on the application, but it is obvious there are safety implications that need to be considered.

One approach is to define how long an output driven by the link is allowed to be uncontrolled, say 2 secs. The originating PLC_A then sends a toggling signal via the link at a slightly shorter period, say 1·5 secs. Inside PLC_B, the true and complement forms of this signal trigger two delay off timers set for 2 seconds. In normal healthy operation both timers should never time out, so both timers energised can be used as a 'link healthy' signal. If the link fails, one timer will time-out (and one stay energised) causing the 'link healthy' signal to de-energise and all link controlled outputs to go to a safe state.

A network also introduces extra delays into the system. These delays obviously depend on the loading on the network, but are typically of the order of 0·2 to 0·5 seconds on proprietary networks,

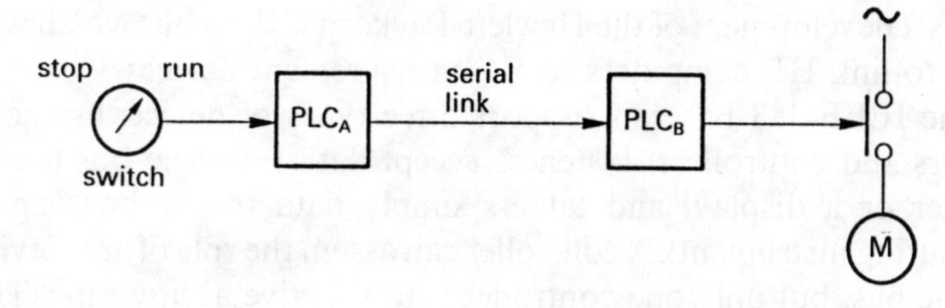

Fig. 6.84 Safety considerations with serial communications.

and possibly a bit slower on Ethernet and MAP.

Noise is a major source of problems, and normally manifests itself as an increase in the delay time introduced by the network (caused by a large number of retries). Because of checking and CRC discussed earlier, noise rarely causes operational problems, and when it does the effect is almost always something not working when requested rather than something starting unexpectedly. Noise prevents signals getting through, it does not usually cause faulty signals to be accepted, but the prudent designer should not rely on this characteristic.

Obvious precautions against noise are separation from power cables, and the use of conduit or trunking (mainly to identify low level signal cables). Cable screens should be continuous and earthed in one, and only one, place. Great care should be taken to prevent screens accidentally grounding inside junction boxes. Fibre optics give almost total freedom from interference.

Most proprietary networks have monitoring facilities and it is worth logging the error rate when a network is first commissioned. These allow regular checks to be made later, and any deterioration noticed before problems start to arise.

6.10. Parallel bus systems

Discussions so far have considered serial based systems. Parallel bus systems are less widely used in industry, largely because of cabling costs and restrictions on bus lengths, but are useful where a group of instruments or controllers can be located locally to a computer.

The commonest parallel bus system is probably the IEEE-488 general purpose parallel interface bus (sometimes called the GPIB standard). This allows the linking of up to 15 devices and a computer with a total transmission length of 20 m. The standard was formalised by the American Institute of Electrical and Electronic Engineers, and is a development of the Hewlett-Packard HP-IB bus which was used to link HP computers to HP measurement devices.

The IEEE-488 bus can support three types of device: listeners, talkers and controllers. Listeners accept data from the bus (e.g. a printer or a display) and talkers supply data to the bus (e.g. a measuring instrument). A controller can assign the role of any device on the bus, but only one controller can be active at any time. The designations listener-talker-controller are really attributes of a unit, and it is possible for a piece of equipment to fill more than one role. A computer, for example, can act as all three.

Signals on the system can be grouped into a bidirectional 8 bit data bus, transfer control, general interface management and grounds/shields. The allocation is shown in Table 6.2.

Signalling is done at TTL levels, with 0 V representing '1' and 3·5 V a '0' (note that this is the inverse of normal logic operation). Open collector drivers are used to allow the bidirectional data bus and control signals such as NDAC and NFRD to operate. The maximum bus length is 20 m, and there must be no more than 2 m between

Table 6.2

Group	*Designation*	*Description*	*Pin*
Data bus	DIO 1	Data Input/Output 1	1
	2	2	2
	3	3	3
	4	4	4
	5	5	13
	6	6	14
	7	7	15
	8	8	16
Transfer control	DAV	Data Valid	6
	NRFD	Not Ready for Data	7
	NDAC	Not Data Accepted	8
Interface management	IFC	Interface Clear	9
	SRQ	Service Request	10
	ATN	Attention	11
	REN	Remote Enable	17
	EOI	End or Identify	5
Grounds/shield	Shield		12
	DAV ground		18
	NRFD ground		19
	NDAC ground		20
	IFC ground		21
	SRQ ground		22
	ATN ground		23
	LOGIC ground		24

Table 6.3 *Control selection with ATN line (bit 7 not used)*

Function	*Bit*	*7*	*6*	*5*	*4*	*3*	*2*	*1*	*0*
Bus Command CCCCC		X	0	0	C	C	C	C	C
Enable Listen Address LLLLL		X	0	1	L	L	L	L	L
Enable Talk Address TTTTT		X	1	0	T	T	T	T	T
Enable Secondary Address SSSSS		X	1	1	S	S	S	S	S

devices. The data bus is used for several purposes. It can carry data to one or more listeners, data from one talker and addresses to enable or disable devices. The system can support 15 devices with 16 additional secondary devices (generally allocated to secondary functions within a unit). These devices are identified by an address with address 31 being used as a 'disable all listeners' address.

The data bus action is determined by the active controller, which uses the ATN line to signal whether the bus is carrying data or control information. When the ATN line is taken low any active talker is disabled and control is selected as shown in Table 6.3.

The major attraction of IEEE-488 is that the user is generally unaware of the actual bus operation, with data transfers being controlled by simple programming instructions. For example, when data is sent to a listener, the command

```
OUTPUT 701; Value
```

is used. This selects the computer (address 700) as Talker, address 01 as Listener and sends the data held in the variable Value. In terms of the bus operation the following steps are performed:

(1) Deselect all Listeners
(2) Select Talker
(3) Select Listener
(4) Perform Data Transfer

These steps are invisible to the user.

Chapter 7
Computer simulation

7.1. Introduction

Many of the analytical techniques described in previous chapters are mathematically straightforward, but laborious to perform by hand. Increasingly computers are being used to analyse control systems. This chapter provides several programs and utilities to show some of the methods used in computer analysis. The programs are not designed for serious analytical use. They have been written to demonstrate ideas and provide a basis on which the interested reader can build and develop further.

The programs have been written in Turbo Pascal Version 6 or Visual Basic. Pascal was chosen for several reasons; mainly the author likes the language, and the resulting programs are easier to read and follow than more efficient languages such as C. Visual Basic offers an attractive front end, and the 'clock' function block seems to be well suited for computer simulation programs.

The programs are deliberately verbose. In many places several statements could have been compacted into a single statement. This has been avoided where the compaction would have hindered the readability. The programs have also been liberally annotated. This verbose code and documentation can obviously be amended by anyone wishing to develop the programs further.

The Turbo Pascal code should be compatible with Version 4 and later (but not Turbo Pascal for Windows). With earlier versions (including the workhorse Version 3) there are minor differences in graphical procedures and functions, DRAW(X1,Y1,X2,Y2) in Version 3 performs a similar function to LINE(X1,Y1,X2,Y2) in Version 4 and later. A more fundamental difference, though, concerns user written library functions.

In early versions of Turbo Pascal (up to and including Version 3), these were written as standard Pascal programs, and linked in with an include statement using the I compiler directive of the form:

```
{$Ig_utils.pas}
```

which includes the user written library program g_utils.pas into the compilation.

From Version 4, library files are written specifically as library units and declared as such by using the word 'unit' rather than 'program' in their headers. The public links to other programs are first declared followed by the private code:

```
unit g_utils {a collection of graphic utilities}
interface
{links defined here}
implementation
{private code here}
end. {of unit}
```

The units are linked into the main program with a USES statement. For example

```
program control; {a program in version 6}
uses crt, g_utils, m_utils;
{followed by the program}
```

There are several library units in this chapter, typical of which are:

g_utils, a collection of graphics routines.
t_utils, a collection of text and miscellaneous utilities performing functions such as input within predefined limits and stopping a program crash if the user enters a letter where a number is expected.
mathutil, a collection of mathematical functions and procedures for solving polynomial equations, and performing arithmetic with complex numbers.

The Turbo Pascal programs are written around a VGA/EGA screen of nominal size 640 pixels wide by 480 pixels high. The standard graphics origin (x = 0, y = 0) in Turbo Pascal is the top left-hand corner of the screen. This severely degrades the readability of graphical programs. Procedures in the g_utils unit scale the screen size and invert the y axis so that x = 0, y = 0 is apparently at the bottom left-hand corner as one would expect when drawing graphs or time relationships.

The programs in Visual Basic were written with Visual Basic Professional Version 3.0. The use of the Professional Version is recommended because of the inclusion of the Gauge and Graph VBXs.

7.2. Printing results

The programs are not much use if the results have to be traced off the screen. MSDOS provides a simple screen dump program called GRAPHICS which should be run from DOS before starting the Pascal programs. The Print-Screen PB can then be used to trigger a screen dump to the printer. The Pascal printouts in this chapter were obtained with this method. Better quality screen dump utilities can be obtained from most shareware companies. These utilities generally give better quality and can handle multicoloured screens. An alternative approach is to use a screen grabber utility which converts the screen to a PCX, TIF or similar format file which can be loaded into a paintbox or desk top publisher program and modified/printed as required.

Printouts in Windows are easier to obtain. Pressing the Print-Screen PB causes a bit mapped copy of the screen to be placed on the clipboard. It can then be pasted into Paintbrush or any normal Windows wordprocessing or DTP program.

7.3. Numerical analysis

All of these simulation programs represent the signals in a control system as numbers. Inherently a computer program works in binary and represents a number with a certain resolution. Turbo Pascal handles so-called 'real' (as opposed to integer) numbers in exponent form, i.e. 3 245 613 is held as $3{\cdot}245613 \times 10^6$.

Numbers in exponent form have an upper and lower limit on their allowable range. A standard real number in Turbo Pascal can only exist in the range $2{\cdot}9 \times 10^{-39}$ to $1{\cdot}7 \times 10^{38}$ with a resolution of 11 to 12 significant digits. This limited resolution causes rounding errors.

Suppose, for example, that we have a quadratic equation to solve. As every GCSE student knows, the formula for solving the equation $ax^2 + bx + c = 0$ is $(-b \pm \sqrt{b^2 - 4ac})/2a$. If, however, b^2 is much greater than 4ac, the two roots are greatly different in magnitude and rounding errors will cause one root to be found with poor accuracy

Consider the equation $x^2 + 10x + 0{\cdot}01 = 0$, a typical equation in process control, and suppose, for illustrative purposes, we have only four significant digits. b^2 is 100, and 4ac is 0·04 giving $\sqrt{b^2 - 4ac}$ to be 9·998. The two solutions are evaluated as −9·999 (correct to four decimal places) and −0.001 (correct to one decimal place and possibly in error by 100%).

The effect of rounding errors should be of great concern in commercial numerical analysis programs. A common technique is individual 'polishing' of the crude results by successive approximation similar to the straddling method described later. Another approach is to recognise, and deal with, the problem of rounding directly.

With the quadratic equation, for example, it is not commonly known that there are two solutions, the common

$$x = (-b \pm \sqrt{b^2 - 4ac})/2a$$

and the far less well known

$$x = 2c/(-b \pm \sqrt{b^2 - 4ac})$$

A more accurate set of results can then be calculated by first evaluating

$$-0{\cdot}5(b + \mathrm{sgn}(b)\sqrt{b^2 - 4ac}) \quad \text{call this q}$$

The two results are then

$$x_1 = q/a \text{ and } x_2 = c/q$$

Both results are accurate to the rounding error in q.

Many languages allow higher resolution. Turbo Pascal supports Double (16 significant places) and Extended (20 significant places) by using the built-in maths co-processor 80 × 87 emulation. This does, however, significantly slow down the programs and only reduces, not eliminates, the problem.

The topic of rounding errors and numerical analysis techniques are discussed in detail in the excellent book *Numerical Recipes in Turbo Pascal, The Art of Scientific Computing* by Press, Flannery *et al.* published by Cambridge Press. They say, with a certain truth, that the handling of rounding errors is practically the entire content of the study of numerical analysis.

7.4. Program DIGSIM

The program DIGSIM draws the closed loop step response for a system described by an open loop transfer function in series with an optional transit delay as shown on Fig. 7.1.

Any reasonable real life system can be described by a transfer function whose numerator power is lower than the denominator power as below:

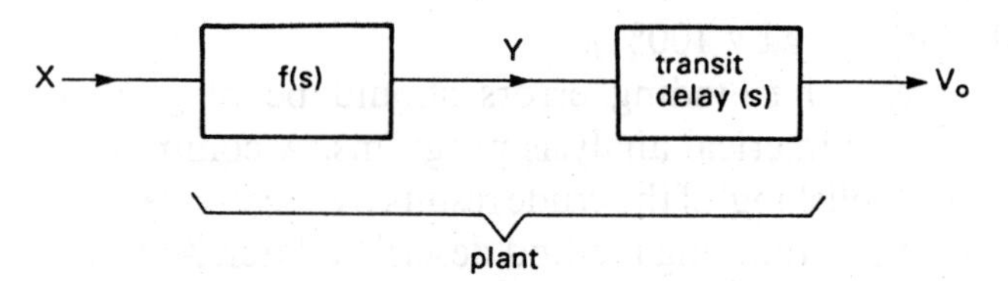

Fig. 7.1 Plant representation for DIGSIM program.

$$\frac{Y}{X} = \frac{A_0 + A_1 s + A_2 s^2 + \ldots A_{n-1} s^{n-1}}{B_0 + B_1 s + B_2 s^2 + \ldots B_{n-1} s^{n-1} + B_n s^n} \tag{7.1}$$

This is described as a minimum phase system, see Section 5.1.

Equation 7.1 can be reorganised into:

$$Y(B_0 + B_1 s + B_2 s + \ldots + B_n s^n) = X(A_0 + A_1 s + \ldots + A_{n-1} s^{n-1})$$

If we divide by $B_n s^n$ we get

$$Y\left(\frac{B_0}{B_n} s^{-n} + \frac{B_1}{B_n} s^{-(n-1)} + \ldots + 1\right)$$

$$= X\left(\frac{A_0}{B_n} s^{-n} + \frac{A_1}{B_n} s^{-(n-1)} + \ldots + \frac{A_{n-1}}{B_n} s^{-1}\right)$$

from which

$$Y = \left(\frac{A_0}{B_n} X - \frac{B_0}{B_n} Y\right)\frac{1}{s^n} + \left(\frac{A_1}{B_n} X - \frac{B_1}{B_n} Y\right)\frac{1}{s^{n-1}}$$

$$+ \ldots + \left(\frac{A_{n-1}}{B_n} X - \frac{B_{n-1}}{B_n} Y\right)\frac{1}{s}$$

The operation of the open loop system can thus be represented by Fig. 7.2. Terms in 1/s represent integration which can be performed numerically as successive summation. The program uses the simple integration algorithm of Fig. 4.25b described in Section 4.4.2. The trapezoid representation of Fig. 4.25c would give better results but make the program more complex.

The program is organised as Fig. 7.3, with the plant response being calculated as Fig. 7.2 at regular time intervals ΔT. The transit delay is simulated by a simple bucket brigade shift register organised as Fig. 7.4. Values are passed down one place per ΔT, the length of the chain being set at (delay time)/ΔT.

To use this program, the open loop transfer function must be

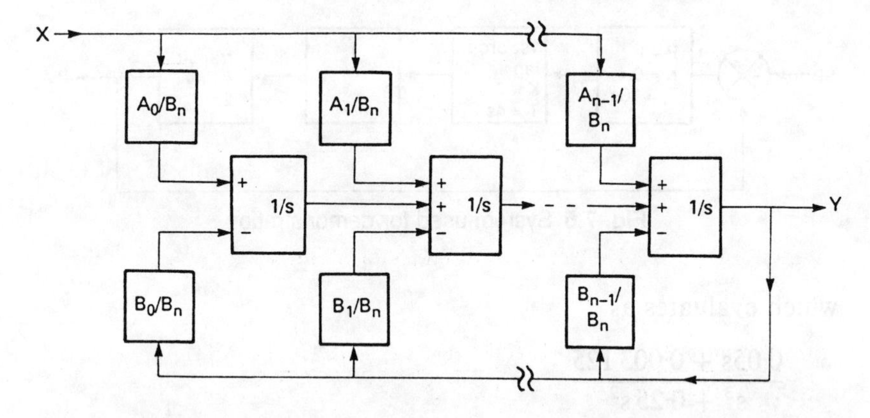

Fig. 7.2 Representation of plant function block.

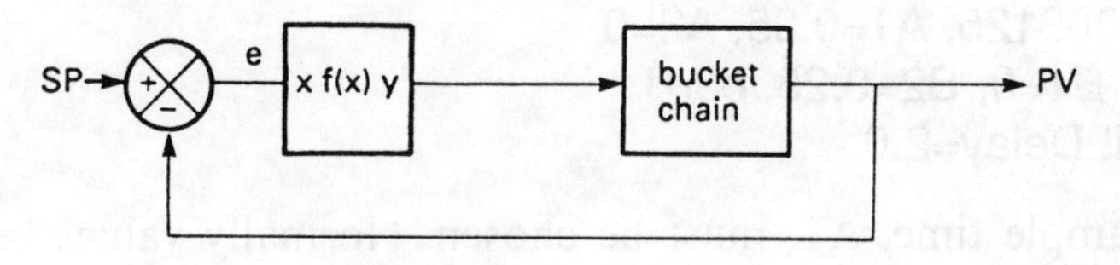

Fig. 7.3 Program organisation.

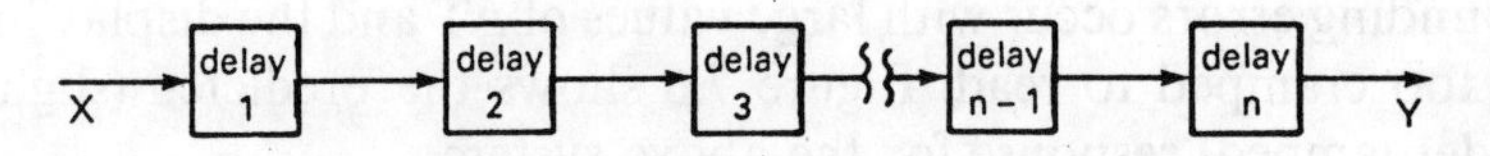

Fig. 7.4 Method used to simulate transit delay.

found. Consider the system in Fig. 7.5 which represents a simple level control with a PI controller and first order lag on the actuator. The sensor imposes a two second transit delay.

The PI controller has the transfer function

$$\frac{K(s + 1/T_i)}{s}$$

The lag has the transfer function

$$\frac{K/T}{s + 1/T}$$

The integral action transfer function is simply 1/s, giving the composite open loop transfer function

$$\frac{0{\cdot}4(s + 1/16) \times 0{\cdot}5 \times 0{\cdot}25}{s \times (s + 1/4) \times s}$$

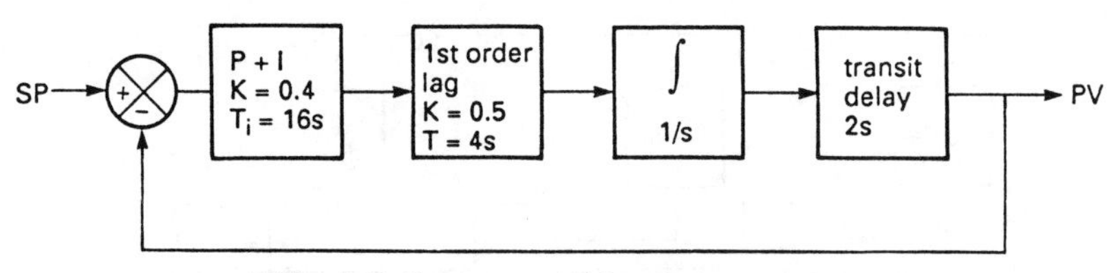

Fig. 7.5 System used for demonstration.

which evaluates as

$$\frac{0{\cdot}05s + 0{\cdot}003\,125}{s^3 + 0{\cdot}25\,s^2}$$

This is input as

A0=0.003125, A1=0.05, A2=0
B0=0, B1=0, B2=0.25, b3=1
Transit Delay=2.0

The sample time, ΔT, must be chosen. Normally values between 0·01 s and 1 s are best. A very short time gives the best accuracy, but the interesting part of the response may not fit on the screen. Rounding errors occur with large values of ΔT and the display may be too cramped to read. Figure 7.6 shows the predicted (slightly underdamped) response for the above system.

7.5. Transfer function formation from individual blocks

The program DIGSIM and the next two programs DISTURB (showing the effect of a disturbance) and ROOTLOKE (which plots root loci) all require the open loop transfer function to be entered in the form of equation 7.1 above. For complex systems this can appear to be a formidable task. It can be simplified, though, by simplifying the block diagram in stages with the equivalent relationships shown in Fig. 7.7

Figure 7.8 shows the stages in simplifying a very disorganised block diagram to the form needed by these three programs

7.6. Program DISTURB

The program DISTURB shows the response of a system to a disturbance. The system is considered to consist of two parts, one

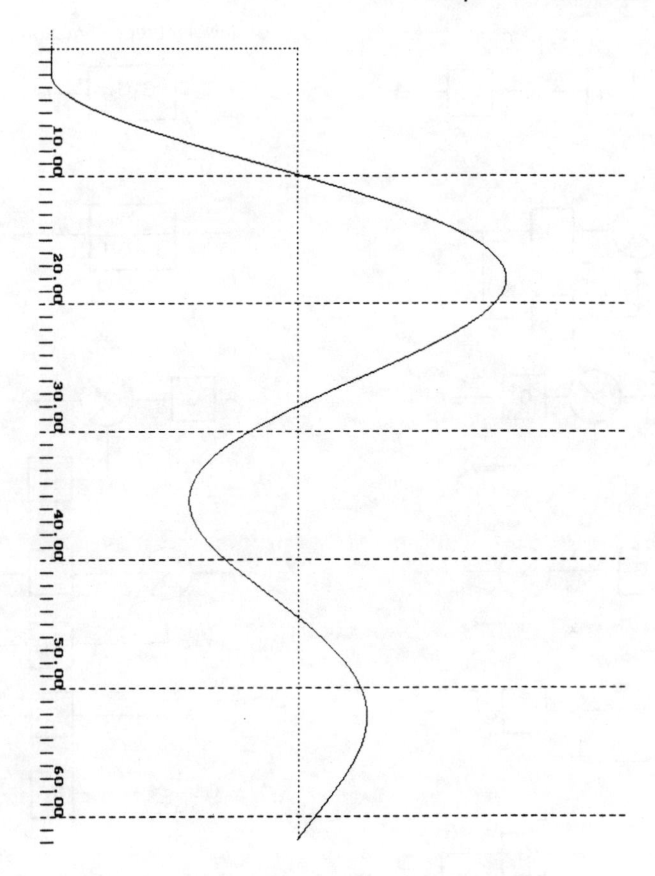

Fig. 7.6 Computer printout (DIGSIM).

before the disturbance and one after the disturbance as shown in Fig. 7.9. Each part is defined by its transfer function and each can have a separate transit delay.

The program operates in a similar way to the program DIGSIM following the steps from equation 7.1 as above.

A typical response is shown on Fig. 7.10.

7.7. Solving polynomial equations

Many process control calculations involve the solving of polynomial equations. A common example is drawing a root locus, an operation performed by the next program ROOTLOKE. In process control we are interested in complex solutions as well as the real roots.

Most school children should be able to solve a quadratic equation

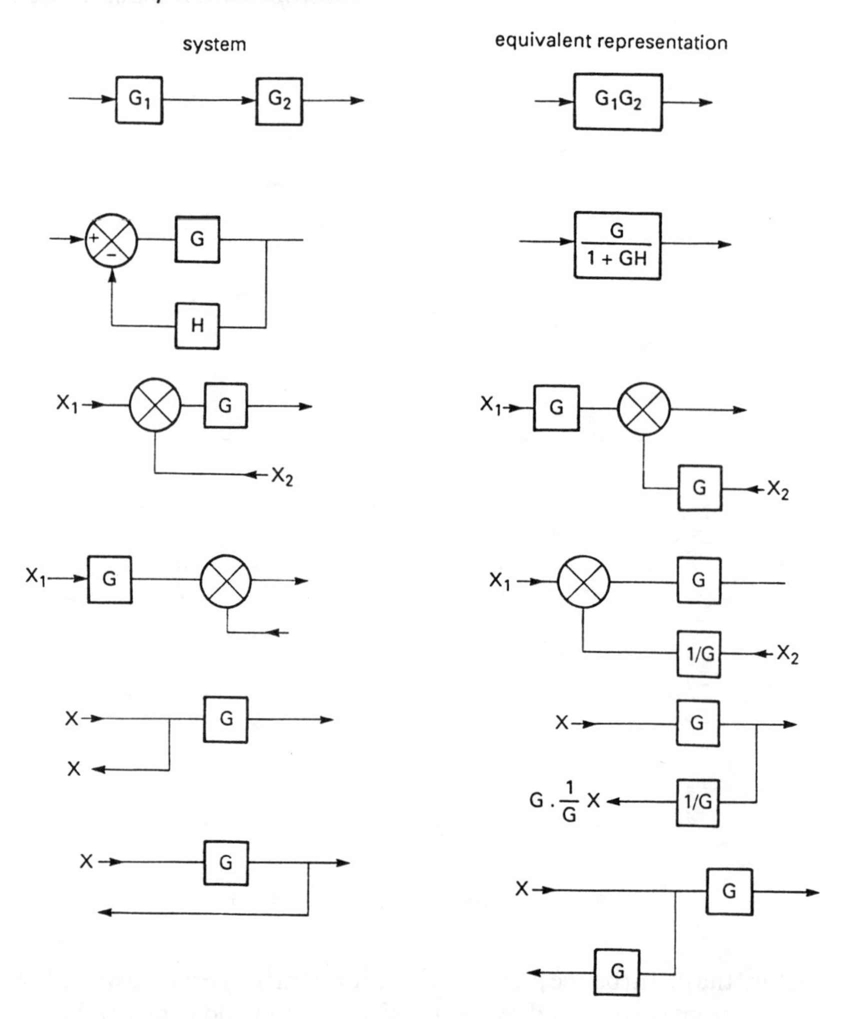

Fig. 7.7 Useful equivalent structures for forming the transfer function.

using the well-known formula $(-b \pm \sqrt{b^2 - 4ac})/2a$. The difficulties caused by rounding errors were described earlier in Section 7.3. The MATHUTIL unit provides a straightforward GCSE procedure in QUADRAD and a more refined version in QUADRAD2.

Cubic equations are more difficult. Any polynomial of odd power has at least one real root. A real root can always be found by straddling and homing in on the root by seeing which half of the straddle range the root lies in by looking for a change of sign of the polynomial between each half. The procedure CUBIC_SOL in the MATHUTIL unit finds the first real root by this method.

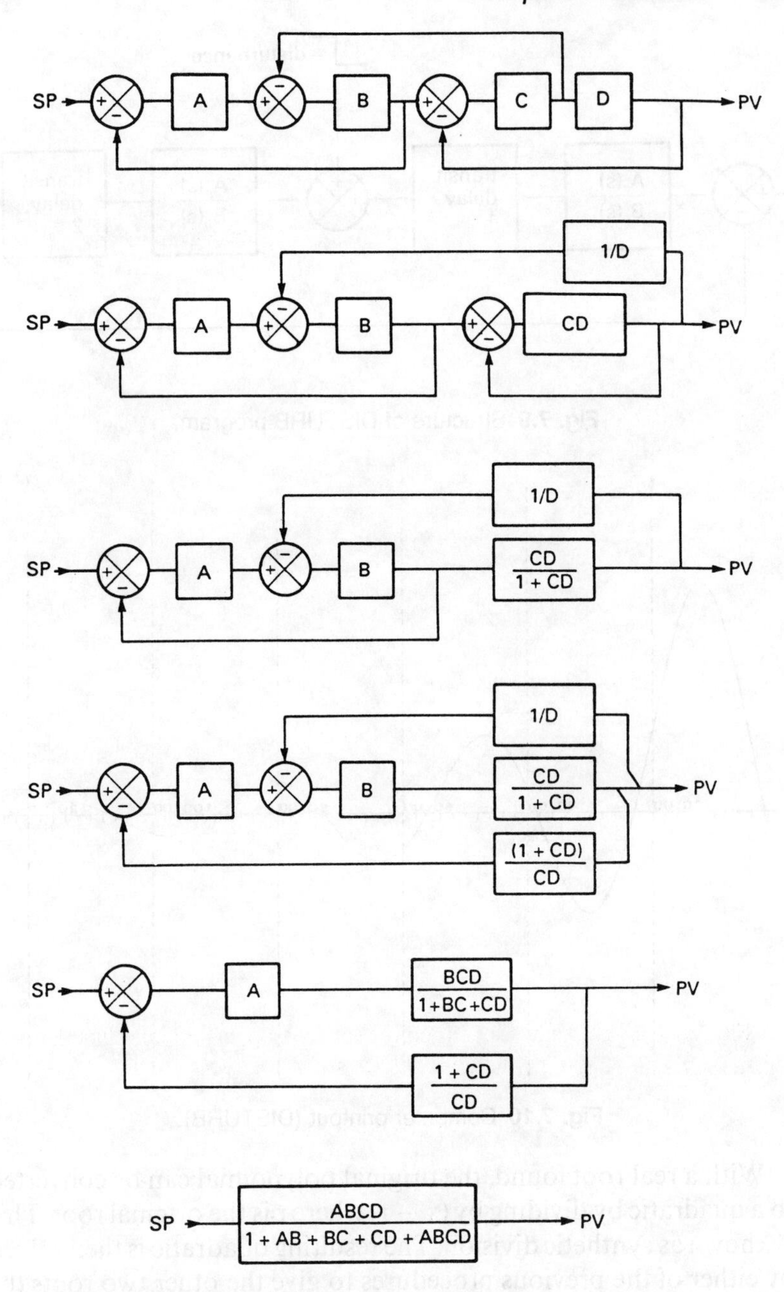

Fig. 7.8 Using equivalent structures to simplify a block diagram.

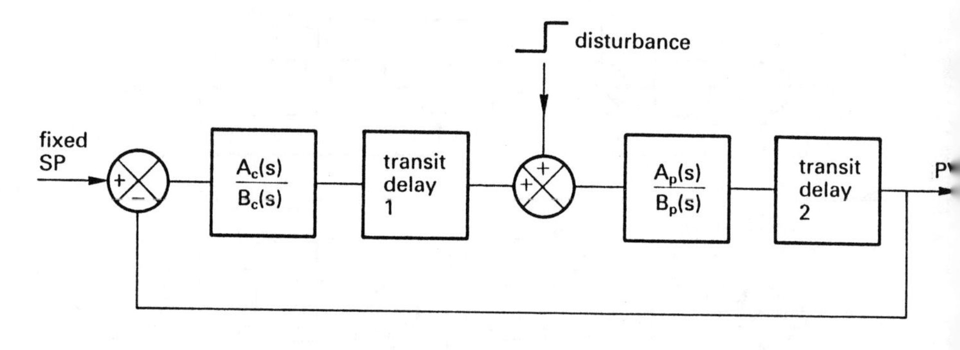

Fig. 7.9 Structure of DISTURB program.

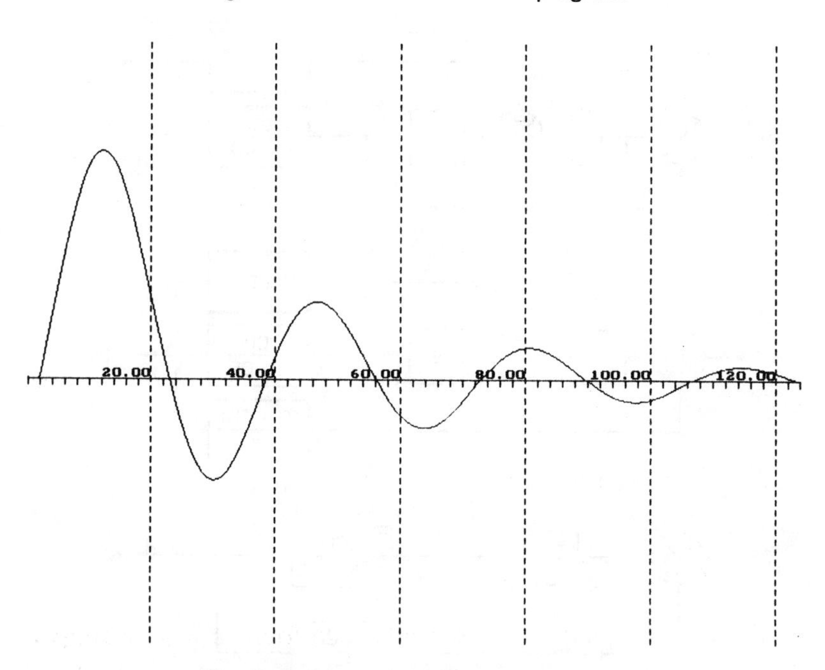

Fig. 7.10 Computer printout (DISTURB).

With a real root found, the original polynomial can be converted to a quadratic by dividing by $(x - r)$ where r is the original root. This is known as synthetic division. The resulting quadratic is then solved by either of the previous procedures to give the other two roots (be they two real or two imaginary).

An alternative, and very interesting, method was given by Francois Viete (in 1615!!) for the equation

$$x^3 + ax^2 + bx + c = 0$$

we first calculate Q and R where

$$Q = \frac{a^2 - 3b}{9}$$

and

$$R = \frac{2a^3 - 9ab + 27c}{54}$$

Next we check that $Q^3 \geq R^2$. If it is, the equation has three real roots. To find these we evaluate θ where

$$\theta = \arccos(R/\sqrt{Q^3})$$

then the three roots are

$$r_1 = -2\sqrt{Q}\cos\left(\frac{\theta}{3}\right) - \frac{a}{3}$$

$$r_2 = -2\sqrt{Q}\cos\left(\frac{\theta + 2\pi}{3}\right) - \frac{a}{3}$$

$$r_3 = -2\sqrt{Q}\cos\left(\frac{\theta + 4\pi}{3}\right) - \frac{a}{3}$$

If $R^2 > Q^3$ there is only one root which is given by the rather horrendous

$$r_1 = -\,\mathrm{sgn}(R)\left[(\sqrt{R^2 - g^3} + |R|)^{1/3} + \frac{Q}{(\sqrt{R^2 - Q^3} + |R|)^{1/3}}\right] - \frac{a}{3}$$

and the other two (complex) roots have to be found by converting the cubic to a quadratic with synthetic division, then solving the resulting quadratic as above.

It is an amazing bit of mathematics for the early seventeenth century, and it does work. The straddle and synthetic division method is easier to code, however, so CUBIC_SOL does not use Viete's method.

A quartic equation can have four real, four complex or two real, two complex roots. Straddling cannot, therefore, be used to find a real root. More cunning techniques have to be used. Consider the equation

$$ax^4 + bx^3 + cx^2 + dx + e = 0$$

By making the substitution

$$y = x + \frac{b}{4a} \tag{7.2}$$

we can convert it to

$$y^4 + Cy^2 + Dy + E = 0$$

i.e. no cubic term. This equation can be further rearranged into two quadratics

$$(y^2 - Ay + B)(y^2 + Ay + G) = 0$$

By multiplying out and equating coefficients we get

$$B + G - A^2 = C$$
$$A(B - G) = D$$
$$BG = E$$

These are three simultaneous equations with three unknowns A, B, G, so the unknowns can be found. It is usually simpler (with a computer) to find the value of A by successive approximation from which B and G follow. With A, B, G found the two quadratics can be solved to give the four roots for y. The four roots for x can then be found by going back to equation 7.2 and using

$$x = y - \frac{b}{4a}$$

for each root.

This operation is performed in the procedure QUARTIC_SOL in the unit MATHUTIL. The procedure is very heavily annotated and by removing the comment braces {} around the written instructions it will describe its operation as it is working. These braces must be in place when the procedure is being used by ROOTLOKE.

A quartic equation is the highest power polynomial that can be solved directly. Beyond a quartic, iterative methods have to be used. The real root of a quintic polynomial, however, can again be found by straddling as described previously for the cubic equation. With a real root found, the quintic equation can be reduced to a quartic by synthetic division by $(x - r)$ where r is the found real root. The four roots of the quartic can then be found by the previous procedure. The procedure QUINTIC_SOL solves a quintic polynomial in this manner.

Polynomials of power higher than a quintic cannot be solved by straightforward methods, and iterative searches for roots have to be made in two dimensions (real and imaginary). The MATHUTIL unit and hence the program ROOTLOKE, therefore only deal with polynomials to order five.

Simple test programs are provided for checking these procedures. These are programs TESTQUAD, TESTCUBC, TESTQUAR and TESTQUIN. With TESTQUAR and TESTQUIN it is interesting to remove the comment braces {} in the procedure QUARTIC_SOL.

7.8. Unit MATHUTIL

The unit MATHUTIL contains a set of useful mathematical utilities including the four polynomial procedures above. The unit also defines the form in which complex numbers are held. A record type complex is defined which has real and imaginary parts. The complex number c1 thus exists with c1.real and c1.imag.

A set of procedures (c_add, c_sub, etc.) handle four function maths for complex numbers. Note that these procedures can be used to turn a complex fraction c1/c2 into a single equivalent complex number c3, a very common operation in process control. Other complex utilities are evaluating absolute value and polar angle, and conversion/combination of real parts to a complex number. Write procedures for displaying complex numbers on the screen are also given.

Other utilities for non-complex numbers are max and min of two numbers, conversion both ways between degrees and radians, tangent (oddly omitted from standard Pascal), log, power x ^ y and a WITHIN test which checks that a number lies within specified limits.

7.9. Program ROOTLOKE

Program ROOTLOKE draws the root loci for a system with a specified transfer function with a denominator of power five or less (see earlier comments on solving polynomials). It cannot handle higher powers or systems with a transit delay.

Consider a system with a transfer function G (representing both plant and the controller, except for the controller gain K). If this is connected with unity feedback as Fig. 7.11 the closed loop response will be

$$C = \frac{KG}{1 + KG} \tag{7.3}$$

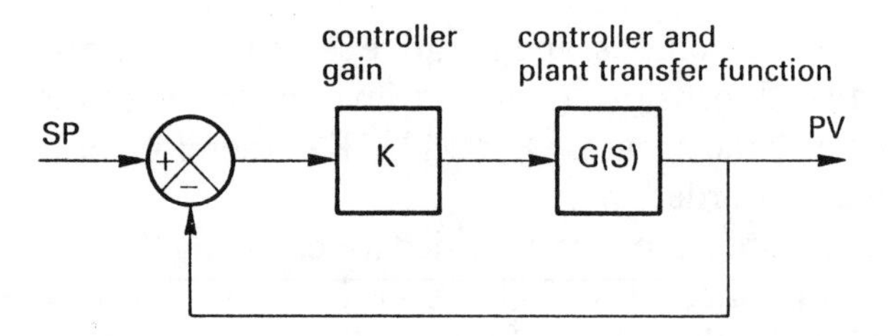

Fig. 7.11 Organisation of program ROOTLOKE.

If G has the form a(s)/b(s) equation 7.3 can be arranged

$$C = \frac{Ka(s)}{b(s) + Ka(s)} \tag{7.4}$$

The open loop poles lie at the solutions of $b(s) = 0$ and the zeros at solutions of $a(s) = 0$. The closed loop poles for gain K lie at the solution of the denominator polynomial of equation 7.4, i.e.

$$b(s) + Ka(s) = 0 \tag{7.5}$$

The root loci are drawn by solving equation 7.5 as K is varied and plotting the (probably complex) solutions onto a representation of the s-plane.

The program requests the polynomials a(s) and b(s) then plots the loci as K is varied from near zero to 30. The s-plane is drawn complete with the asymptotes and the $b = 0{\cdot}5$ line (useful for choosing the gain which gives an approximate quarter amplitude decay). The program then allows the user to specify a smaller, or larger, gain range for closer examination. The program ends when the high and low gains are the same.

Figures 7.12 and 7.13 show typical root loci.

7.10. Unit SETPID and program TUNETEST

Section 4.6.6 described a simple and quick closed loop tuning method which required the user to perform a step change in set point and record the overshoots and time as shown on Fig. 7.14. The controller is used in proportional only mode.

This test is coded in the unit SETPID and can be used with the program TUNETEST. The calculation is performed in a unit as this test is also incorporated into the autotuner function of the later program SIMULATE.

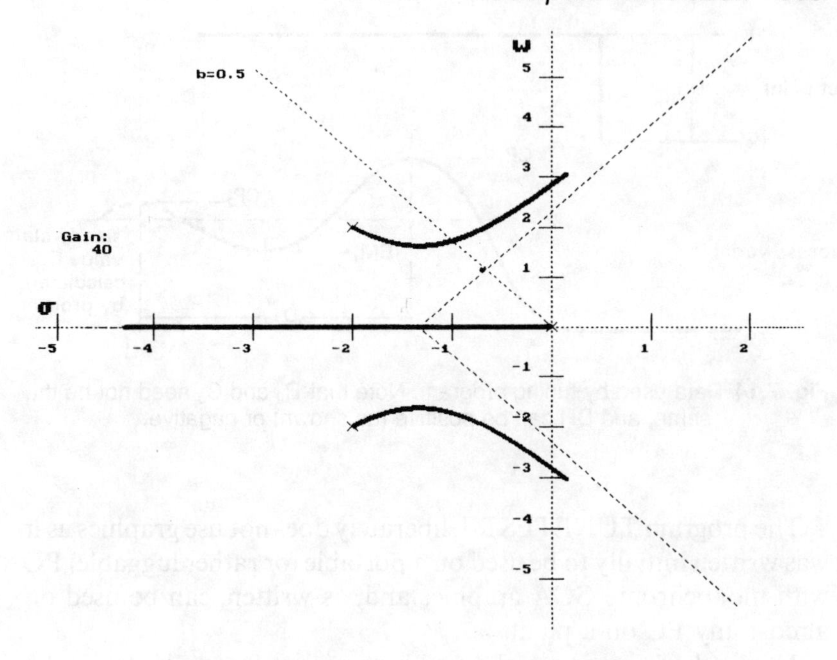

Fig. 7.12 Computer printout (ROOTLOKE).

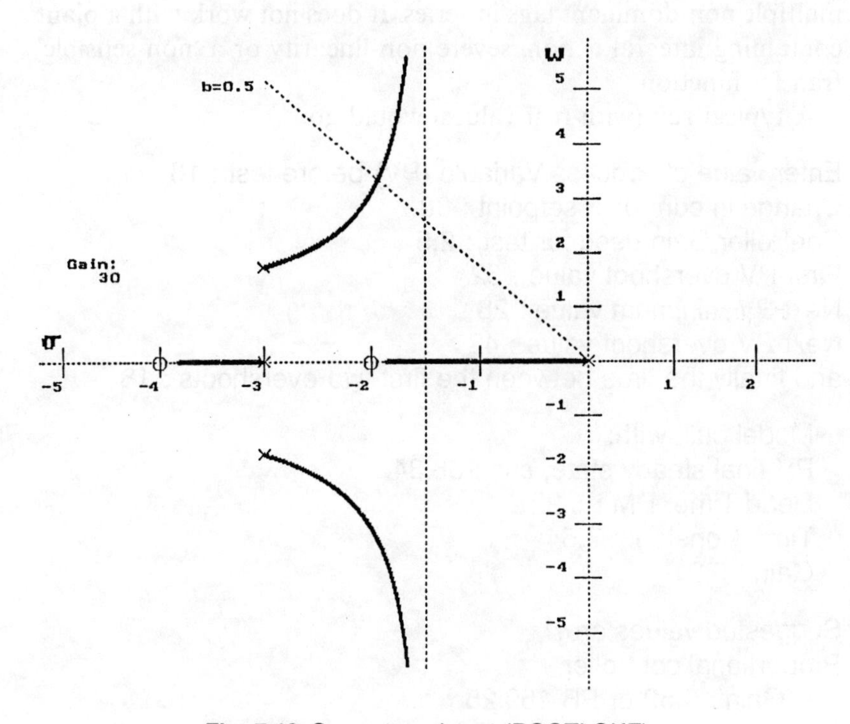

Fig. 7.13 Computer printout (ROOTLOKE).

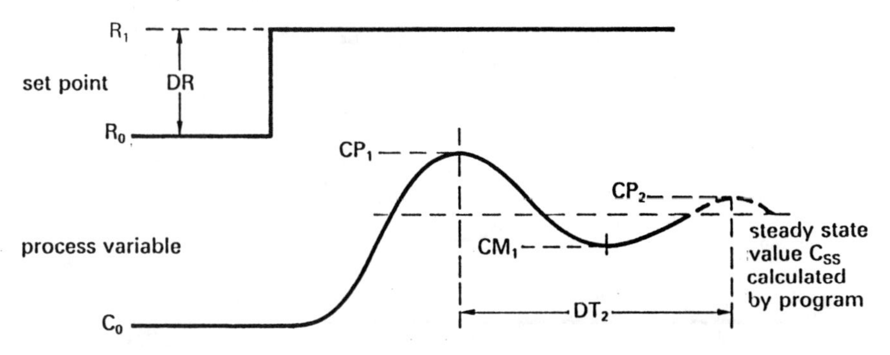

Fig. 7.14 Data used by tuning program. Note that R_0 and C_0 need not be the same, and DR can be positive (as shown) or negative.

The program TUNETEST deliberately does not use graphics as it was written initially to be used on a portable (or rather luggable) PC with monochrome CGA graphics and, as written, can be used on almost any PC on a plant.

The method assumes a plant with dominant first order lag and a transit delay. The transit delay can be a real delay or arising from multiple non-dominant lags in series. It does not work with a plant containing integral action, severe non-linearity or a 'non sensible' transfer function.

A typical run (with real values) would go:

Enter value of Process Variable (PV) before test : 16
Change in controller setpoint : 30
Controller Gain used for test : 2.5
First PV overshoot value : 52
Next PV minimum value : 28
Next PV overshoot value : 42
and finally the time between the first two overshoots : 18

Model built with:
PV final steady state, css : 36.84
Dead Time, DM : 6.92
Time Constant : 7.54
Gain : 0.91

Suggested values are:
Proportional controller
Gain : 0.59 or PB 169.25
PI controller

Gain : 1.03 or PB 97.44
Ti : 10.56

PID controller

Gain : 1.62 or PB 61.84
Ti : 8.33
Td : 2.64

Note that the units of time for T_i and T_d will be the same as the units used for the entered time between the first two overshoots. If the entered time was 18 minutes, the T_i recommended above for a PI controller was 10·56 minutes.

7.11. Other units

7.11.1. T_UTILS

This small unit provides mainly text based utilities, including Read_ln procedures for integers and real numbers. In standard Turbo Pascal entering a bad number (e.g. hitting an alpha key rather than a number key) will cause the program to crash with a not very informative message. These procedures simply erase bad input data and let you try again. An allowable data range is also specified when these procedures are called, so, for example, the calling program can ask for a value of T_i in an allowable range 0·1 to 100.

The operation of the other three procedures is given by their names: pause, pauseprompt and beep.

7.11.2. G_UTILS

One of the problems with Turbo Pascal in mathematical applications is the graphics origin at the top left of the screen. This makes graph drawing rather awkward. The G_UTILS procedures provide the normal plot, move and circle operations on a more sensible screen layout with the graphics origin at bottom left.

Procedure setgraph sets up the graphics card, obtaining the screen dimensions in xmax and ymax. The screen is assumed to be 640 pixels wide by 480 pixels high, and xscale, yscale are calculated scale factors if this assumption does not apply.

Horizline and vertline draw absolute horizontal and vertical lines. The normal plot function can give a one pixel kink which looks odd.

The gwrite procedures write a string at a given position on the graphics screen (with origin bottom left). Three versions give text justified left, right or centred on the specified position.

7.11.3. DB_PHI

This unit calculates the gain (linear and in db) and phase shift (in degrees) for a variety of blocks at a specified frequency (in radians/sec). These are used for drawing Bode/Nyquist diagrams and Nichols charts in the program SIMULATE.

The blocks covered are

Simple gain
First order
Second order
PID controller (also used for PI controller with $T_d = 0$)
Integral action
Transit delay
Lead

For each, the block characteristics are provided by the calling program (e.g. gain, T_i, T_d for a PID controller) along with the frequency and the procedure returns the linear gain, the gain in db and the phase shift.

7.11.4. CALC_STEP

This unit calculates the response of the blocks listed above to a step change over a short time Δt (denoted by dt in the program). These use difference equations as an approximation to differential equations as described in Section 4.4.2.

The step response of the lead block is limited to prevent the program blowing up. Real life lead blocks operate in a similar way, so this approximation is true to life.

The procedures take an input signal (xin), Δt (dt), the input and output signals at the previous time called and give the new output signal (yout). With the integral block, for example

$$Y_{out} = Y_{prev} + \text{gain} \times x_{in} \times dt$$

7.12. Program SIMULATE

The program SIMULATE analyses a system specified as a series of common blocks. Not surprisingly the blocks available are:

Controllers (P, PI, PID)
First order lag
Second order
Integral action
Transit delay (one off only)
Lead

Lead–lag blocks of both types can be made by having a lead block followed by a first order lag block.

The user specifies the number of blocks required. As written the program can handle ten blocks which should be adequate for simulating real systems. Remember most plants can be approximated by a controller, one or two dominant blocks and a possible transit delay.

The user then specifies the characteristics of each block (e.g. gain and T_c for a first order lag). With all blocks specified, Nichols charts, Bode diagrams, Nyquist diagrams and step/disturbance response can be drawn. The method used for drawing step/disturbance responses is based on difference equations rather than the transfer function method used earlier in programs DIGSIM and DISTURB. It is interesting to plot the response of a system by both methods and see if they agree.

An autotune function is provided. This works in two distinct ways. The first method performs a Ziegler Nichols experiment and finds the critical gain and frequency (i.e. where the phase shift is -180 degrees). The critical gain and frequency are then plugged into the equations in Section 4.6.2.

The second method is based on the closed loop test of Section 4.6.6 (using the unit SETPID). The controller block is temporarily set to P only control with a gain margin which should give a damped oscillatory response. The step response is then calculated and the required values read off and passed to SETPID.

The efficiency of this tuning method can be tested by building a system with purely a first order lag and transit delay (which is what the tuning method assumes) then calling autotune and seeing if SETPID finds the values entered.

Do not be surprised if this second autotune returns error messages, it cannot handle integral action and gets very finicky if the system is complex.

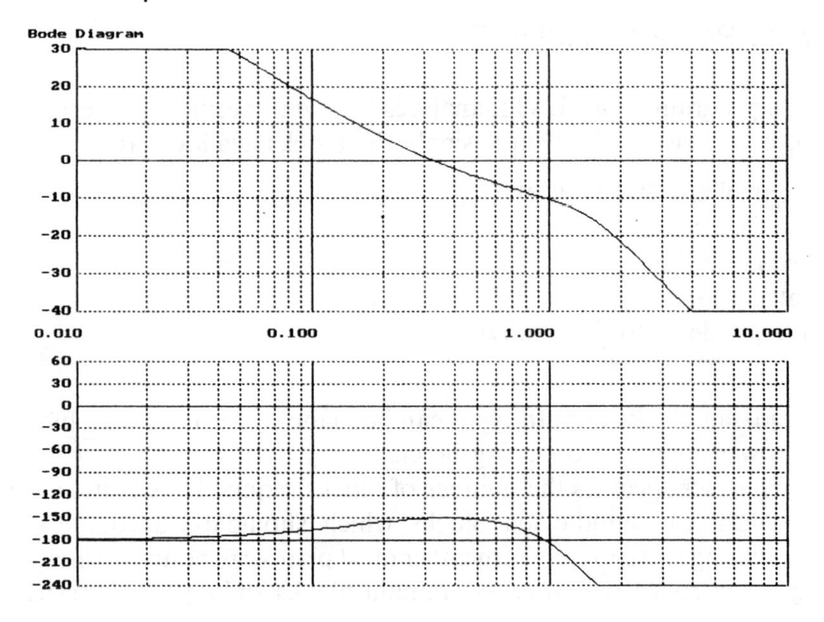

Fig. 7.15 Computer printout (Bode).

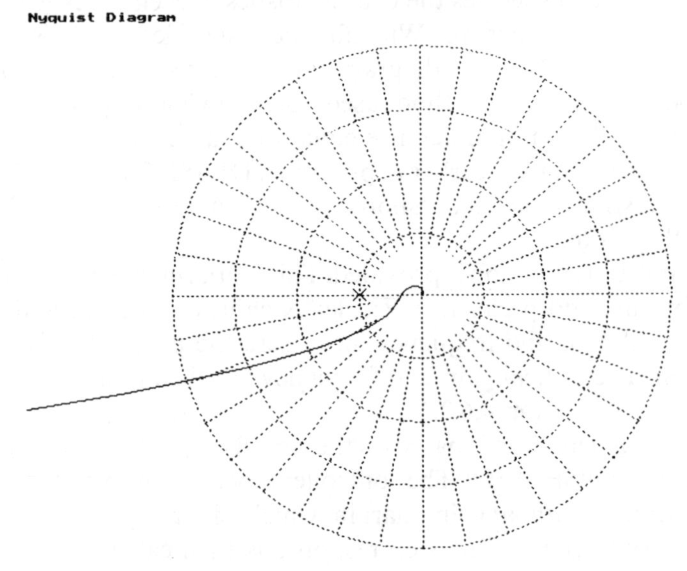

Fig. 7.16 Computer printout (Nyquist).

Unlike the other programs, SIMULATE has facilities for editing, adding and deleting blocks, so different controller types can be tried, and the characteristics of the plant can be changed.

Typical printouts for Bode, Nyquist, Nichols, step and disturbance are given on Figs 7.15–7.19.

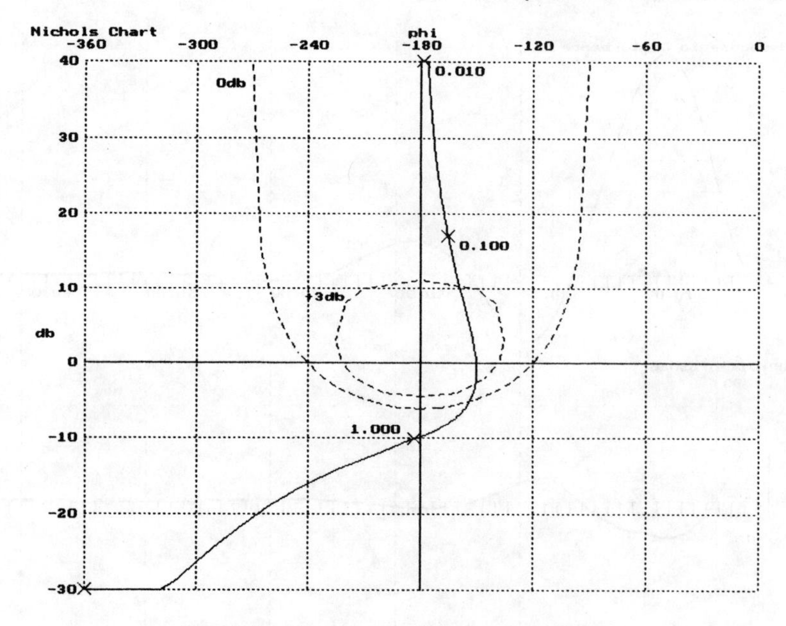

Fig. 7.17 Computer printout (NICHOLS).

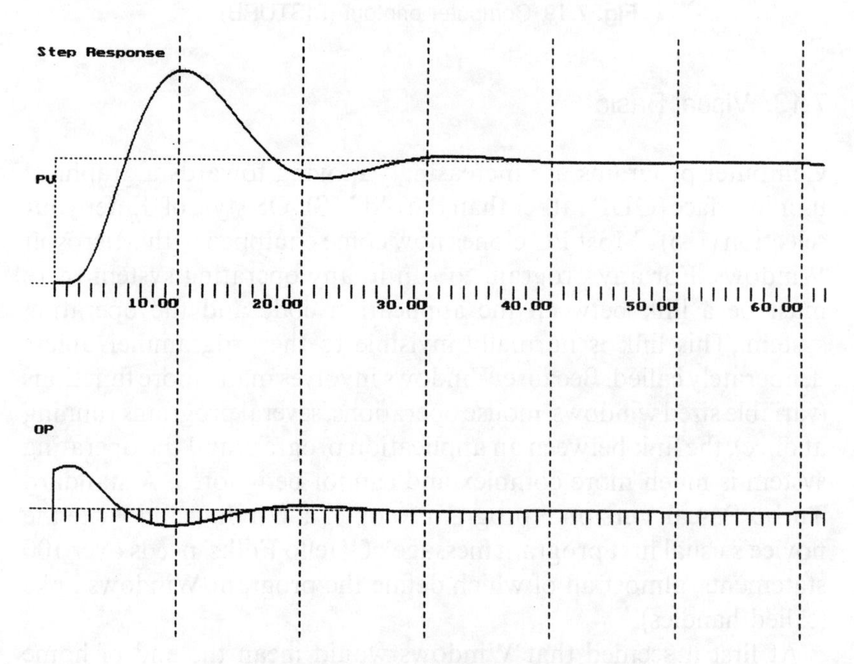

Fig. 7.18 Computer printout (STEP).

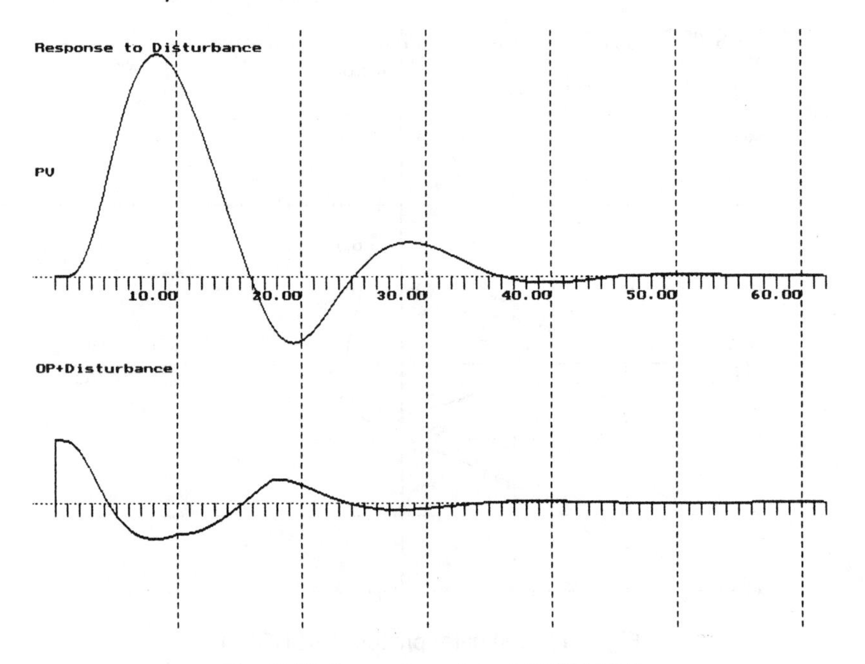

Fig. 7.19 Computer printout (DISTURB).

7.13. Visual Basic

Computer programs are increasingly moving towards a graphical user interface (GUI) rather than the old MSDOS style of 'Enter your selection (1–5)'. Most PC clones now come equipped with Microsoft Windows. For any program to run in any operating system there must be a link between the application code and the operating system. This link is normally invisible to the programmer unless deliberately called. Because Windows involves many more functions (variable sized windows, mouse operations, several programs running at once), the link between an application program and the operating system is much more complex and cannot be ignored. A standard Turbo Pascal Version 6 program to open a Window and write the novice's usual first program message of 'Hello Folks' needs over 100 statements, almost all of which define the program/Windows links (called handles).

At first it seemed that Windows would mean the end of home programming for all but the most dedicated, but Microsoft released a version of Basic for Windows which almost allows the designer to forget about the operating system. This language is known as Visual Basic.

The language starts with forms. These are blank windows onto which the standard Windows controls (slide bars, text boxes, graphics, mouse operated buttons, etc.) can be picked from a toolbar, sized and placed. Unlike the Pascal programs in the previous sections, Visual Basic is event driven, i.e. the Basic program is written to respond to changes caused by actions such as the user clicking on boxes with the mouse, or altering the position of a scroll bar.

Code is written for each action that the user can perform. If there are two buttons marked respectively Auto and Manual, the code for each button is written which is activated when the button is clicked with the mouse. Slightly simplified the code for these objects could be:

```
proc cmdAutopb.click
  begin
    automode=true {internal variables}
    manmode=false
    txtMode.caption="Automatic" {message in a text box}
  end

proc cmdManpb.click
  begin
    automode=false
    manmode=true
    txtMode.caption="Manual"
  end
```

The forms are designed with Pick & Place from objects on a toolbar. The objects' size, colour, etc. are then specified and the code for each written.

Visual Basic is sold in two forms; a cheap and cheerful simple version, and the so-called Professional Version. The latter is well worth the extra money for control system work because among the additional facilities are circular scale and bargraph meters, and a powerful utility for drawing graphs.

This book is about process control, not programming, so it is not intended to devote much space to describing Visual Basic. There are many excellent books and courses on the language.

Both of the programs below operate on a single form, and are built around the operation of the timer control. This is an (invisible) object which enables code to be run at regular intervals. An interval (in milliseconds) is specified for the timer, and the assigned code is run once per timer interval. The two programs use a timer interval of 0·1 secs.

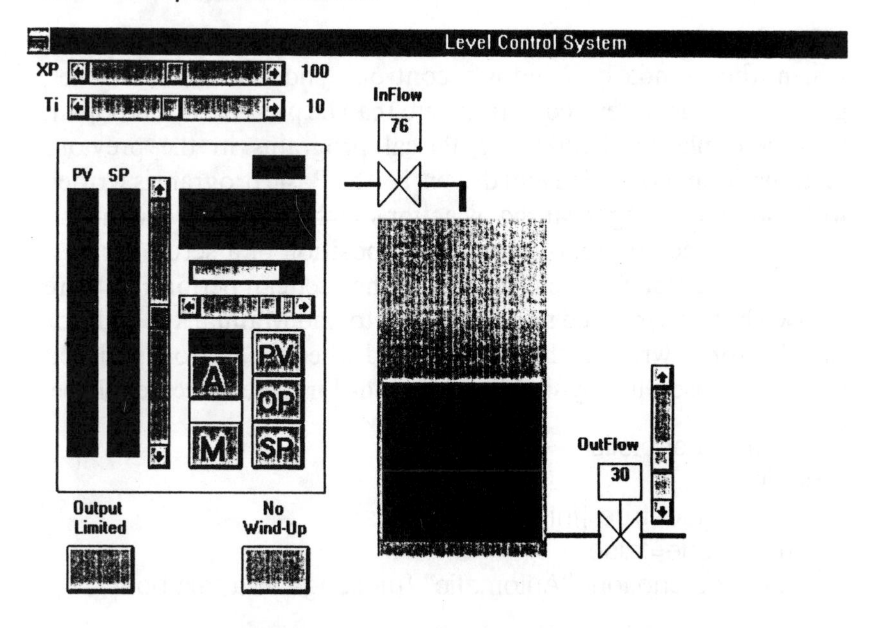

Fig. 7.20 Computer printout (ScreenDump CONTROL).

The programs operate as difference equations with the interval time ΔT replacing dt in the differential equations describing the control algorithm and process. Essentially at each interval the programs perform the tasks below in order:

read the state of all the variables from the scroll bars and the process model
follow the control algorithm
calculate the response from the plant model
update the screen indication.

7.14. Program LINLEVEL

Both the Visual Basic programs aim to provide real time simulation of a control system. LINLEVEL, shown on Fig. 7.20, is a linear level control system with a PI controller. The controller setup (PB and T_I) can be changed along with the outflow and set point.

The controller layout is based on a Eurotherm Controls 6360 controller, although the control algorithm is written by the author. The controller can be switched with bumpless transfer between automatic and manual modes. The set point is changed with the vertical scroll bar alongside the controller bargraphs. The controller output in manual mode is set by the horizontal scroll bar below the controller digital display.

The PV, SP and controller output can be displayed on the digital display as well as the bargraphs. The display is selected by clicking on the relevant button.

The controller has protection against integral windup. The effect of removing the integral windup protection can be seen by clicking on the integral action button to the lower right of the controller.

The controller output behaves like a real life instrument, and is limited to the range 0–100%. The controller can be made absolutely linear by clicking on the left-hand button below the controller. This allows the controller to have infinite positive and negative output range, but this does have the anomalous effect that the controller causes liquid to be sucked out of the tank!

The controller settings of XP and T_i can be varied on the scroll bars above the controller.

7.15. Program NON_LIN

This program simulates the action of a position servo with deadband control. The set point is shown on the left-hand gauge, and the plant position on the right-hand gauge. The plant will drive forward or reverse dependent on the error. The plant has a first order lag between demanded speed and actual speed, so it accelerates and decelerates in an exponential manner.

The deadband is set by the lower horizontal scroll bar, and the maximum speed by the vertical scroll bar. Note that the plant will only attain near maximum speeds on long movements.

7.16. Program disk

The disk contains the program source files (*.PAS for the Pascal programs and units) and compiled files (executable *.EXE for the Pascal programs and *.TPU for the Turbo Pascal units). The executable files can be run from the floppy disk and do not need Turbo Pascal to be present on the machine. The *.BGI files are video drivers and must exist in the same disk/directory as the executable files.

The Visual Basic programs exist in their own directories LINLEVEL and NON_LIN on the floppy, along with an executable file called (in

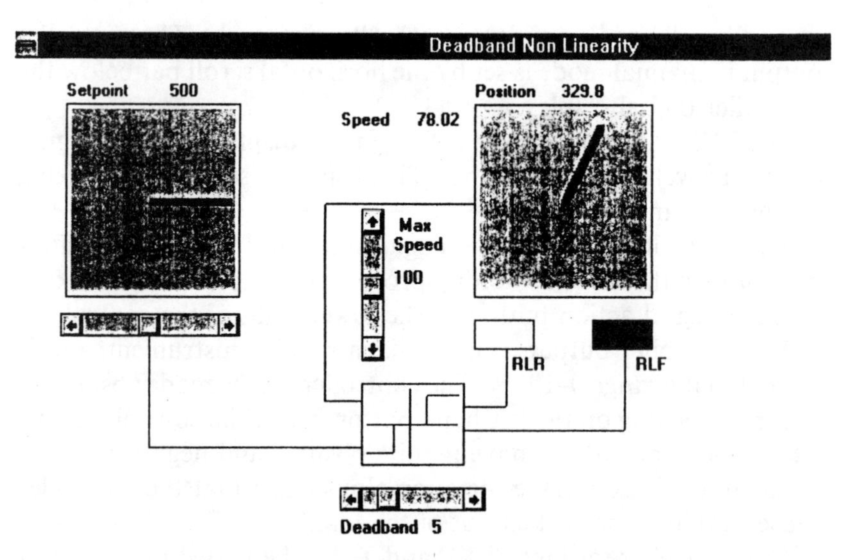

Fig. 7.21 Computer printout (ScreenDump NON_LIN).

each case) SETUP. The file SETUP was created with the Visual Basic Setup Wizard. If these SETUP files are run from Program Manager in Windows, they will create executable Windows programs which can be run without Visual Basic being present.

The Visual Basic programs are also provided on the disk as *.TXT, *.MAK, *.FRM and *.EXE files along with the VBX files utilised by the programs. These can be used with Visual Basic. Any late comments are included in a README.TXT file on the disk.

Appendix A
Complex numbers

The solution of quadratic polynomial equations often leads to results of the form

$$x = a \pm b\sqrt{-1}$$

The $\sqrt{-1}$ cannot be evaluated in real terms, but if an operator j is defined as $\sqrt{-1}$ the above expression can be expressed as

$$x = a \pm jb$$

Such numbers are called 'complex numbers' and have a real part (a) and an imaginary part (jb). In pure mathematics 'i' is used rather than 'j', but in engineering 'i' could be confused with electrical current.

Complex numbers can be visualised on an Argand diagram as Fig. A.1.

An alternative representation is shown in Fig. A.2, where the point P (which is a + jb) is expressed in terms of its magnitude μ and angle ϕ. By simple geometry and trigonometry

$$\mu = \sqrt{a^2 + b^2}$$

$$\phi = \tan^{-1}(b/a)$$

A complex number can thus be expressed in polar form as $\mu\angle\phi$. Given a complex number in polar form, the cartesian form can be obtained by

$$a = \mu \cos \phi$$

$$b = \mu \sin \phi$$

which gives the common representation of a complex number as

$$Z = \mu(\cos \phi + j \sin \phi)$$

Complex numbers can be added or subtracted by dealing with real

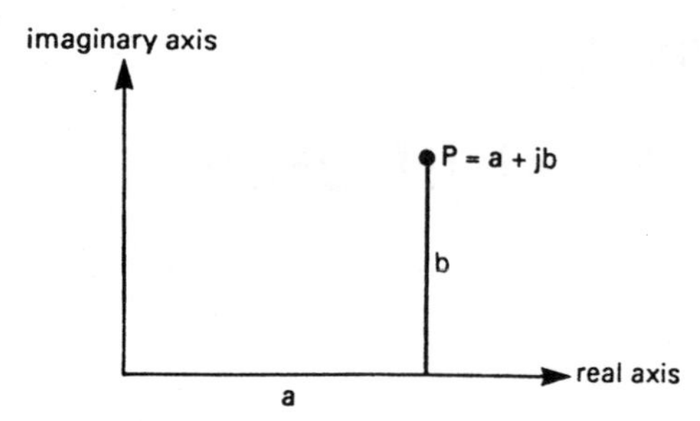

Fig. A.1 A complex a+jb.

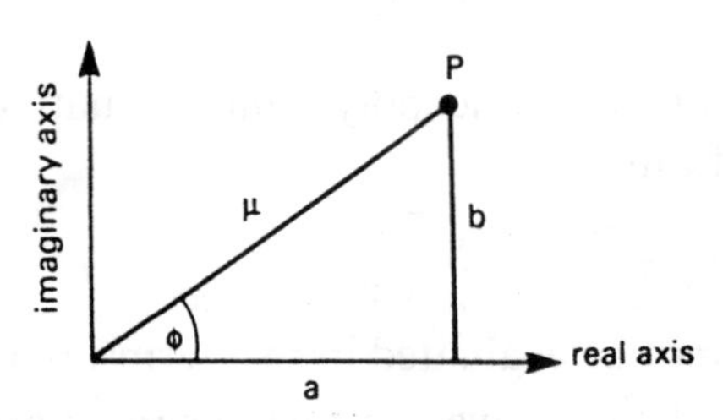

Fig. A.2 A complex number in terms of its amplitude μ and phase shift ϕ.

and imaginary parts separately, i.e.

$$(a + jb) + (c + jd) = (a + c) + j(b + d)$$

$$(a + jb) - (c + jd) = (a - c) + j(b - d)$$

Addition and subtraction are best performed in cartesian form. Multiplication is achieved by noting that $j^2 = -1$, e.g.

$$\begin{aligned}(a + jb)(c + jd) &= ac + jad + jbc + j^2bd \\ &= (ac - bd) + j(ad + bc)\end{aligned}$$

In polar form

$$(\mu_1 \angle \phi_1)(\mu_2 \angle \phi_2) = \mu_1\mu_2 \angle (\phi_1 + \phi_2)$$

Division in cartesian form is achieved via the complex conjugate. If a complex number has the form $(a + jb)$, the complex conjugate is $(a - jb)$. The product of a complex number and its conjugate always gives a real result

$$\begin{aligned}(a + jb)(a - jb) &= a^2 - jab + jab - j^2b^2 \\ &= a^2 + b^2 \text{ since } j^2 = -1\end{aligned}$$

A division of the form $1/(a + jb)$ can therefore be evaluated

$$\frac{1}{a + jb} = \frac{1}{a + jb} \cdot \frac{a - jb}{a - jb}$$

$$= \frac{a - jb}{a^2 + b^2}$$

$$= \left(\frac{a}{a^2 + b^2}\right) - j\left(\frac{b}{a^2 + b^2}\right)$$

In polar form

$$\frac{\mu_1 \angle \phi_1}{\mu_2 \angle \phi_2} = \frac{\mu_1}{\mu_2} \angle (\phi_1 - \phi_2)$$

Complex numbers in process control often occur in the form

$$\frac{1}{(a + jb)} \cdot \frac{1}{(c + jd)} \cdot \frac{1}{(e + jf)}$$

These can be evaluated by multiplying out the denominator and noting that

$$j^2 = -1$$
$$j^3 = -j$$
$$j^4 = 1 \qquad \text{etc.}$$

which will give a form $1/(A + jB)$ which can be evaluated as shown above.

A complex number can also be expressed in exponential form. Mathematical functions can be expressed as a power series, e.g.

$$e^x = 1 + x + \frac{x^2}{2!} + \frac{x^3}{3!} + \frac{x^4}{4!} + \ldots$$

$$\sin x = x - \frac{x^3}{3!} + \frac{x^5}{5!} - \frac{x^7}{7!} + \ldots$$

$$\cos x = 1 - \frac{x^2}{2!} + \frac{x^4}{4!} - \frac{x^6}{6!} + \ldots$$

The complex number $e^{j\varphi}$ is thus

$$e^{j\varphi} = 1 + (j\phi) + \frac{(j\phi)^2}{2!} + \frac{(j\phi)^3}{3!} + \frac{(j\phi)^4}{4!} + \ldots$$

Noting that $j^2 = -1$, $j^3 = -j$, $j^4 = 1$ etc.

$$e^{j\varphi} = 1 + (j\phi) - \frac{\phi^2}{2!} - j\frac{\phi^3}{3!} + \frac{\phi^4}{4!} + j\frac{\phi^5}{5!} \ldots$$

$$= \cos \phi + j \sin \phi$$

The complex number $\mu \angle \phi$ can thus be expressed in exponential form as $\mu e^{j\varphi}$.

A pure time delay can also be expressed in exponential form. Since $\phi = -\omega T$ where T is the transit delay time, we have

$$\frac{X_o}{X_i} = \mu e^{-j\omega T}$$

Complex numbers in exponential form can be easily multiplied or divided

$$\mu_1 e^{j\varphi_1} \times \mu_2 e^{j\varphi_2} = \mu_1 \mu_2 e^{j(\varphi_1 + \varphi_2)}$$

$$\frac{\mu_1 e^{j\varphi_1}}{\mu_2 e^{j\varphi_2}} = \frac{\mu_1}{\mu_2} e^{j(\varphi_1 - \varphi_2)}$$

Equations involving complex numbers should be dealt with as separate real and imaginary parts. For example, if

$a + jb = c + jd$
then $a = c$
and $b = d$

Appendix B
Trigonometrical relationships

Basic trigonometrical ratios (sin θ, cos θ, tan θ) are shown on Fig. B.1.
By simple reorganisation and Pythagoras

$$\tan\theta = \frac{\sin\ \theta}{\cos\ \theta}$$

$$\text{and } \cos^2\theta + \sin^2\theta = 1$$

Figure B.2 shows the sign of the basic ratios for various angles. Compound angles can be evaluated as below:

$$\sin(A + B) = \sin A \cos B + \cos A \sin B$$

$$\sin(A - B) = \sin A \cos B - \cos A \sin B$$

$$\cos(A + B) = \cos A \cos B - \sin A \sin B$$

$$\cos(A - B) = \cos A \cos B + \sin A \sin B$$

$$\tan(A + B) = \frac{\tan A + \tan B}{1 - \tan A \tan B}$$

$$\tan(A - B) = \frac{\tan A - \tan B}{1 + \tan A \tan B}$$

$$\sin(2A) = 2 \sin A \cos A$$

$$\begin{aligned} \cos(2A) &= \cos^2 A - \sin^2 A \text{ and since } \cos^2\theta + \sin^2\theta = 1 \\ &= 1 - 2\sin^2 A \\ &= 2\cos^2 A - 1 \end{aligned}$$

$$\sin A \cos B = \frac{1}{2}(\sin(A + B) + \sin(A - B))$$

$$\sin A \cos B = \frac{1}{2}(\cos(A + B) + \cos(A - B))$$

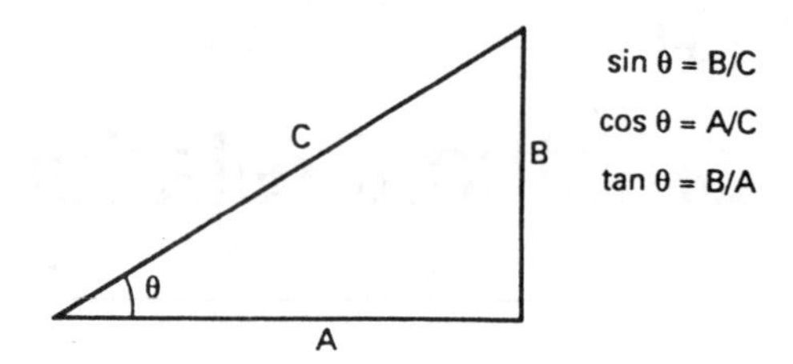

Fig. B1 Definition of trigonometrical ratios.

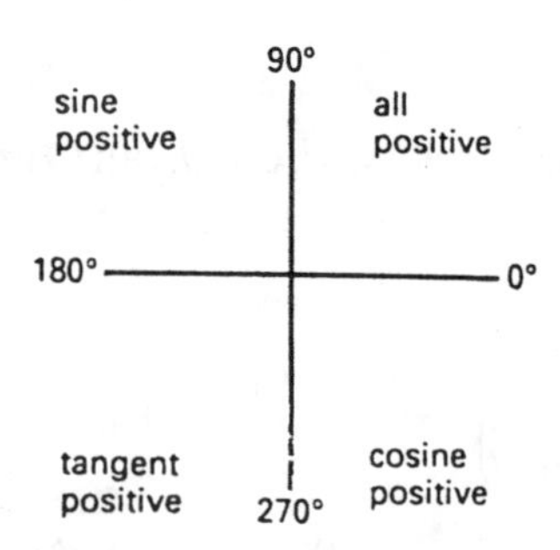

Fig. B.2 The sign of trigonometrical ratios.

$$\sin A \sin B = \frac{1}{2}(\cos(A - B) - \cos(A + B))$$

$$\sin A + \sin B = 2 \sin\left(\frac{A + B}{2}\right)\cos\left(\frac{A - B}{2}\right)$$

$$\sin A - \sin B = 2 \cos\left(\frac{A + B}{2}\right)\sin\left(\frac{A - B}{2}\right)$$

$$\cos A + \cos B = 2 \cos\left(\frac{A + B}{2}\right)\cos\left(\frac{A - B}{2}\right)$$

$$\cos A - \cos B = -2 \sin\left(\frac{A + B}{2}\right)\sin\left(\frac{A - B}{2}\right)$$

If $\tan(\theta/2)$ is expressed as t, then

$$\sin \theta = \frac{2t}{1 + t^2}$$

$$\cos \theta = \frac{1 - t^2}{1 + t^2}$$

$$\tan \theta = \frac{2t}{1 - t^2}$$

List of derivatives

y	$\frac{dy}{dt}$
sin x	cos x
cos x	−sin x
tan x	$\sec^2 x$

where sec $\theta = c/a$ in Fig. B.1.

Index